Interactions Between Agroecosystems *and* Rural Communities

Advances in Agroecology
Series Editor: Clive A. Edwards

Soil Ecology in Sustainable Agricultural Systems,
  Lijbert Brussaard and Ronald Ferrera-Cerrato
Biodiversity in Agroecosystems,
  Wanda Williams Collins and Calvin O. Qualset
Agroforestry in Sustainable Agricultural Systems,
  Louise E. Buck, James P. Lassoie, and Erick C.M. Fernandes
Agroecosystem Sustainability: Developing Practical Strategies,
  Stephen R. Gliessman
Structure and Function in Agroecosystem Design and Management,
  Masae Shiyomi and Hiroshi Koizumi

Advisory Board

Editor-in-Chief
Clive A. Edwards
  The Ohio State University, Columbus, OH

Editorial Board
Miguel Altieri
  University of California, Berkeley, CA
Lijbert Brussaard
  Agricultural University, Wageningen, The Netherlands
David Coleman
  University of Georgia, Athens, GA
D.A. Crossley, Jr.
  University of Georgia, Athens, GA
Adel El-Titi
  Stuttgart, Germany
Charles A. Francis
  University of Nebraska, Lincoln, NE
Stephen R. Gliessman
  University of California, Santa Cruz
Thurman Grove
  North Carolina State University, Raleigh, NC
Maurizio Paoletti
  University of Padova, Padova, Italy
David Pimentel
  Cornell University, Ithaca, NY
Masae Shiyomi
  Ibaraki University, Mito, Japan
Sir Colin R.W. Spedding
  Berkshire, England
Moham K. Wali
  The Ohio State University, Columbus, OH

Interactions Between Agroecosystems *and* Rural Communities

Edited by
Cornelia Flora, Ph.D.

CRC Press
Boca Raton   London   New York   Washington, D.C.

Cover photograph courtesy of André Muelhaupt of Basel, Switzerland.

Library of Congress Cataloging-in-Publication Data

Interactions between agroecosystems and rural communities / edited by Cornelia Flora.
    p. cm.— (Advances in agroecology)
    Includes bibliographical references and index (p.).
    ISBN 0-8493-0917-4 (alk. paper)
    1. Agricultural ecology. 2. Agricultural systems. 3. Rural development. I. Flora,
Cornelia Butler, 1943- II. Series.

S589.7 .I5685 2000
306.3′49—dc21   00-050723

Visit our Web site at www.crcpress.com

Preface

The idea for this book began over ten years ago through conversations with Clive Edwards, editor of an agroecosystems series of books. It was clear to both of us that the human intersect with the environment is mediated by social forces that are international, national, and local. But at that point there was little research to demonstrate the nature of the relationships between human institutions and ecosystems.

Seven years later, innovative research was underway. Building on the farming systems tradition of research, which looked at farmers as actors with socio-economic constraints and values, the importance of communities—of interest and communities of place—in determining individual actions had become much more widely recognized. It was time to revisit the project. Encouraged by Professor Edwards, I cast a very wide net to discover who was doing work that would help define these important interactions. Rather than survey articles, I sought original research from researchers and practitioners of agroecosystem development and management. Thus the book contains chapters that combine the disciplines of forestry, economics, plant breeding, agronomy, anthropology, sociology, geography, soil science, agricultural education, history, landscape planning, landscape ecology, agricultural engineering, and human ecology.

The authors have lived and worked in many parts of the globe, which has given them a great appreciation for the specificity of people and places. Despite that specificity, a number of important conclusions can be drawn in terms of the impact of monoculture on people and land and the exploitative nature of an export-only strategy.

MULTIFUNCTIONALITY

Interactions between agroecosystems and rural human communities change landscapes and livelihoods, as John Soluri explains in Chapter 3. These interactions are highly interactive and in constant flux. Further, the interactions are contested by those who seek alternative uses of the ecosystem, the labor it supports, and the wealth it generates. The interactions are, of course, constrained by pre-existing ecosystems and the cultural constructs of the human communities involved on what is possible and proper. Thus most of the authors link particular communities to particular places at particular historical moments.

Communities work best to develop and sustain healthy agroecosystems under conditions of widespread participation, transparency, and widespread community benefits. When control of agroecosystems and rural communities is highly concentrated, when all decisions are made behind closed doors and there is no clear and regular accountability, and when only a few profit from what is viewed as a common legacy, the entire agroecosystem rapidly deteriorates, as Klooster makes clear in Chapter 5 and Mayda illustrates in Chapter 4. Unequal power, both within rural communities and between rural communities and outside market actors, leads to both social and environmental deterioration.

The articles demonstrate the applicability of the principles of evolutionary biology. As new crops are introduced and have a year or two of success, pests quickly invade and radically reduce yield. When pesticides are used to overcome these pests, the pests evolve resistant strains and the land becomes unusable for that crop. Melons in Mexico (Magdalena Barros Nock in Chapter 6) and bananas in Honduras (John Soluri in Chapter 3) serve as examples of that type of evolution. Both chapters illustrate the impacts on the local ecologies and populations.

When agricultural production is approached as a linear process, identifying a problem, implementing a counter measure, which natural pests and ecosystems respond to in ways that create a new problem, a technological treadmill is created. This is shown in the case of application of nutrients in the chapters by Monsen (Chapter 12) and by Flora, McIsaac, Gasteyer, and Kroma (Chapter 9). Only a systems approach, which rethinks the relation sustainability is not the same as stable.

So, too, do institutions evolve to overcome the constraints imposed by policy aimed to protect the environment and social equity. Klooster (Chapter 5) and Bendini (Chapter 8) illustrate these principles. Because institutions have purposes other than survival, they can evolve in ways that enhance the environment and promote social equity through their positive impact on a agroecosystems as illustrated by Herzog and Oetmann (Chapter 7), Mountjoy (Chapter 15), Lightfoot, Fernandez, Noble, Ramírez, Groot, Fernandez-Baca, Shao, Muro, Okelabo, Mugenyi, Bekalo, Rianga, and Obare (Chapter 10), Rule, Szymanski, and Colletti (Chapter 13), and Flora et al. (Chapter 9), and Barón and Barkin (Chapter 14).

These institutions can be a part of the market sector, the state sector, or civil society. The best combine institutional collaboration among all three sectors, as shown by Mountjoy and by Rule et al. But institutional institutions that take agroecosystems into account seem more complicated than technological silver bullets. Our willingness as a society to invest in institutions and systems as well as in specific technologies will determine the future health of both agroecosystems and rural human communities.

Acknowledgments

This book would not have been possible without the editorial and formatting assistance of Pam Cooper and the graphics assistance of Kristi Hetland, both of the North Central Regional Center for Rural Development. Their hard work, dedication, and flexibility made this a more enjoyable experience for the authors and editors alike. A special note of gratitude goes to John Sulzycki, Senior Editor at CRC Press, who stuck with us through delays and formatting difficulties.

About the Contributors

Maria De Lourdes Barón and **David Barkin** are, respectively, Professor of Rural Development at the Universidad Autónoma de Chapingo, Morelia Campus, and Professor of Economics at the Universidad Autónoma Metropolitana, Xochimilco Campus, in Mexico.

Isaac Bekalo is with the International Institute for Rural Reconstruction (IIRR), Nairobi, Kenya.

Monica Bendini, sociologist, is a professor in the Department of Social and Political Sciences, researcher and coordinator of the Social Agrarian Studies Group, and director of the postgraduate program on Sociology of Latin American Agriculture at the National University of Comahue, Argentina. She has authored publications on peasantry in Andean areas and on wage labor in production chains. She has co-edited an international book on globalization, work, and environment, and research books on sexual division of agrarian work, rural seasonal migrations, and agro-industrial changes and flexibility of labor.

Gary Bentrup is a landscape planner at the U.S. Department of Agriculture (USDA) National Agroforestry Center in Lincoln, Nebraska. With Craig Johnson and Dick Rol, he co-authored a Natural Resources Conservation Service publication entitled *Conservation Corridor Planning at the Landscape Level: Managing for Wildlife Habitat* (2000).

Lorna Michael Butler holds the Henry A. Wallace Endowed Chair for Sustainable Agriculture, Iowa State University, and is Professor in the Departments of Sociology and Anthropology. She conducts research and education on the human dimensions of sustainable food, agriculture, and community systems, with a particular interest in rural–urban linkages and sustainable business development. Her work includes participatory plant breeding, farming systems research in Africa and the Middle East, community and international development and urban–peri-urban agriculture. She has authored *The Sondeo: A Rapid Reconnaissance Approach for Situational Assessment* (1995), and is co-author of *Focus Groups: A Tool for Understanding Community Perceptions and Experiences* (1995) and *Human Diversity, Community, and Viable Food and Agricultural Systems* in *Exploring the Role of Diversity in Sustainable Agriculture,* edited by R. Olson, C. A. Francis, and S. Kaffka (1995).

Richard Carkner is a professor and extension economist with Washington State University (WSU). He recently completed a sabbatical with the FAO in Rome looking at urban agriculture and food security. His current emphasis is on agriculture in urbanizing counties. He is on the executive committee of WSU's Food and Farm Connections team, whose mission is to enhance sustainable community food and farm systems through education, research, and partnerships. He currently is studying marketing in the Pacific Northwest.

Joe P. Colletti is Associate Professor of Forestry: Economics and Quantitative Methods. He has been president of the Association for Temperate Agroforestry (AFTA) and a co-leader of the Leopold Center for Sustainable Agriculture's Agroecology Issue Team since 1992. He has written several articles on the economics of riparian buffers applied to agricultural landscapes and multipurpose shelterbelts. He was lead principal investigator of an agroforestry project with the Winnebago Tribe of Nebraska and learned to "think like a mountain" from his Winnebago friends.

Maria E. Fernandez works on farmer participatory research, focusing on gender and agrobiodiversity management issues. She has worked for Food and Agriculture Organization (FAO), Center for Research and Information on Low-External Input and Sustainable Agriculture (ILEIA), and Centro Internacional de Agricultura Tropical (CIAT), and is a member of ISG and Grupo Yanapai. She currently is an independent consultant and visiting professor at the Universidad Nacional Agraria La Molina in Peru.

Edith Fernández-Baca is a member of Grupo Yanapai, a nongovernmental organization involved in participatory-action research in the Central Andes of Peru. She also is a member of the International Support Group (ISG). A Doctor in Veterinary Medicine, she has been working for the past 7 years with rural communities in natural resource management and sustainable agriculture. She currently is engaged in graduate studies in rural sociology at Iowa State University and is working as a research associate in the North Central Regional Center for Rural Development.

Cornelia Butler Flora is Professor of Sociology and Director of the North Central Regional Center for Rural Development (NCRCRD), located at Iowa State University. One of four regional centers in the United States, the NCRCRD combines research and outreach for rural development and covers the 12 North Central states. She held the Endowed Chair in Agricultural Systems at the University of Minnesota from October 1999 to May 2000. Previously, she was head of the Sociology Department at Virginia Polytechnic Institute and State University, a University Distinguished Professor at Kansas State University, and a program officer for the Ford Foundation. A past president of the Rural Sociological Society, she is author and editor of a number of recent books, including *Agroecosystems and Rural Communities, Rural Communities: Legacy and Change, Rural Policies for the 1990s*, and *Sustainable Agriculture in Temperate Zones*.

Her current research addresses alternative strategies of community development and measurement of community change in the United States and Latin America. Her Bachelor of Arts degree is from the University of California at Berkeley in 1965, and her M.S. (1966) and Ph.D. (1970) degrees are from Cornell University, where she received the 1994 Outstanding Alumni Award from the College of Agriculture and Life Science. She was president of the Board of Directors of the Henry A. Wallace Institute of Alternative Agriculture, and now serves on the Board of Directors of Winrock International. She currently is on the Board of Directors of the Heartland

Institute for Community Leadership, the Board on Agriculture and Natural Resources of the National Research Council of the National Academy of Science, and the Northwest Area Foundation.

Charles A. Francis is Professor of Agronomy and Extension Cropping Systems Specialist at University of Nebraska, Lincoln. He is editor or co-editor of *Multiple Cropping Systems, Sustainable Agriculture in Temperate Zones, Exploring the Role of Diversity in Sustainable Agriculture,* and *Crop Improvement for Sustainable Agriculture.* He is on the editorial boards of three journals, and has written numerous articles on multiple cropping, sustainable farming systems, on-farm research methods, and methods for teaching agroecology and integrated crop/animal systems. He currently is active with colleagues in developing the new Nordic M.Sc. degree in agroecology.

Stephen Gasteyer is a Ph.D. candidate in sociology at Iowa University, focusing on community development and natural resources management. He has worked in the United States, Africa, and the Middle East on natural resources, agriculture, and development issues. He has co-authored articles, chapters, and reports on local indicators of environmental quality, participation in water quality protection, power, and perceptions of landscape change, integrated pest management, biodiversity, and sustainable agriculture and community development. He currently is working on his dissertation on community organization to protect water quality.

Annemarie Groot is with the International Support Group (ISG), Amersfoort, Netherlands.

Felix Herzog is a senior researcher at the Swiss Federal Research Station for Agroecology and Agriculture. He leads a group that evaluates the effect of the Swiss agri-environmental program on biodiversity and on the quality of ground and surface water with respect to excess nutrients. He has worked for 15 years in landscape ecology and agroecology with experience in West Africa, Germany, and Switzerland. He has published his work in numerous articles in journals and books.

Daniel James Klooster is Assistant Professor of Geography at Florida State University in Tallahassee, Florida. His publications address community forestry through the lenses of common property theory, resistance, development theory, and the implications for global environmental change. In his current research, he continues to examine how local institutions of forest management can help to guide a sustainable relationship between people and forests.

Margaret Kroma is an assistant professor in the Department of Education at Cornell University. She teaches courses in international extension education and adult education. Her research focuses on participatory technology development, gender, and participatory processes for research and extension.

Clive Lightfoot divides his time between chairmanship of the International Support Group (an international nongovernment organization that provides support to local groups in learning approaches to ecologically sound agriculture) and teaching in systems thinking and agroecosystem analysis with the International Center for Development Oriented Agriculture (ICRA) based in Montpellier, France. He has authored many publications on agroecosystems analysis and participatory technology development over a research career of more than 20 years in tropical agricultural development.

Chris Mayda is an assistant professor of Geography at Eastern Michigan University. She is a dedicated technician working on the web and in documentary digital video on cultural landscape subjects, especially the rural and agricultural world. She publishes an online cultural geography journal, www.Geographing.com, and continues her Canadian border studies for future publication while writing her book on the changing geography of pigs. Really!

Gregory McIsaac is an assistant professor in the Department of Natural Resources and Environmental Sciences at the University of Illinois, where he teaches courses in watershed hydrology and ecosystem management. His research addresses various aspects of land management and water quality.

Wayne Monsen is the program coordinator for the grant, loan, and whole-farm planning programs for the Energy and Sustainable Agriculture Program at the Minnesota Department of Agriculture. A former farmer, he brings a broad depth of farming experience in production systems and agricultural issues. He networks with farmers, rural communities, agency staff, extension educators, researchers and nonprofit organizations to adopt sustainable farming practices and systems, implement whole-farm planning, and encourage citizen leadership.

Daniel C. Mountjoy is Area Resource Conservationist for the USDA—Natural Resources Conservation Service in Salinas, California. He develops technical and educational outreach programs for minority farmers along California's central coast to assist them with conserving natural resources. His work provides him with the opportunity to improve communication between farmers and multiple public agencies. He earned a doctorate in human ecology from the University of California Davis for his studies in conservation adoption by Japanese, Mexican, and Anglo strawberry farmers.

Anthony Mugenyi is with Development Support Services (DSS), Soroti, Uganda.

Grace Muro is with the Tanzania Food and Nutrition Center (TFNC), Dar es Salaam, Tanzania.

Reg Noble is with the International Support Group (ISG), Amersfoort, Netherlands.

Magdalena Barros Nock is a Social Anthropologist with a PhD. in Development Studies from the Institute of Social Studies in the Hague, Netherlands. She has done research on the fruit and vegetable business, family enterprises, transnational social networks, and migration.

Lynette Obare is with the Forest Action Network (FAN), Nairobi, Kenya.

Anja Oetmann is an agricultural scientist who has done research on grassland ecologic systems, with a special focus on the effect of different ecologic and management conditions on morphologic and genetic diversity of populations. Since 1994, she has been responsible for public information services and political advice in the field of plant genetic resources for food and agriculture (PGRFA), whose main objective is the national implementation of the Global Plan of Action on PGRFA that was adopted at the 4th International Technical Conference on Plant Genetic Resources (ITCPGR) 1996 in Leipzig, Germany.

Simon Okelabo is with Development Support Servvices (DDS), Soroti, Uganda.

Ricardo Ramirez is Secretary of the International Support Group (ISG), which links local experience in agroecosystem management. His research includes multistakeholder coordination for natural resource management, and he is finishing his Ph.D. through the University of Guelph's interdisciplinary program on Rural Studies for Sustainable Communities. His thesis title is "The Human Side of Information and Communication Technologies for Rural and Remote Community Development."

Andrew Rianga is with the Kisii Network for Ecological Agricultural Development (KNEAD), Kisii, Kenya.

Lita C. Rule is Associate Professor of Forestry at Iowa State University. She teaches courses on forestry economics, policy, and administration. She has worked with natural resource researchers and other scientists in the United States and countries abroad such as Mexico, South Africa, the Philippines, and International Center for Research in Agro-Forestry (ICRAF), on agroforestry and forestry production systems, and has just started work on a collaborative research and teaching capacity development program in South Africa. She has authored publications on agroforestry and forestry production systerms, mostly on the socioeconomic aspects.

Michele M. Schoeneberger is Research Project Leader and Supervisory Soil Scientist at the USDA Forest Service/National Resource Conservation Service (FS/NRCS) National Agroforestry Center, part of the Rocky Mountain Research Station located in Lincoln, Nebraska. Her research interest in long-term productivity has involved her in projects ranging from forest harvesting impacts to atmospheric pollutant effects on vegetation, and now has her looking at better ways to design and integrate tree-based buffer practices into the landscape to better meet landowner and societal objectives.

Francis Shao is with Food, Agriculture, and Natural Resources Management Research Consultants (FANRM), Dar es Salaam, Tanzania.

John Soluri is an Assistant Professor of History at Carnegie Mellon University, where he teaches courses on Latin American and environmental history and policy. His research on banana production has appeared in the *Hispanic American Historical Review*. He currently is writing a book-length manuscript about the ecosocial history of banana production in Honduras.

Marcella B. Szymanski is Assistant Extension Professor of Forest and Natural Resource Economics at the University of Kentucky. She has trained with the Praxis Group, an international training center for participatory development in Canada and has a broad background that includes natural resource assessment, forest economics, agroforestry, and participatory appraisal methods. She has a passion for participatory development and has worked closely with native peoples in the United States and Kenya.

Table of Contents

Chapter 1
Introduction... 1
Cornelia Butler Flora

Chapter 2
Shifting Agroecosystems and Communities... 5
Cornelia Butler Flora

Chapter 3
Altered Landscapes and Transformed Livelihoods: Banana Companies,
Panama Disease, and Rural Communities on the North Coast of
Honduras, 1880–1950... 15
John Soluri

Chapter 4
Community Culture and the Evolution of Hog Production: Eastern
and Western Oklahoma... 33
Chris Mayda

Chapter 5
Forest Conservation and Degradation in a Subsubsistence Agricultural
System: Community and Forestry in Mexico.. 53
Daniel James Klooster

Chapter 6
Community Fruits and Vegetables for Export: The Impact on Two
Mexican Ecosystems and Rural Communities...................................... 69
Magdalena Barros Nock

Chapter 7
Communities of Interest and Agro-Ecosystem Restoration:
Streuobst in Europe... 85
Felix Herzog and Anja Oetmann

Chapter 8
Transhumant Communities and Agroecosystems in Patagonia.......... 103
Monica Bendini

Chapter 9
Farm–Community Entrepreneurial Partnerships in the Midwest........................ 115
Cornelia Butler Flora, Gregory McIsaac, Stephen Gasteyer, and Margaret Kroma

Chapter 10
A Learning Approach to Community Agroecosystem Management................. 131
Clive Lightfoot, Maria Fernandez, Reg Noble, Ricardo Ramírez,
Annemarie Groot, Edith Fernandez-Baca, Francis Shao, Grace Muro,
Simon Okelabo, Anthony Mugenyi, Isaac Bekalo, Andrew Rianga, and
Lynette Obare

Chapter 11
Bridges to Sustainability: Links between Agriculture, Community
and Ecosystems... 157
Lorna Michael Butler and Richard Carkner

Chapter 12
Rural Community Leadership in the Lake Benton Watershed......................... 175
Wayne Monsen

Chapter 13
The Winnebago Tribe's Agroforestry Project: Linking
Indigenous Knowledge, Resource Management Planning, and
Community Development.. 187
Lita C. Rule, Marcella B. Szymanski, and Joe P. Colletti

Chapter 14
Innovation in Indigenous Production Systems to Maintain Tradition..................... 211
Maria de Lourdes Barón and David Barkin

Chapter 15
Ethnicity, Multiple Communities, and the Promotion of Conservation:
Strawberries in California... 221
Daniel C. Mountjoy

Chapter 16
Ecobelts: Reconnecting Agriculture and Communities.. 239
Michele M. Schoeneberger, Gary Bentrup, and Charles A. Francis

Chapter 17
Afterword: An Optimistic Future Scenario... 261
Charles A. Francis

Index... 269

1 Introduction

Cornelia Butler Flora

The agroecosystem concept involves multiple levels of organization: an individual organism, a population (groups of similar organisms), communities (groups of different organisms), and an ecosystem, which links the abiotic factors of an environment and the communities of populations and organisms that occur in a specific area (Gliessman, 1998).

Human communities can be understood in the same way as comprising individuals, a population (a group of similar or related individuals), communities (interactions among individuals and populations for mutual support), and coalitions of communities that work together or oppose each other on the basis of interests or values. Like agroecosystems, communities of place and of interest have emergent qualities that are more than the sum of the parts. The interactions between populations and communities of interest and of place can enhance or detract from system sustainability.

Sustainability is an emergent quality of the interactions between communities of interest and of place that includes a healthy ecosystem, vital economies, and social equity. The President's Council on Sustainability refers to these as the "3 E's" (The President's Council on Sustainable Development, 1999).

The chapters in *Interactions between Agroecosystems and Rural Communities* all address these three components of sustainability, and they do so from an interdisciplinary perspective. The authors include historians, geographers, sociologists, anthropologists, agronomists, soil scientists, foresters, and agricultural engineers. All of the authors are scholars. Many are also practitioners, seeking to bring about sustainable rural communities embedded in healthy agroecosystems.

Chapter 2 lays out a framework for analyzing the systems levels that impact and are impacted by human action on the land to grow food and fiber. Chapters 3 through 7 describe how specific rural communities have attempted to vary their agroecosystems in response to changing external conditions and internal community dynamics.

John Soluri shows how the entry of a new commodity—bananas—into the world market led to an economic boom then an agroecosystem bust in rural communities in Honduras. Decisions made outside the community—that only a certain color, shape, and size of fruit, the Gros Michel, was a banana, despite the wide variety of species present in Honduras—increased the monoculture-devastation cycle. Soluri documents the response of local communities, government authorities, and transnational corporations to the agroecosystem changes.

Chris Mayda demonstrates the intersection between culture and agroecosystem. She analyzes two distinct hog-producing communities in Oklahoma, one with a tradition of small farm hog production and one in which a large-scale industrial hog production was introduced. The relation of the two communities to the pork buyers—transnational corporations in both cases—was quite different, and the impacts on the local communities in terms of ecosystem health and social and economic well-being diverged. Mayda convincingly shows that agroecosystems emerge in response to cultural as well as biophysical context.

Daniel Klooster examines how community organizational structure impacts an agroforestry ecosystem in Mexico. Building on Mayda's theme of cultural context, he shows how the Mexican government and community power structures made it difficult actually to implement sustainable agroforestry programs in mountain communities. In a situation of corruption and timber poaching, community members outside the power structure organized to demand transparency, accountability, and equity in access to forest resources.

Magdalena Barros Nock examines another effort by rural communities in Mexico to become integrated into international vegetable markets. Despite patient attempts at organization and market development, the monoculture necessary to meet the brokers' demands led to pest infestations that made continuing production impossible. Moreover, dependence on a single broker to transport the fruit across the border further increased the vulnerability of the farmers, despite government investment in irrigation and community investment in packing sheds. The development of local production and handling was based on community organizations and structures.

Felix Herzog and Anja Oetmann show the intersection of community culture, government policy, and agroecosystems in discussing the perseverance of the *streuobst* in central Europe. These diverse, dispersed orchards, after being nearly wiped out by intensive fruit production in the 1950s through the 1970s, gained recognition and support by communities of interest and of place. Policies were put into place that rewarded the ecosystem services, particularly biodiversity and landscapes, that the *streuobst* provide. Furthermore, new markets were developed for the heirloom and site-specific fruits and juices they produce.

Monica Bendini analyzes the intersection of nomadic herding communities in the Patagonian region of Argentina with changing world markets, government programs, and ecosystem health. Community organization of both indigenous and creole herders provided basic control over the commons that worked well in times of high demand and continuous access to land. But a shifting political agenda has gradually excluded the herders from the most productive pastures and increased the pressure on arid and high mountain pastures. The community organizational structures are hard pressed to respond to the agroecosystem changes imposed by policy and overgrazing.

Cornelia Butler Flora, Gregory McIsaac, Stephen Gasteyer, and Margaret Kroma examine the intersection of market and community in Illinois by looking at entrepreneurial partnerships between farms and communities. Their analysis makes it clear that institutions off the farm must change if practices on the farm can change, and that when these changes can take place in rural communities, both farm and townspeople benefit.

The remaining chapters offer a look at planned interventions in agroecosystems to increase sustainability. These chapters have a strong emphasis on process rather than technology. Community-agroecosystem actions to increase sustainability are "abnormal": they generally do not just happen. Specific interventions can bring about system change in the layered systems to support the various dimensions of sustainability.

An international, interdisciplinary team made up of Clive Lightfoot, Maria Fernandez, Reg Noble, Ricardo Ramirez, Annemarie Groot, Edith Fernandez-Baca, Francis Shao, Grace Muro, Simon Okelabo, and Anthony Mugenyi present a synthesis and analysis of their projects in Africa, Asia, and Latin America that addressed community-based agroecosystems. The process they present involves phases of involvement and involvement tools built on a vision of the future and assessment of the resources available for working toward that future, negotiating access to those resources, implementing them, and reflecting on movement toward the desired future conditions.

Lorna Michal Butler and Richard Carkner analyze efforts at linking agriculture, community, and environment in North America. By using the concept of the *food-shed,* they demonstrate how different agroecologies can be enhanced through closer ties between producers and consumers by utilizing organizations in rural communities. They focus on sustainability connectors linking agroecosystems to communities.

Wayne Monson examines the process of linking farmers, a rural community, and agroecosystem deterioration in Minnesota. Multiple uses of the agroecosystem, both crop production and tourism, require new alliances as well as new practices on the farmers' field. Facilitated processes for vision and attention to inclusiveness build social capital between farmers and rural communities that made it possible for all sectors to be involved in positive action for ecosystem health.

Lita C. Rule, Marcella B. Szymanski, and Joe Colletti analyze a process of building rural community and ecosystem health simultaneously. They describe the development of the Winnebago Tribe's agroforestry project and the processes used to help the tribal community balance issues of culture, agroecosystem health, and economic vitality. The tools they use to help determine balance in agroecosystem management also build community social capital.

Maria de Lourdes Barón and David Barkin examine a process of combining indigenous knowledge and cutting-edge science to build on community resources for sustainability. By analyzing resources currently existing in a rural community's agroecosystem—hogs and avocados—they introduce a project to grow low-fat hogs that fit the community cultural structure and emerging market demands.

Daniel C. Mountjoy analyzes strawberry production in the foothills of California. The agroecosystems in such situations can be very destructive of the environment, workers, and communities. By systematically addressing communities of interest and of place—and particularly by understanding the cultural diversity among strawberry farmers—technical assistance and community organization can be oriented toward more sustainable agricultural practices.

Michele M. Schoeneberger, Charles Francis, and Gary Bentrup examine an innovation in community connections: ecobelts, discussing their contributions to agroecosystems and communities once they are in place. In addition, they demonstrate that the process of their participatory design can be important, for building community around a concept of sustainability.

Finally, Charles Francis, describes life in a utopian community with a sustainable agroecosystem. He makes the point that agroecosystems are important not only for rural areas, but are continued into urban zones. Sustainable agroecosystems are consistent with high technology, although not necessarily high use manufactured inputs. However, they require that attention be paid to community institutions and resources to make them happen.

REFERENCES

Gliessman, S. R., *Agroecology: Ecological Processes in Sustainable Agriculture,* Ann Arbor Press, Chelsea, Michigan, 1998, 18.

The President's Council on Sustainable Development, *Towards a Sustainable America: Advancy Prosperity, Opportunity, and a Healthy Environment for the 21st Century,* U.S. Government Printing Office, Washington, DC, 1999. Available: http://www.whitehouse.gov/PCSD.

2 Shifting Agroecosystems and Communities

Cornelia Butler Flora

CONTENTS

Introduction 5
Market, State, and Civil Society 6
Resources, Rural Communities, and Agroecosystems 9
Forms of Capital 9
Shifting toward Sustainable Agroecosystems and Rural Communities 11
References 12

INTRODUCTION

The intersection of people and the landscape transforms both. Urbanization can be seen as humans separating themselves from nature through built roads; electric, utility, and telephone systems; water works; skywalks; buildings; and underground tunnels that link and house the activities of people who are as removed as possible from the intrusions of the natural world. Nature is overcome, except when earthquake, flood, or storm severs the constructed linkages and destroys the barriers between people and nature (Harvey, 1985).

Suburbanization is, in part, a reaction to removal from nature, taking people back to interaction, albeit very controlled, with plants, animals, and the natural landscape. The remaining separation of suburbanites from the local ecosystem is indicated by the ubiquity of turf grass and lawns throughout the United States, requiring huge inputs of water, chemicals, and machines to survive. But community norms support the sanctity of the English garden, with communities writing and enforcing covenants and laws against actions by homeowners that deviate from the agreed-on aesthetics of idealized place.

Agriculture and forestry represent attempts of human communities to use the perceived potential of a local landscape to extract value and maintain human communities: rural, suburban, and urban. The systems by which agriculture and forestry transform the ecosystem into an agroecosystem are varied. Some systems deplete the natural capital of place, whereas others replenish it. The agroecosystems that emerge are not simply natural outgrowths of humans and landscapes with productive potential, but the product of human communities mediated by culture and technology.

5

Communities of interest (which can be global) and communities of place (which must be local) both influence agroecosystems. Agroecosystems are constantly changing in response to natural events (precipitation, drought, global warming) and anthropogenic phenomena. Furthermore, these events can be proximate (immediately at hand) or distant, making linear causality all the more difficult to identify. Indeed, by considering agroecosystems and communities as systems, we assume multiple causality and feedback mechanisms. We also assume multiple systems levels: field, farm, community, watershed, state, nation and the globe. Although most approaches cannot look at all these systems levels, it is important to look at more than one simultaneously.

Communities of place are a useful lens for viewing changes in agroecosystems. Communities of place, however, do not have neat, tight barriers, but influence and are influenced by other parts of society.

Communities are shaped by three sets of institutions: the market, the state, and civil society. These institutions occur at various levels: the community, the county, the region within a state, the state, the region within the nation, the nation, and the international community. These institutions influence the resources available to local access and control. Local resources, in turn, influence what happens to land and people. The way local resources are consumed, stored, or invested shapes local institutions of market, state, and civil society.

MARKET, STATE, AND CIVIL SOCIETY

The institutional spheres of market, state, and civil society, which overlap in different ways at different times and places, all are critical for communities, rural and urban, to flourish (Tester, 1992; Zijderveld, 1999).

Markets are the many firms and institutions that exchange goods and services at a profit. When there is competition and a free flow of information, markets are incredibly efficient at distributing goods and services to those who can pay. They are not particularly efficient at distributing goods and services to those who cannot pay or at protecting the environment. The market is highly dynamic, with much competition and the constant entrance and exit of firms.

Market institutions are present at the local, state, national, and transnational levels. These institutions sometimes compete, sometimes collaborate, and are integrated forward and backward to differing degrees. The purpose of market institutions is to make a profit for their owners. Sometimes the owners are individuals or families. Sometimes owners are stockholders, who tend to evaluate firms on the basis of two things: how much profit they have generated in the last quarter and their market value. When either of these is viewed as unsatisfactory, owners seek to change the hired managers. Consolidation, competition, and cooperation among market firms suggest a very dynamic sphere. Farms, cooperatives, and transnational firms all are part of the market sector, with direct impacts on agroecosystems and the communities they surround. Increasingly around the world, landowners are not the land managers, nor do they live in the community of place. This separation of ownership and management, whether through rental agreements or management contracts, puts more emphasis on financial capital and less on natural and social capital.

The state makes markets possible. Markets need fairly stable conditions in which to operate. Markets need contracts that are enforceable through an effective administrative and judiciary system. They need a reliable money supply. They need to know that rules put into place by the legislative system apply to all, and that that the rules will be administered in a universal way: the same rules will be applied to everyone. Therefore, the state, which is government from international levels down to the national, state, and local levels, is critical to the market. The state, however, has the additional responsibility of providing for the public welfare. Governments are simultaneously a foundational force underlying environmental destruction and the principal agent of sustainability reforms (Buttel, 1998).

Different state agencies within a given level and at different levels of the state often are at odds. The state, like the market, is a dynamic, contested sector. In the United States, governors and legislators disagree. Very often, local levels of government, particularly counties and small cities, feel imposed on by the state or federal governments, particularly as they deal with unfunded mandates and land use regulation. Therefore, within the state sphere, which sets the rules and conditions for the market and the safety net for its citizens, the terrain is very contested. In particular, there is disagreement on where the authority lies to determine the rules and regulations that influence agroecosystems.

The state includes local, state, national, and international government institutions, including the three branches of government: the legislative (which makes the laws and allocates resources), the administrative (which implements the laws and distributes the resources), and the judicial (which imposes sanctions on those who do not follow the laws). The state provides the rules under which market operates so that the common good is served and firms at the same time are profitable. The state also provides a safety net for people and protects natural resources deemed to be in the common good.

The state is a highly contentious sphere, in which state governments disagree with the national government, the legislative branch contests with the administrative branch, and even within institutions, different bureaucracies or agencies seek to gain or maintain hegemony, influence, and budget. The purpose of the state under capitalism is to make sure that making a profit also serves the common good. Elected officials often are judged by the degree to which they serve the common good. However, definition of the common good is almost always contested. This is particularly true when it comes to agroecosystems. Are they to yield the highest possible profit? Is it their responsibility to provide clean air and clean water, a habitat for birds and animals, or scenic views for visiting tourists? How do elected officials decide?

Civil society determines the common good. Groups in society, formal and informal, join together around common interests or values. Through their organized activity, they influence the market and the state. The faith community, including churches, synagogues, and mosques; the National Rifle Association; and antigun groups all are part of civil society. Also included are the Sierra Club and Ducks Forever, as are parent–teacher organizations and Rotary clubs. These organizations articulate shared interests and values in a variety of ways as they interact with the market and the state. The strength of civil society is related to agricultural sustainability in the United

States (Lyson and Barham, 1998). Tolbert et al. (1998) found that the strength of civil society is related to local capitalism and positive community outcomes. The presence of civil society organizations with environmental and social justice concerns increasingly means attention is paid to agroecosystems and how they are changing.

Civil society influences the market by forming consumer groups, which can engage in boycotts and information campaigns. By boycotting certain products, such as those containing transgenic organisms, civil society has an impact on what is farmed and how it is farmed. Civil society influences the state by bringing lawsuits against agroecosystems seen as polluting the local environment (influencing the judicial branch of government), by forging legislation such as taxes on greenhouse gases (influencing the legislative branch of government), and by urging that particular laws related to farm labor or nonpoint pollution be enforced (influencing the administrative branch of government).

Civil society generally exerts influence on the basis of deeply held values or desired future conditions. Groups in civil society, both formal and informal, form around these shared future conditions and their mental/causal models of how the world works. Individuals relate to civil society when they become participants or members. Groups in civil society often are in hot dispute. Because this is where the definition of the "collective conscience" is negotiated, groups struggle to gain participants and to co-opt other groups. The dynamism of this sector influences both the market and the state. It also influences agroecosystems and rural communities directly and indirectly through its influence on the market and the state. The innovations in informatics have created global civil society organizations that influence the shifts in agroecosystems and rural communities (Smith, 1998).

As individuals, we have roles in each sphere. We are part of the market as individual producers and consumers. As individual citizens, our roles in the state sphere involve rights (e.g., as voting and running for public office) and responsibilities (e.g., paying taxes and following laws). As individuals, we can become parts of interest or value groups; many Americans are members of several such groups.

Each of these spheres is not one dimensional, yet there are tendencies for one to dominate another. In the former Soviet Union, the state dominated the society to the point that the market was small and did not work very well, and civil society was nonexistent. Many agroecosystems became degraded. There were no environmental or farm groups to protect them outside the state, whose vision for the future was production oriented. With the fall of the Berlin Wall and the end of the Soviet Union, the belief was that free markets alone would transform the newly reformed countries into nations similar to the capitalist democracies of Western Europe, North America, and the Pacific, including Japan, Australia and New Zealand. The state was very inefficient in distributing goods and services in the old centrally planned economies. The external advisors and the nationalist reformers forgot the need for a strong state to enforce contracts and the need for organized groups of citizens in civil society to be outraged by corruption and able to pressure the state to put more rules on the market. As a result, "mafia" capitalism reigns in the former Soviet Union. Agroecosystems continue to be degraded. Individuals are upset, and organized corrective action still is not in place (Rose, 1998). A lack of a civil society and the absence of a strong state have resulted in a market dominated purely by power, which includes control of the means of violence, a power reserved for and protected by the state in more stable parts of the world.

Having an imbalance in favor of the market also is problematic. When things are judged as moral because they are profitable and the state therefore acts only to increase profitability, the other important roles of the state—protecting those in society who cannot protect themselves and protecting the environment—fall by the wayside. In situations wherein the market dominated, as with banana production in Honduras (see Chapter 3), both rural communities and the agroecosystem suffered.

RESOURCES, RURAL COMMUNITIES, AND AGROECOSYSTEMS*

All communities have multiple resources. These resources can be consumed and therefore not available to current or future generations, stored for future but not current generations, or invested for the present and future needs of all generations (World Commission on Environment & Development, 1987). When resources are used to create new resources, they are called *capital*. This is close to Daly's (1968) definition of capital: matter capable of trapping energy and using it for human purposes. In this definition, both biologic organisms and the environment are included. Folke and Berkes (1998) are even more explicit, defining capital as "a stock of resources with value embedded in its ability to produce a flow of benefits (see page 21)." This is different from the business usage of the term "capital," which simply refers to money. Sociologists find it useful to distinguish different forms of capital: human, social, natural, and financial/built. Sustainability means investing in the forms of capital that do not deplete other forms (Flora, 1998). Thus, a strategy to maximize short-term financial capital through intensive irrigated agriculture can decrease natural capital through salinization of the soil and loss of wetlands. In another instance, the use of children as field labor can increase financial capital by decreasing labor costs while lowering human capital as the children leave school to work.

The way agriculturists and communities handle their resources, both individually and collectively, depends on their collective vision for the future (Engel, 1997; Maarleveld and Dangbégnon, 1999). In Chapter 10, Lightfoot and colleagues stress the importance of collective visioning. That vision can be sustainable or not. If the vision has a single end, such as "more is better," then sustainability is impaired. Sustainability requires choice and balance, which give resiliency to change over time. Resilience means that a community has conserved opportunities for innovation and renewal (Folke and Berkes, 1998).

FORMS OF CAPITAL

Four forms of capital are used as a model for analyzing the interaction of people with the agroecosystem (Pretty, 1998). It is critical to name these resources in order to act

---

* Portions of this section are drawn from Flora, 1999.

toward them. George Herbert Mead (1934) argued that you must name something to have consciousness of it. Without consciousness of resources, we cannot act to sustain them. Not only will unknown species become extinct, but key forms of human relationships can vanish unless we are aware of what they are. Ignorance does not always lead to the disappearance of species or social forms, but only awareness will enable action to sustain them. Therefore, knowledge of these various stocks within a system is critical. Only by explicit awareness, through naming, can we act toward these stocks in ways to enhance them as capital.

The four forms of capital (Figure 2.1) in the order that they best address community development are (1) human capital, (2) social capital, (3) financial/built capital, and (4) natural capital. The author's work with communities has shown that one must deal first with the people (human capital) and the relationships (social capital) before one can mobilize these to enhance natural capital. Starting with natural capital often leads to conflict that detracts from all forms of capital (Paulson, 1998). Successful actions to enhance natural capital start from human capital (recognizing the skills, knowledge, and abilities of local people) and build social capital (increased communication and networks and increased initiative and responsibility). These activities are performed around environmental issues, usually in terms of place, but are critical predecessors of actions to improve natural capital (Cortner and Moote, 1999; Pomeroy and Beck, 1999).

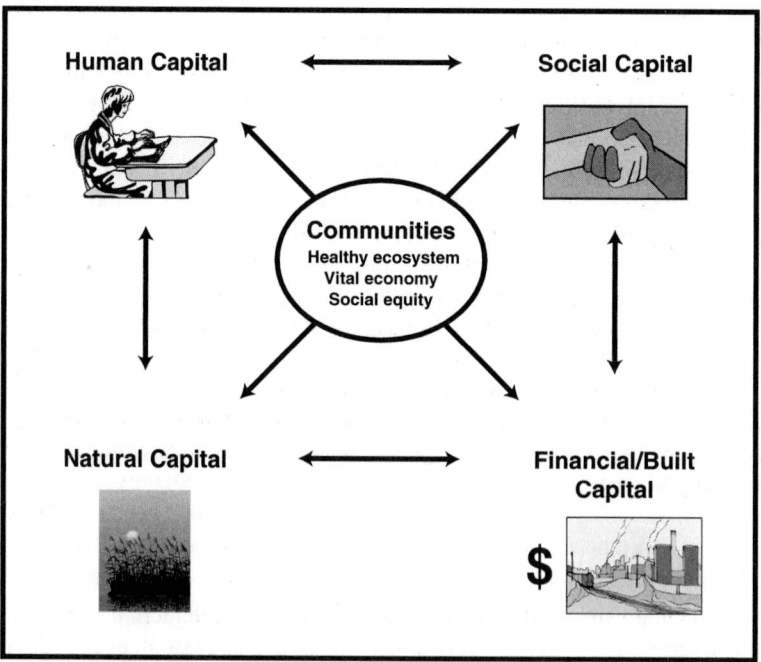

FIGURE 2.1 Community Capitals and System Sustainability

SHIFTING TOWARD SUSTAINABLE AGROECOSYSTEMS AND RURAL COMMUNITIES

Healthy agroecosystems with multiple community benefits are more likely to be sustainable than agroecosystems which enhance only one of the capitals. The responsible stewardship of natural resources sustains businesses and families in communities over the long term. Finding the common ground among people who have emotional, symbolic, or economic identification with a place, whether they live there or not, is essential to making decisions about development and resource use that will enable communities and their resource base to survive and thrive. Rural communities that plan and act in concert with the natural systems in which they are located can encourage farmers to do the same (Flora, 1995). Colonizers' treatment of agroecosystems in the United States, Latin America, Africa, and Asia often was aimed at increasing short-term outputs, with no awareness or concern about depletion of the very resources that produced the bounty.

Vital economies deploy financial, natural, and human resources to create, maintain, and improve local livelihoods. A diverse economic base, including a mixture of natural resource–based activities, helps maintain services, businesses, and households when markets fluctuate. In rural communities with vital economies, community residents move toward self-sufficiency and prosperity; local businesses modernize and find new markets; local ownership of homes and businesses increases; and local people and financial institutions invest in the community (Flora et al., 1997).

Social equity does not mean that all individuals, populations, and communities have the same things or are similar. It does mean that all have opportunities for participation in market activities as producers and consumers, for participation in the state sector (government at all levels) with corresponding rights and obligations, and for participation in civil society through gathering together with those who share their interests and values. It means that the skills, abilities, and knowledge of the individual are recognized and enhanced, that there is transparent access to information and ease of association within and among communities, and that there are options for innovation and responsibility. Rural communities that achieve healthy ecosystems and vital economies find that these aspects of social equity are a critical base for the other system functions (Flora et al.,1996).

Diversity varies among rural communities and networks of communities. Generally, communities are like agroecosystems, in that those that are more resilient are more diverse. Natural capital shifts with global warming and human intervention. Financial capital shifts rapidly in an increasingly global economy (Flora, 1990). Human capital becomes ever more mobile, and social capital, both bridging and bonding social capital (Narayan, 1999), can be created and destroyed through shifts in the other capitals, through the new forms of networking, and through the digital, biotechnology, and lifestyle revolutions of the 21st century.

Privileging one form of capital over another can destroy rural communities and agroecosystems. When either the state or the market dominates a place, civil society can be destroyed. Increasing globalization and industrialization in the market, decentralization and privatization in the state, and polarization and engagement in civil society can redirect resources in ways that are beneficial or detrimental to rural communities and the agroecosystems they support. Understanding these interrelationships can inform the different ways that human communities organize to benefit both agroecosystems and rural human communities.

REFERENCES

Buttel, F. H., Some observations on states, world orders, and the politics of sustainability, *Organ. Environ.,* 11, 261, 1998.

Cortner, H. J. and Moote, M. A., *The Politics of Ecosystem Management,* Island Press, Washington, D.C., 1999.

Daly, H. E., On economics as a life science, *J. Polit. Econ.,* 76, 392, 1968.

Engel, P. G., *The Social Organization of Innovation: A Focus on Stakeholder Interaction,* Royal Tropical Institute, The Netherlands, 1997.

Flora, C. B., Rural peoples in a global economy, *Rural Sociol.,* 55, 157, 1990.

Flora, C. B., Social capital and sustainability: agriculture and communities in the Great Plains and the corn belt, *Res. Rural Sociol. Dev.: Res. Annu.,* 6, 227, 1995.

Flora, C. B., Sustainable production and consumption patterns: community capitals, in *The Brundtland Commission's Report Ten Years,* Softing, G. B., Benneh, G., Hindar, K., Walloe, L., and Wijkman, A., Eds., Scandinavian University Press, Oslo, 1998, 115.

Flora, C. B., Sustainability of Human Communities in Prairie Grasslands, *Great Plains Res.,* 9, 397, 1999.

Flora, C. B., Flora, J. L., and Wade, K., Measuring success and empowerment, in *Community Visioning Programs: Practices and Principles,* Waltzer, N., Ed., Greenwood Press, Westport, CT, 1996, 57.

Flora, C. B., Goddard-Clark, K., Kinsley, M., Luther, V., Wall, M., Odell, S., Topolsky, J., and Ratner, S., *Working Toward Community Goals: Helping Communities Succeed,* North Central Regional Center for Rural Development, Ames, IA, 1997. Available: http://www.ncrcrd.iastate.edu/Community_Success/how.html. Accessed: Sept. 16, 2000

Folke, C. and Berkes, F., *Understanding the Dynamics of Ecosystem-Institution Linkages for Building Resilience,* Biejer Discussion Paper Series No. 112, Beijer International Institute of Ecological Economics, Stockholm, Sweden, 1998.

Harvey, D., *Consciousness and the Urban Experience: Studies in the History and Theory of Capitalist Urbanization,* John Hopkins University Press, Baltimore, 1985.

Lyson, T. A. and Barham, E., Civil society and agricultural sustainability, *Soc. Sci. Q.,* 79, 554, 1998.

Maarleveld, M. and Dangbégnon, C., Managing natural resources: a social learning perspective, *Agric. Hum. Values,* 16, 267, 1999.

Mead, G. H., *Mind, Self and Society from the Standpoint of a Social Behaviorist,* University of Chicago Press, Chicago, 1934.

Narayan, D., *Bonds and Bridges—Social Capital and Poverty,* Poverty Research Working Paper No. 2167, World Bank, Washington, D.C., 1999.

Paulson, D., Collaborative management of public rangeland in Wyoming: lessons in comanagement, *Prof. Geogr.,* 50, 301, 1998.

Pomeroy, C. and Beck, J., Experiment in fisheries comanagement: evidence from Big Creek, *Society Nat. Resour.,* 12, 719, 1999.

Pretty, J., *The Living Land: Agriculture, Food and Community Regeneration in Rural Europe,* Earthscan Publications, London, 1998.

Rose, R., *Getting Things Done in an Anti-Modern Society: Social Capital Networks in Russia,* Work Bank Social Capital Initiative Working Paper No. 6, 1998. http://www.inform.umd.edu/EdRes/Colleges/BSOS/Depts/IRIS/zzzmisc/soccap7.html. Accessed: Sept. 22, 2000.

Smith, J., Global civil society? transnational social movement organizations and social capital, *Ame. Behavi. Scientist,* 42, 93, 1998.

Tester, K., *Civil Society,* Routledge, New York, 1992.

Tolbert, C. M., Lyson, T. A., and Irwin, M. D., Local capitalism, civic engagement, and socioeconomic well-being, *Soc. Forces,* 77, 401, 1998.

World Commission on Environment and Development, *Our Common Future,* Oxford University Press, New York, 1987.

Zijderveld, A. C., *The Waning of the Welfare State: The End of Comprehensive State Succor,* Transaction Publishers, New Brunswick, NJ, 1999.

3 Altered Landscapes and Transformed Livelihoods: Banana Companies, Panama Disease, and Rural Communities on the North Coast of Honduras, 1880–1950

John Soluri

CONTENTS

Introduction 15
A New Livelihood on the Caribbean Littoral of Honduras 17
Monocultures, Plant Pathogens, and International Markets 20
Claiming Spaces, Making Places: The Municipality of Sonaguera, Colón 22
Altered Landscapes and Transformed Livelihoods 24
Conclusion 30

INTRODUCTION

In late May 1931, some 90 residents of Omoa, an export banana-growing community located on the Caribbean coast of Honduras, sent a petition to Honduran President Vicente Mejía Colindres expressing their anger and frustration over the United Fruit Company's decision to abandon the area:

> There are more than 500 laborers who are losing their daily work and along with it their ability to provide for their families. Many years of struggling, patient labor, perseverance, and cooperation with the company are going for naught simply due to an order, as if the labor of an entire community were not worth even the tiniest consideration.[1]

---

[1] Mayor Samuel E. García, et al., Omoa to President Mejía Colindres, Tegucigalpa, May 31, 1931, National Archive of Honduras (NAH) Folder, Correspondence from the province of Cortés.

0-8493-0917-4/01/$0.00+$.50
© 2001 by CRC Press LLC

15

The situation in Omoa was hardly unique. During the first half of the 20th century, U.S. banana companies operating in Central America shifted their production centers to such an extent that one observer likened the industry to "the shadow of a fleeting protean cloud passing over the lands of the Caribbean."[2] Unlike a shadow, however, the space-invading banana industry radically transformed both the landscapes and livelihoods of the regions through which it passed.

This chapter offers an agroecologic interpretation of the banana industry's uneven historical development along the Caribbean lowlands of Honduras and explores the cross-cutting effects of shifting production patterns on local livelihoods. The author argues that the export banana trade helped to create overlapping communities of interest and communities of place that were highly unstable because of their connection to a changing agroecosystem shaped by disease dynamics in monocultures, international markets, and the power of U.S. banana companies.

The author's interest in telling this story is twofold. The primary goal is to illustrate, via a historical case study, the relationships between changing tropical landscapes and livelihoods. Although the link between agroecological change and social transformations in rural areas may seem intuitive, the number of diachronic studies remains small, particularly in the context of Latin America.[3] Second, this chapter stresses the need to view communities and agroecosystems as ecosocial processes that are fluid, turbulent, and subject to change as they articulate and re-articulate themselves with other ecosystems, social structures, and discourses. Given that communities and agroecosystems are not static, they are best examined from the vantage point of particular places and times. This appeal for specificity is not merely academic. In a world of dynamic agroecosystems and fragmented social identities, policy makers must recognize that pathways to sustainable rural development are both contested and, paradoxically, shifting.

After sketching the historical and agroecologic contexts in which export banana growing has taken place in Honduras, the author focuses on three communities of place that were transformed by the banana trade during the late 19th and early 20th centuries. The community of interest on which the author primarily focuses is that of small-scale banana cultivators, a category that incorporates a considerable diversity of people. "Small-scale", of course, is a relative term. Some noncompany growers accumulated considerable amounts of wealth through their participation in the banana trade. Many more struggled to get by, often forging livelihoods that combined wage work on company plantations with small-scale production of crops for both

---

[2] Charles Kepner, *Social Aspects of the Banana Industry,* Vanguard Press, New York, 1936.

[3] In the North American context, important works include William Cronon, *Changes in the Land,* Hill and Wang, New York, 1983; Carolyn Merchant, *Ecological Revolutions,* Chapel Hill University of North Carolina Press, 1989; Richard White, *Land Use, Environment, and Social Change,* University of Washington Press, Seattle, 1980; and Donald Worster *Dust Bowl,* Oxford University Press, New York, 1979. For Latin America, see Elinor G. K. Melville, *A Plague of Sheep,* Cambridge University Press, New York, 1994; Douglas L. Murray, *Cultivating Crisis,* University of Texas Press, Austin, 1994; Michael Painter and William Durham, Eds., *The Social Causes of Environmental Destruction in Latin America,* University of Michigan Press, Ann Arbor, 1995; Robert G. Williams, *Export Agriculture and the Crisis in Central America,* University of North Carolina, Chapel Hill, 1986; and Angus Wright, *The Death of Ramón Gonzales: the Modern Agricultural Dilemma,* University of Texas Press, Austin, 1990.

export and local markets. Space limitations do not permit a full portrait of these worker-cultivators. Important questions related to gender, ethnicity, and kinship networks—questions central to community formation—go largely unexplored here. Their absence serves as a reminder to readers that what follows is an interpretation of a history that can never be definitively told.

A NEW LIVELIHOOD ON THE CARIBBEAN LITTORAL OF HONDURAS

The first export banana growers in Honduras were residents of the Bay Islands. Situated some 50 km off the Honduran coastline, the Bay Islands did not become subject to Honduran rule until an 1859 treaty definitively ended a century-long effort by Great Britain to control the small archipelago. By the mid-1860s, the Bay Islands' predominantly English-speaking inhabitants were selling bananas and coconuts to schooners bound for New Orleans. The fruit trade prospered until the 1880s, when a combination of changing government policies, declining yields, and competition from newly established mainland farms led to a sharp drop in production.

As the century came to a close, Bay Island exports had all but collapsed as shippers increasingly plied the waters of the Honduran coastline, purchasing bananas from growers in communities such as Omoa, Puerto Cortés, El Porvenir, La Ceiba, and Balfate. In 1899, a government survey recorded the presence of more than 1,000 export banana growers in seven mainland municipalities.[4] The vast majority of the production took place on a small scale; with 85% of the growers having 14 hectares of land or less in banana plantings, and more than 40% planting fewer than 4 hectares. All told, more than 10,000 hectares of banana plants yielded approximately 3 million stems of fruit. Both the amount of land in production and the exports continued to rise steadily as the first decade of the 20th century opened.

Bananas (*Musa* species) thrived in the region's warm, humid climate.[5] Between 1880 and 1910, mainland banana production took place primarily along the narrow coastal plain in close proximity to either navigable inland waterways or the National Railroad that ran from Puerto Cortés to a point south of San Pedro Sula. This spatial distribution can be explained largely by the exigencies of production and marketing. Export bananas prospered in the well-drained, nutrient-rich alluvial soils found on the banks of the numerous rivers that coursed through the region's lowlands. Once harvested, the highly perishable fruit, cut green, had to be transported rapidly yet

---

[4] Honduras, Junta Registradora, "Datos relativos a las fincas de bananos," July 1899. Manuscript. NAH. The author found this manuscript while working in the archive in 1995. Because of its deteriorated state, archive staff intended to remove the manuscript from the general collection. The author possesses a photocopy of the document.

[5] Average annual rainfall can vary considerably along the coast. Data collected near the port of Tela during the 1920s indicate that annual rainfall ranged from 127 to 172 inches. The heaviest rains generally fell between October and January before giving way to a dry season in March, April, and May. Average monthly temperatures varied little throughout the year (low 70s F in the rainy season to high 70s F in the dry season). Paul C. Standley, The Flora of Lancetilla, *Field Museum of Natural History: Botany,* 10, 8–49, 1931.

gently to a port where it would be sold to an exporter. Thus, before the building of extensive railroad networks, growers tended to concentrate near the coastline.

Fragmentary evidence from various banana-growing communities indicates that arable land was relatively abundant in the late 19th century, either in the form of *ejido* (communal) lands held by local municipalities or national lands whose distribution was controlled by the central government. Other factors, in addition to the relative availability of land, helped to make export banana cultivation a viable livelihood for growers with limited capital. Labor inputs, as described in the late 19th century, consisted of clearing forest growth (often accomplished through burning), planting, and a single cycle of light weeding. Harvesting the heavy, easily bruised banana stems from plants that stood 8 to 10 m high was by no means easy, but it was less arduous than the harvesting processes associated with other tropical export crops such as coffee and sugarcane. Finally, the self-propagating plants did not require annual reseeding and potentially yielded marketable fruit throughout the year, meaning that growers may have been able to avoid taking on large amounts of debt while awaiting the harvest.[6] Reports of plant pests and diseases are notably absent during the first decades (1880–1910) of mainland banana production in Honduras.

Export banana production was not risk free. Losses resulting from wind damage were considerable, and flooding or prolonged droughts periodically saddled growers with heavy losses. Tensions between growers and shippers surfaced during the trade's early years (before the formation of the United Fruit Company), because shippers frequently rejected harvested fruit that they judged to be damaged, undersized, or overripe. Having one's fruit rejected was a shared experience that led producers to form associations in order to gain greater leverage in their negotiations with shippers.[7] The formation and dissolution of grower organizations suggest that the early years of the banana trade were no golden age in terms of stability or equity. On the other hand, a growing market for fruit in the United States, combined with the availability of land and low start-up costs made banana growing an attractive livelihood in an economy that offered few other possibilities for accumulating capital.

However, economic structures and biophysical conditions cannot entirely explain the emergence of the export banana industry in Honduras. The late 19th century witnessed the conjuncture of expanding consumer markets in the United States and the rise of liberal governments in Central America that embraced direct foreign investments and export markets as crucial for the modernization of the region's economies. For example, Honduran President Marco Aurelio Soto, during his tenure in office (1876–1883), initiated a number of projects designed to build infrastructure and promote foreign investment in mining and agriculture. However, before resources could be transferred to individual or corporate entities, they had to be "nationalized." In other words, the state needed to establish its authority over

---

[6] Other export crops such as coffee, coconuts, and cacao required greater initial capital investments and did not yield marketable fruits for several years.

[7] John Soluri, *Landscape and Livelihood: an Agroecological History of Export Banana Growing in Honduras. 1870–1975*, Ph.D. dissertation, University of Michigan, 1998, Ch. 1; and Charles A. Brand, *The Background of Capitalistic Underdevelopment: Honduras to 1913*, Ph.D. dissertation, University of Pittsburgh, 1972.

regions such as the Caribbean lowlands that were both geographically and culturally distant from the Tegucigalpa-based liberal elites. Honduran political leaders portrayed the humid tropical forests as "green deserts" awaiting the light of civilization. Such a view marginalized several distinct ethnic groups (e.g., Garifuna, Jicaque, and Sumo) that inhabited the region. These peoples, who tended to shun sedentary agriculture as a way of life, were increasingly marginalized in the plans of Honduran liberals, who envisioned a nation of culturally homogeneous mestizo farmers.[8] Thus, the rise of banana production on the Honduran mainland during the late 19th century is best understood in the context of expanding export markets and the growing power of liberal state institutions and ideologies.

The arrival of heavily capitalized U.S. fruit companies in Honduras during the first decades of the 20th century dramatically accelerated the growth of the banana industry. North American investors received a number of extremely generous concessions granted by Honduran leaders who hoped to stimulate railroad-building projects deemed vital to facilitate trade and communication between the Caribbean coast and Tegucigalpa. The concessions provided the legal means by which the fruit companies laid claim to and exploited the region's resources. In return for the construction of railways, bridges, piers, and telegraph lines, the fruit companies were ceded the rights to vast amounts of the nation's soil, timber, water, and mineral resources, along with numerous tax and duty exemptions and the right to import laborers.[9]

Expressing its official approval for a 1912 concession granted to a subsidiary of the United Fruit Company, the Honduran Congress declared that "the more railroads that exist, the more we will cultivate and export."[10] If bananas were what the congressional proponents of railroad building had in mind, they must have been satisfied with the results. During the years 1914–1918, U.S. imports of Honduran fruit increased from 8.4 million to 11.1 million stems.[11] Export totals nearly tripled over the next ten years. In 1929, a record 29 million stems left Honduran shores, making the country the world's leading exporter of bananas. Most of this increased production was from fruit company plantations, but noncompany growers continued to supply between one fourth and one third of total banana exports from Honduras between 1921 and 1940.

---

[8] On the relationship between liberals and national identity in Honduras, see Darío Euraque, The Banana Enclave, Nationalism and Mestizaje in Honduras, 1910s–1930s, in Aviva Chomsky and Aldo Lauria-Santiago, Eds., *Identity and Struggle at the Margins of the Nation-State,* Duke University Press, Durham, 1998, 151–168.

[9] Mario Argueta, *Bananos y Política: Samuel Zemurray y la Cuyamel Fruit Company,* Editorial Universitaria, Tegucigalpa, 1989; Enrique Flores Valeriano, *La Explotación Bananera en Honduras,* Editorial Universitaria, Tegucigalpa, 1987; and Antonio Murga Frassinetti, *Enclave y sociedad,* Editorial Universitaria, Tegucigalpa, 1978.

[10] Honduras, *La Gaceta* no. 3998, July 29, 1912.

[11] C. F. Marbut and Hugh H. Bennett, "Informe de los terrenos y la agricultura de la región cubierta por el estudio económico de la expedición Guatemalteco-Hondureña," *Estudio Economico de la zona fronteriza entre Guatemala y Honduras practicado durante los meses de mayo y junio de 1919 bajo la superintendencia de la Sociedad Geográfica Americana, para el departamento de Estado de los Estados Unidos de America* (trans., J. E. M. Alonso) NAH loose mimeograph.

A sharp increase in population accompanied the dramatic expansion of banana production. The population of the principal banana-growing departments tripled between 1910 and 1935, climbing from 65,000 to nearly 200,000 inhabitants.[12] Women and men migrated to the North Coast from the Honduran highlands, El Salvador, and Jamaica. Smaller numbers came from other Latin American and Caribbean countries, the Middle East, and the United States. These immigrants helped to transform port cities such as La Ceiba, Tela, and Puerto Cortés into multilingual, culturally diverse places.

Many people also formed communities in banana company work camps and the numerous villages that surrounded the plantations.[13] During the first half of the 20th century, the banana companies hired field laborers via a decentralized system of contractors and subcontractors. Contracts tended to be of short duration, and laborers could be summarily fired by *mandadores* (field bosses). Field hands, for their part, often expressed their discontent with working conditions by quitting and moving to another plantation.

One result of this system of labor organization was an ongoing internal migration on the North Coast. Thus, the populations in banana zone communities were extremely transitory. Beginning in the early 1920s, these patterns of migration were reinforced by the fruit companies' abandonment of lands in reaction to the arrival of a plant pathogen that infected export banana plantations.

MONOCULTURES, PLANT PATHOGENS, AND INTERNATIONAL MARKETS

Humans were not the only life forms drawn to the expanding banana plantations. Panama disease (*Fusarium oxysporum* f. *cubense*) may have reached the export banana zones of Honduras as early as 1910. The soil-borne fungus invaded the roots of susceptible plants, passed through the rhizome and vascular system of the host plant, and eventually reached the leaf petioles before killing the plant. Spores released from decaying plant tissues germinated on coming into contact with the roots of neighboring plants, thereby spreading the disease in a radial pattern. Unfortunately for banana growers, the variety on which the U.S. market had been created, the Gros Michel, was highly susceptible to the pathogen.[14]

Panama disease epidemics in Central America and the Caribbean coincided with the rapid expansion of the export banana industry and were linked to agroecological changes driven by local and international forces. The growth of banana plantations, railroad lines, and human settlements was accompanied by forest fragmentation, the

---

[12] Honduras, Dirección General de Estadísticas y Censos, *Honduras en cifras,* Tegucigalpa, 1965.

[13] The lives of workers on the North Coast of Honduras have been best captured by a number of Honduran novelists including Ramón Amaya Amador, *Prisión Verde,* Editorial Universitaria, Tegucigalpa, 1990; Marcos Carías Reyes, *Trópico,* Editorial Universitaria, Tegucigalpa, 1990; and Pacas Navas de Miralda, *Barro,* Editorial Guaymuras, Tegucigalpa 1992 [1951].

[14] For a comprehensive review of scientific understandings of Panama disease through the early 1960s, see Robert H. Stover, Fusarial Wilt (Panama Disease) of the Banana and other Musa Species, *Phytopathological Paper* no. 4, Commonwealth Mycological Institute, 1962.

loss of wetlands, and altered hydrologic systems. When workers cleared lowland forests and replaced them with banana plants, they transformed agroecosystems characterized by an extremely high diversity of flora and low population densities of individual species into high-density monocultures of extremely limited inter- and intraspecific diversity. The result was a new agroecosystem highly conducive to the rapid spread of plant pathogens. The dense plantings of Gros Michel clones presented few barriers, physical or genetic, capable of slowing the spread of the soil fungus.[15] Large-scale transformations of the landscape, including extensive deforestation, draining of wetlands, installation of irrigation systems, and building of railroad networks that transported people, animals, machines, and bananas from one plantation to another, facilitated the pathogen's spread between banana-growing districts.

Market structures and the aesthetic sensibilities of retailers and consumers in the United States indirectly contributed to the spread of Panama disease. The U.S. market favored the Gros Michel because of its bruise-resisting peel, symmetric bunch shape, consistent ripening color, and pleasing flavor. Throughout the 1920s and beyond, both fruit companies and small holders attempted to breed disease-resistant varieties that also possessed the aesthetic qualities of the Gros Michel. These initial efforts ended in failure, in part because of the banana plant's physiology. Varieties of edible *Musa* lack seeds in the fruit pulp, a characteristic that made it difficult to crossbreed the Gros Michel. Early breeding projects were tedious and yielded unpredictable results.[16]

Biologic stumbling blocks were not the only challenge involved in breeding a new export banana. Wholesalers and retailers in distant marketplaces were reluctant to adopt a variety whose shape, flavor, and color did not meet their very particular notions about what constituted a banana. The myriad varieties of *Musa* found in tropical regions did not exist in the minds of most U.S. wholesalers, retailers, and consumers, whose image of the banana was restricted to the Gros Michel. As one United Fruit Company scientist assessing the set of hybrids yielded from a 1929 breeding project wrote: "In no case is their quality equal to the fruits that are generally recognized by the public as "bananas."[17] Thus, Panama disease was not a "natural" phenomenon, but rather the product of an agroecosystem designed to maximize production of a very particular kind of banana.

Reluctant to replace the prized Gros Michel with another variety, the highly capitalized and politically powerful U.S. fruit companies adopted a strategy of shifting plantation agriculture, in which individual farms and entire regions were abandoned in favor of soils that had not been intensively cultivated with the Gros Michel. When production declined in a particular area, the fruit companies asked for, and generally

---

[15] Fungal diseases capable of multiple cycles of infection in a single cropping season frequently reveal a positive correlation between plant density and disease incidence. See Christopher C. Mundt, Disease dynamics in agroecosystems, in Carroll, C. Ron, John Vandermeer, and Peter Rosset Eds., *Agroecology*, McGraw-Hill, New York, 1990, 277.

[16] For a more detailed discussion of fruit company breeding projects, see Soluri, "Landscape and Livelihood" Ph.D. diss., Univ. of Michigan, 1998. Chapter 3.

[17] J. H. Perman, Banana Breeding, *United Fruit Company Research Department Bulletin 21*, October 14, 1929, p. 1.

received, amendments to their concessions that allowed them to reroute their tracks through new lands. This practice was facilitated by natural resource subsidies given to the companies that enabled them to externalize a significant portion of the costs incurred by relocation. By 1940, the fruit companies had all but ceased growing export bananas along the Caribbean littoral in favor of soils located inland along the Sula and Aguán valleys.

CLAIMING SPACES, MAKING PLACES: THE MUNICIPALITY OF SONAGUERA, COLÓN

The shifting banana industry gave rise to communities of place throughout the coastal lowlands. Indeed the "North Coast" referred less to a fixed geographic region than to a dynamic social space shaped by export banana growing. For example, as late as 1907, people residing in Sonaguera did not consider their community to be a part of the North Coast.[18] At that time, the founders of Standard Fruit, the Vacarro brothers, were directing railroad building to the west of La Ceiba. However, by 1919, soil exhaustion and Panama disease compelled Standard Fruit to abandon many of its original farms and prompted the company's engineers to redirect their gaze eastward toward Sonaguera and the Aguán valley beyond. Almost simultaneously, the Truxillo Railroad Company, a United Fruit subsidiary, began constructing a railroad toward Sonaguera from the east.

Livestock raising was the principal livelihood in Sonaguera throughout much of its recorded history. Ranchers continued to enjoy a privileged status as late as 1923, when the municipal council declared all of the community's ejidal lands to be "mixed zones," in which livestock could range freely. Farmers were required to fence their fields.[19] However, 1 year later, local cultivators began to challenge the power of the ranchers. Inés Lanza presented a petition on behalf of 100 residents, calling for the establishment of an agricultural zone in the ejidal lands of Sonaguera.[20] The petition argued that, in anticipation of the "foreign companies" arrival, agricultural expansion should be vigorously supported. The problem, the petitioners explained, was that they lacked the necessary funds to erect fences capable of restricting the growing number of free-ranging livestock.

In 1925, Sonaguera Mayor R. Martínez reported that the Standard Fruit Company had lodged complaints about damages inflicted on the company's new plantations by cattle belonging to residents of Sonaguera.[21] The municipality responded by establishing an agricultural belt around ejido lands, which retained their designation as a mixed zone.[22] Cattle ranchers were given 3 months to corral

---

[18] In 1907, the Sonagueran municipal council gave its support to a railroad project in the hope that the railroad would put an end to the "constant emigration of our sons to the *costa norte* where banana cultivation is flourishing." Sonaguera Municipal Acts, June 30, 1907.

[19] Sonaguera Municipal Acts, March 15, 1923.

[20] Sonaguera Municipal Acts, November 15, 1924.

[21] Sonaguera Municipal Acts, September 21, 1925.

[22] Sonaguera Municipal Acts, September 21, 1925.

their cattle and prevent further grazing on agricultural lands. One year later, some 40 livestock owners requested permission to erect a barb-wire fence around the entire mixed zone to avoid damaging the farms of the "foreign companies" whose operations now encircled municipal grazing lands.[23] The request was granted, but in subsequent years, ranchers continued to complain about having to pay "severe fines" because their cattle entered the farms of the Truxillo Railroad and Standard Fruit Company.[24] Contending that they lacked the financial resources to fence their pastures, ranchers called on the national government to oblige the fruit companies to fence plantations that shared a border with Sonaguera's ejidal lands.[25] But the assault on free grazing did not stop. In 1930, a group of Sonaguerans called for the rezoning of ejidal lands from mixed to agricultural, citing the familiar problem of crop damages caused by wandering livestock.[26] The municipal council agreed to redesignate all of the ejidal lands as an agricultural zone. Ranchers were given time to fence their pasturelands. Two weeks after this petition, the provincial governor approved the new zoning scheme for Sonaguera.[27]

Therefore, less than 10 years after declaring that grazing would not be restricted on ejidal lands, Sonaguera's municipal council reversed its zoning policies. All pastureland was to be fenced at the ranchers' expense. The zoning battles that transpired during the 1920s reflected the transformations taking place in the local Sonagueran political economy. Anticipating the economic opportunities that would accompany the arrival of the fruit companies and railroads, some residents tried to create a landscape conducive to banana cultivation. However, area cattle ranchers succeeded in beating back zoning changes until the Standard Fruit Company began establishing plantations in the region. The presence of the powerful company, whose banana plants fared no better against marauding cattle than those of local growers, provided the political leverage needed by Sonagueran cultivators to advance their interests. To judge by the history of Sonaguera, a dynamic interplay between international capital and local actors shaped or reshaped communities along the North Coast in complex ways that at times gave rise to a new set of regional elites.[28]

By 1930, some 70% of the bananas exported by the Standard Fruit Company came from Sonaguera. A place once considered to be on the periphery of the North Coast had become one of the region's key production centers. But the prosperity

---

[23] Sonaguera Municipal Acts, March 1, 1926.

[24] Nicolas M. Robles, Petrona Ocampo, et al., Sonaguera to Minister of Development, Tegucigalpa, February 15, 1927; NAH leg. Notas varias, 1927.

[25] Minister of Development, Tegucigalpa, to Nicolás Robles, Petrona Ocampo, etc., February 17, 1927; NAH leg. Notas varias, 1927.

[26] Sonaguera Municipal Acts, February 1, 1930.

[27] Sonaguera Municipal Acts, March 1, 1930.

[28] In his classic 1969 study, *Interpretación del desarrollo centroamericano*, published in 1969, Edelberto Torres Rivas stressed the role of national elites in the formation of the Central American banana industry. However, he tended to de-emphasize the ways in which the industry helped to give rise to local and regional elites. Recently, a Honduran historian has argued that the banana industry gave rise to a new entrepreneurial class in San Pedro Sula that controlled local and regional politics throughout much of the 20th century. See, Darío Euraque, *Reinterpreting the Banana Republic: Region and State in Honduras, 1870–1972*, University of North Carolina Press, Chapel Hill, 1996.

brought by the railroads was fleeting because of the fruit companies' practice of shifting plantation agriculture. The dollars that flowed into Sonaguera flowed out of other communities in the wake of abandonments. In towns all along the Caribbean coastline, residents were left with the vexing task of creating new livelihoods in altered landscapes.

ALTERED LANDSCAPES AND TRANSFORMED LIVELIHOODS

The fruit companies' response to Panama disease, abandoning infected soils in favor of new ones, was not an option for the majority of *poquiteros,* or small-scale growers, who lacked the capital resources and government concessions that made relocation economically feasible. Some growers tried to plant disease-free Gros Michel rootstock, whereas others experimented with disease-resistant varieties. In 1927, the Atlántida Chamber of Commerce, "clearly convinced that the fall of commerce . . . is rooted in the destruction of the banana farms," proposed importing planting stock from Colombia and Mexico that was believed to possess "greater resistance than does the worn-out planting stock available here."[29] At about this same time, the Standard Fruit Company initiated a program to supply growers with disease-free planting materials. However, these initial attempts to replant infected soils were unsuccessful.[30]

Poquiteros also experimented with disease-resistant varieties such as Lacatan. Defending a group of squatters accused of occupying Standard Fruit Company lands near La Masica, Jacobo P. Munguía admitted that the occupied lands belonged to the company, explaining that the squatters wanted to collaborate, not fight, with Standard: "They want to grow that resistant banana, and if the company finds a market for it, they will sell their output to the company with pleasure."[31] Munguía presented a petition addressed to the President of Honduras bearing the names of 108 individuals who wanted to "work independently" and sought permission to plant 7 hectares each of Lacatan banana plants. However, the squatters' hopes of selling their output were never realized because the variety was not widely accepted by U.S. markets.[32] As a result, the fruit companies grew and purchased bananas only in areas capable of supporting Gros Michel production. Elsewhere, communities were compelled to find new livelihoods after the companies moved on.

---

[29] Ernesto Crespo, Secretary and A. Lopez Villa, President, Atlántida Chamber of Commerce, La Ceiba, to the Minister of Development, August 18, 1927; NAH.

[30] To further stimulate small-scale production, Standard Fruit Company offered higher purchase prices for 1929. See N. R. Park, "Review of Commerce and Industries in La Ceiba for the quarter ended December, 31; 1928," January 19, 1929 United States National Archive U.S. Foreign Agricultural Service, "Narrative Reports, Honduras, 1904–1939," Entry 5, Box 343, Folder, "Fruits."

[31] Jacobo P. Munguía, Esparta, to President Miguel Paz Barahona, Tegucigalpa, May 16, 1927; NAH leg. Correspondencia particular, 1921.

[32] On problems with marketing the Lacatan variety, see *Revista del archivo y de la biblioteca nacional de Honduras*, No. XII (June 1931) p. 434; and Sr. Ordóñez P., La Ceiba to Assistant Secretary of Development, Public Works, Agriculture and Labor, Tegucigalpa, July 3, 1926; Archivo de la Gobernación de Atlántida leg. Libro copiador de cartas 1926.

One of the first places to be abandoned entirely by a major fruit company was the municipality of Omoa. Bananas had been the community's primary export since the late 19th century. At least 140 persons cultivated approximately 1,200 hectares of bananas in 1899.[33] Twelve years later, U.S. citizen Samuel Zemurray took over a railroad concession from another U.S. investor and formed the Cuyamel Fruit Company. Zemurray directed the construction of a railroad from Omoa toward the Guatemalan border. As was the case elsewhere on the North Coast, railroad expansion was accompanied by an increase in export banana production. By 1919, the Cuyamel had felled "large areas" of forest to establish plantations. A "considerable number" of noncompany growers had cleared additional areas reportedly ranging in size from "a few acres to a few thousand acres."[34] All told, Omoa had at least 6,900 hectares of banana farms in 1920.

However, as early as 1925, reports from both Honduran and U.S. government officials noted that Panama disease had reduced banana output in the Omoa region "to an almost negligible quantity."[35] By 1931, the banana plants that once covered the landscape were "seldom found," and secondary growth (*guamiles*) began to replace cultivated fields.[36] Meanwhile, Zemurray had been investing heavily in banana and sugar cane production in the Sula valley, building railroads and establishing plantations along both the Ulúa and Chamelecón rivers. In 1929, Zemurray sold his company to the United Fruit Company. Two years later, company officials announced that they were abandoning their Omoa plantations.

When area residents received this news, they held an open meeting, at which they resolved to ask the national government to intervene to prevent the "death of the only activity that provides a livelihood for the people." Mayor Samuel García pointed out that the suspension of banana growing would likely put an end to local railroad traffic. Area fruit growers admitted that the region no longer produced "what it had in the past," but, defending their fruit as being as good as that cultivated elsewhere in Cortés, they hoped that the Cuyamel would continue to buy their bananas.[37]

In 1932, the company began dismantling its railroad. Area growers made desperate appeals for help in finding a means to transport their produce.[38] For example, Orellano Rodriguez complained that he stood to lose about $1,500.[39] Rodriguez, along with 25 other growers, had new banana plantings located along a portion of the

---

[33] Honduras, Junta Registradora, "Datos relativos a las fincas del banano," 1899.

[34] C. F. Marbut and Hugh H. Bennett, "Informe de los terrenos y la agricultura de la región cubierta por el estudio económico de la expedición Guatemalteco-Hondureña," HNA, Tegucigalpa p. 159.

[35] U.S. Consul at Puerto Cortés, Ray Fox, February 9, 1927, "Excerpt from review of commerce and industry for the year 1926," p. 6. United States Foreign Agricultural Service, "Narrative Reports, 1904–1939," Entry 5, Box 343, Folder, "Fruits." Also see J. Antonio Ynestroza, San Pedro Sula, to Minister of Development, March 10, 1925, NAH leg. Correspondencia telegráfica-Cortés, 1925.

[36] Alonzo Valenzuela, "Informe de la inspección de Omoa y Cuyamel," July 29, 1933; NAH leg. Ministro de Fomento, Informes a varias secciones, 1931–1932.

[37] The letter made no reference to Panama disease.

[38] See editions of *El Pueblo*, May 2, 1932, p. 1; and May 7, 1932, p. 1.

[39] Orellano Rodriguez apparently was a grower of some means. In 1926, he cultivated more than 28 hectares of national land in bananas and forage grass. NAH leg. Datos estádisticos del departamento de Cortés, 1926.

Cuyamel's railroad. They feared that with the removal of the line, their investments would be lost. The national government responded to Rodriguez's plea for help by explaining that the same concessions that gave the fruit companies access to the North Coast's resource base also gave them the liberty to remove infrastructure.

In 1933–1934, of the 6 million stems of fruit harvested in the province of Cortés, less than 30,000 were grown in Omoa. A government official who inspected the community described it as "desolate and dead."[40] He added, however, that the "greater part" of Omoa's 4,000 inhabitants continued to live in the area. Some found work dismantling railroads and buildings for the fruit company. Others cultivated grains and raised livestock on a small scale. Trading and communication were carried out using canoes and mules. The government inspector predicted that there would be insufficient amounts of cargo and passenger traffic to sustain any kind of rail service. However, some residents indicated that they had never viewed the company bridges and railroads as a necessity. Another member of the survey team noted that the removal of an iron bridge over the Cuyamel River would not disrupt local livelihoods, because it had exclusively served railcars.[41] The official reasoned that a government-provided truck could cross the river during the dry season, and that a wooden bridge could be built at little cost to facilitate crossing during the rainy part of the year. Area residents reportedly were "content" with the idea and willing to accept the bridge's removal.

The reported indifference of some local residents toward the removal of transportation infrastructure contradicted the opinions of banana growers such as Orellano Rodriguez who viewed the railroad as essential for their economic survival. The inspectors may have downplayed the importance of rail service to justify their recommendations that the government forego a costly overhaul of the railway in favor of less expensive alternatives.[42] On the other hand, the report may have reflected the extent to which company railroads served the particular needs of a community of interest (i.e., banana growers) more than the general welfare of a community of place.

If the Cuyamel Fruit Company could remove its railroad, it could not do the same with the area's soil and water resources. For both unemployed workers and ex-banana growers, farming abandoned lands was a viable option. The company leased parcels of land to individuals living in the area, a policy that United Fruit Company maintained after its buyout of the Cuyamel.[43] When a rumor circulated in 1933 that the national government was going to acquire the company properties on which they farmed, the inhabitants of Cuyamel wrote to President Tiburcio Carías, requesting

---

[40] Alonso Valenzuela, to Minister of Development, "Report of an inspection of the Omoa and Cuyamel region," July 29, 1933; NAH leg., Ministro de Fomento, Informes a varias secciones, 1931–1932.

[41] Pascual Torres, San Pedro Sula, to Minister of Government, Abraham Williams, Tegucigalpa, July 24, 1933; NAH leg. Correspondencia de las gobernaciones políticas July–September, 1933.

[42] Valenzuela estimated that restoring rail service between Omoa and Cuyamel would cost more than $100,000. He proposed an alternative transportation system that combined marine travel, tramlines, and roads. For details, see Alonso Valenzuela, to Minister of Development, July 29, 1933; NAH leg. Ministro de Fomento, Informes, 1931–1932.

[43] In 1933, the company leased land at the rate of $2 for 10 hectares. Cruz Calix, President of the Cuyamel Development Committee, to President Tiburcio Carías, July 24, 1933; NAH leg. Secretaria de Fomento, Agricultura y Trabajo, Correspondencia de juntas de fomento.

that the government relinquish lands and houses to the residents who had been using them.[44] At that time, former banana lands were supporting annual food crops in addition to fruit trees, plantains, pineapple, and sugar cane.[45] The following year, the Cuyamel Development Committee was created to manage the buildings and properties abandoned by the company.[46]

The Development Committee soon came under criticism for its disorganization and corruption. A government official reported in 1937 that most of the buildings left by the company were badly deteriorated. The lone truck was missing several parts, including tires, and former company lands had become the "spoils" of the Development Committee.[47] The same official also accused the committee of tax evasion and fraud for its failure to pay for the use of national lands and timber resources.[48] In addition, former company houses had been dismantled and sold for lumber. Others were simply given away. One individual was even accused of selling furniture and other objects left by the Cuyamel Fruit Company.

Officials in abandoned banana zones located throughout the North Coast reported similar conditions of acute economic decline. In the La Ceiba area, a "majority" of the former banana growers established small neighborhood stores (*truchas*) that provided "scant earnings" but were considered to be less risky than agriculture.[49] Some ex-cultivators abandoned their farms and migrated out of the region.[50] Still others converted their fields from bananas to crops for local use including corn, beans, rice, plantains and pineapples. The situation varied little in the province of Colón after the Truxillo Railroad Company abandoned several thousand hectares of banana farms and subsequently removed extensive sections of its rail system.[51] As the company completed its pullout from the region in 1941, Governor Carlos Pineda wrote that the department suffered from "complete economic prostration."[52] A "great part" of the region's unemployed inhabitants had left the area, and "every week" families were leaving Trujillo for other parts of Honduras. The decline in shipping led to a collapse of commerce. Several of Trujillo's commercial establishments closed, and the

---

[44] Calix expressed concern that if the government repossessed the lands, wealthy parties would wrest control of them, presumably via public auctions.

[45] United Fruit's subsidiary, the Tela Railroad Company, also adopted a policy of leasing abandoned lands at about this same time. In 1935, Honduran families reportedly were growing rice, corn, beans, vegetables, plantains, and other crops. See Cornelio Mejía, La Ceiba, "Informe de la Gobernador Política del departamento de Atlántida, 1934–1935" n.d., p. 29; NAH leg. Informe de la Gobernador, Atlántida, 1934–1935.

[46] Manuel Paniagua, Cuyamel to Minister of Development, December 10, 1934; NAH leg. Alcaldías municipales, 1934.

[47] Castañeda, San Pedro Sula, to Minister of Government, Justice, and Social Services, September 20, 1937; NAH leg. Correspondencia de las gobernaciones políticas, September–October, 1937.

[48] See correspondence between Castañeda and the Minister of Government, October 15, 1937; and November 16, 1937; NAH leg. Correspondencia de las gobernaciones políticas, September–October 1937.

[49] Ordóñez described local commerce as "paralyzed" on account of there being "as many retailers as consumers." Ordóñez to Assistant Secretary of Development July 3, 1926.

[50] Honduras, *Revista del archivo y de la biblioteca nacional de Honduras*, n. 12, 1931, p. 434.

[51] A. Miralda, Governor of Atlántida, "Certificación: Asunto de Mesapa-Tela Rail-road Co," August 30, 1931; NAH leg. Correspondencia del departamento de Atlántida, 1931.

[52] Carlos L. Pineda, Governor of Colón, Trujillo, to Minister of Development, Tegucigalpa, January 28, 1941; NAH leg. Notas varias, 1932.

general level of poverty was "alarming." Individuals who had turned to farming encountered difficulties marketing their products because of reduced regional demand and high freight and shipping fees.

These fragmentary portraits indicate that the primary effects of abandonment were twofold. First, local and regional economies all but collapsed as a result of massive layoffs, out-migrations, drying up of government tax revenues, and a slowdown in commercial activity. Second, efforts to reinvigorate local economies often confronted serious obstacles with transporting goods to regional markets because of the fruit companies, removal of railroad lines or cutbacks in the number of trains servicing abandoned areas.[53] For example, in the Omoa area, former employees of the Cuyamel Fruit Company gained access to land for a nominal sum, but the loss of train service rendered export banana production all but impossible and compelled growers to plant a range of crops for household use and sale on local markets.

On some occasions, attempts of the fruit companies to remove transportation infrastructure encountered opposition from farmers who, much like the growers in Omoa, feared being left in economic isolation. In 1931, the residents of Mezapa, a village situated in the municipality of Tela, began "creating difficulties" for workers attempting to remove 2 km of a branch line operated by the Tela Railroad Company. Both the mayor of Tela and Atlántida Governor Adolfo Miralda arrived in Mezapa on August 23, 1931, where they met with more than 60 members of the community. Miralda read an official statement that reaffirmed both the Tela Railroad Company's right to remove its branch lines and the government's resolve "to protect the rights of the company."[54] The Governor then acknowledged the railroad's importance to the village, but explained that he could not compel the company to leave the line intact.

The villagers in turn expressed their desire that a series of bridges that spanned several local creeks be maintained in good repair to facilitate the traffic of people and animals during the rainy season. They also called for the rebuilding of a damaged bridge that spanned the Naranjo River. Mezapa residents claimed that before the arrival of the company, the Naranjo River generally "was dry" and "crossed with great ease." However, the Tela Railroad Company had channeled the river and created a network of drainage ditches that combined the flows of several other creeks into the Naranjo. As a result, the river was described by locals in 1931 as "very deep and dangerous," particularly during the rainy season when the swollen waterway carried trees and other debris downstream. Finally, noting that the Mezapa River supplied the drinking water for some company labor camps located in the vicinity, the villagers requested that, as "an act of justice," four water spigots be installed for the community's use.

The following day, the mayor of Tela reported that the villagers had agreed to permit the removal of the railroad in return for a promise that the company would leave the bridges intact and administer repairs as needed.[55] The Governor also

---

[53] For an instance of a fruit company cutting back on train service, see Pineda to Minister of Development, Tegucigalpa, January 28, 1941.

[54] Miralda, "Certificación: Asunto de Mesapa-Tela Rail-road Co," August 30, 1931.

[55] M. Orellano, Tela, to Minister of Government, Tegucigalpa, August 24, 1931; NAH leg. Correspondencia Telegráfica, Atlántida, 1931.

promised to engage the Tela Railroad Company in talks regarding the installation of a drinking water system for the community and the construction of a bridge over the Naranjo River.[56] Less than 1 week after the meeting, Governor Miralda received a message from the assistant mayor of Mezapa reporting that he had ordered the company's workers to stop taking up the rails. The local official justified his action by stating that the company had failed to complete "the construction of the Naranjo River Bridge."[57] Miralda's reply to Torres was firm: the village could not demand a new bridge because one already was in place. The company, Miralda added, had agreed to maintain the existing bridge, but not to build a new one.

However, the tone of Miralda's letter to his Tegucigalpa-based superior was very different. Miralda explained that many of Mezapa's 400 inhabitants made a living by selling food and other products to plantation workers in the nearby municipality of El Progreso. Between Mezapa and these markets lay the Naranjo River and numerous creeks that during the wet season could be crossed only via bridges. Therefore, the bridges were central to the community's economic well-being. Miralda requested that the national government oblige the company to rebuild the bridge over the Naranjo River, noting that "the villagers have justice on their side because the problem has resulted from the channeling [of waterways] that the company has done in that jurisdiction."

Unfortunately, the historical record does not indicate whether the bridge was rebuilt. Nevertheless, the case of Mezapa shows the dynamic intersection between agroecologic change and the North Coast communities. The Tela Railroad Company's arrival in the region created new ways to earn a living while transforming the landscape, including the local waterways. When Panama disease reached the Mezapa area and reduced the profitability of growing bananas, the company pulled out, removing the infrastructure that it had placed there. However, the Naranjo River and the area's drainage basin remained altered, prompting the villagers to impede the removal of the railroad in order to ensure that the company would maintain a series of local bridges essential to the twice-transformed local economy. Mezapa residents did not need the railroad to get their products to market, but they did need a bridge capable of withstanding the increased flow of the channelized river.

The Mezapa incident underscores the complexities of the historical relationship among local agroecologic change, rural communities, and international capital. The bridge over the Naranjo River can be seen as a symbol of the fruit company's engineering wizardry and the benefits it brought to the residents of Mezapa by providing infrastructure that created access to both regional and international markets. The company's production practices, however, transformed both the region's soil and water resources. The changes were manifested in two distinct, but historically linked, ways: seasonal flooding of the Naranjo River and a decline in banana production because of Panama disease. These new dynamics then triggered another series of linked social processes that included the Tela Railroad Company's abandonment of the area, Mezapa residents' search for new livelihoods, and, subsequently, new his-

---

[56] These were the terms of the agreement as reported by Miralda.

[57] Ciriaco Torres, Sta. Rosa del Norte (Mezapa), to Governor of Atlántida, La Ceiba, August 28, 1931 transcribed in A. Miralda, La Ceiba to Ministry of Government, Tegucigalpa, August 30, 1931; NAH leg. Correspondencia recibida de las gobernaciones políticas, 1931.

torical meanings for local bridges. Viewed in this context, the bridge, trembling as it gets pounded by water-driven tree trunks, represents the instability of complex agroecosystems shaped by historical processes operating at local and international levels.

CONCLUSION

Historical processes operating at local and international scales have shaped the interactions between rural communities and agroecosystems on the North Coast of Honduras. Tensions between growers and shippers surfaced from the earliest days of the trade. These tensions resulted from risks associated with the manner in which a tropical agroecosystem articulated with capitalistic economies. The banana was both the product of a dynamic agroecosystem and a commodity bought and sold on volatile markets. Trading bananas for profit required the "rationalization" of both production and marketing. This led to an all-too-familiar set of ecosocial transformations: production shifted from small-scale units to large-scale plantations owned by heavily capitalized U.S. banana companies. Definitions of "quality" became increasingly rigid, compelling growers to adopt relatively uniform management practices. The simultaneous expansion and vertical integration of banana production accelerated both the rate and scale of agroecologic changes. By the mid-1920s, banana farms covered some 50,000 to 75,000 hectares of the region's lowlands, and railroads stretched from the Guatemalan border to Mosquitia.

Export banana production created livelihoods for thousands of people who migrated to the North Coast from places throughout Central America and the Caribbean. These immigrants formed communities characterized by their cultural diversity.[58] Working peoples represented a majority in these rural communities, but the vertical integration of the banana industry did not result in a fully proletarianized rural workforce. In places such as Sonaguera, the expansion of fruit company railroads actually helped to give rise to a community of small-scale banana growers. Thus, communities of place and communities of interest often, but not always, overlapped.[59]

North Coast banana production varied over both time and space. This pattern of uneven development resulted from the interaction of local agroecologic changes and international markets. Large-scale deforestation, the building of extensive railroad networks, and the channelization of surface waters created an agroecosystem highly conducive to plant disease epidemics. Therefore, when Panama disease invaded the North Coast in 1910s, it followed the wide ecological swath cut by the rapidly expanding banana industry.

---

[58] This was particularly true of the port cities of Puerto Cortés, Tela, La Ceiba and Trujillo. These communities had considerable populations of Garifuna, West Indians, and North Americans in addition to Spanish-speaking immigrants from Central America.

[59] Studies carried out in other banana growing areas suggest that "communities" of working peoples were divided in multiple ways. See for example, Phillipe Bourgois, *Ethnicity at Work: Divided Labor on a Central American Banana Plantation*, Johns Hopkins University Press, Baltimore, 1989.

The transformation of regional landscapes, however, was not the only process that shaped export banana production. The structures and aesthetic sensibilities of U.S. marketplaces impeded the fruit companies' conversion to disease-resistant varieties. Consequently, the highly capitalized and heavily subsidized companies rerouted their railroads to find pathogen-free soils. This strategy enabled the companies to maintain production levels, but ultimately served to disperse the pathogen still further, as the cycle of invading people, plants, and pathogens repeated itself. The social costs of abandonment were considerable: local economies collapsed because of wholesale layoffs, tax bases shriveled, and commercial activity declined.

Threatened with the loss of their livelihoods, people living in areas abandoned by the fruit companies responded in multiple ways. Small-scale growers, forced to give up the banana trade after removal of the railroads that linked their farms to export markets, converted their fields to nonexport crops. Unemployed workers migrated to active banana zones or squatted on lands abandoned by the fruit companies. At times, local residents took collective action to preserve their communities. The examples of Omoa and Mezapa suggest that these actions succeeded in compelling government officials to intervene on behalf of struggling communities, but ultimately the risks created by changing landscapes were borne disproportionately by working peoples.

The evidence presented here cautions against conflating communities of interest with communities of place. Paradoxically, a shared sense of community may have been most strongly felt at moments when outside forces threatened to sever the strained ties that linked local livelihoods.[60] Both agroecosystems and communities are in flux and constantly remaking themselves. The ecosocial processes that helped to create the North Coast accelerated rates of agroecologic change and heightened the instability of communities formed by mobile capital, migrant labor, and dynamic biologic processes.

The export banana industry in Honduras did not become geographically stable until the mid 20th century, when changing political, economic, and agroecologic contexts rendered the practice of shifting production locations less viable than planting disease-resistant varieties. By that time, the number of small-scale banana producers had diminished considerably, primarily because of the increasingly capital-intensive

---

[60] This "relational" aspect of community identity can be applied to multiple scales. For example, the hegemony of the U.S. banana companies contributed to the rise of working class nationalism on the North Coast. Worker-cultivators confronting the loss of their livelihoods often based their appeals for justice on their status as *"hijos de la patria."* By asserting their identities as loyal Hondurans, working men imagined new communities that transcended physical space, but that also imposed gendered and racial boundary markers in the process of defining just who could make legitimate claims to membership in the "Honduran" community. See, Darío Euraque, "The Banana Enclave, Nationalism, and Mestizaje in Honduras, 1910s–1930s," *Identity and Struggle at the Margins of the Nation-State,* Durham: Duke University Press, 1998, 151.

[61] In the mid-1930s, another plant pathogen arrived on the North Coast. United Fruit Company scientists devised an effective but expensive control system, the costs of which were beyond the vast majority of non-company growers. As the number of inputs (fertilizers, irrigation, insecticides, and fungicides) continued to rise in the 1940s, growers with limited capital were forced to leave the trade. By midcentury, company plantations produced almost all of the bananas exported from Honduras. For further discussion of inputs and noncompany growers, see Soluri, "Landscape and Livelihood," Chapters 6 and 7.

character of production.[61] The 1950s witnessed a series of transformations in the Central American banana industry. In Honduras, a massive general strike led by banana workers in 1954 resulted in the formation of powerful union and campesino movements in the 1960s and 1970s that pressured the fruit companies to raise wages and relinquish lands.[62] The companies responded by shedding much of their newly unionized workforce through a combination of technological innovations, crop diversification, and "associate grower" programs that created a new group of contract growers.

Entering the 21st century, export banana production remains critical to the North Coast's economy, but it is no longer the region's most dynamic sector. The Sula valley, historically the nation's most important banana zone, is home to a large and expanding *maquiladora* industry that employs thousands of wage workers, mostly young women, in a variety of assembly and manufacturing operations. Banana plantations and cattle pastures are being turned into industrial parks. Not surprisingly then, changing landscapes and livelihoods continue to transform the meanings of place on the North Coast while giving rise to new communities of interest.

---

[62] Concerning the 1954 strike and its impacts on Honduran society, see Mario Argueta, *La gran huelga bananera: los 69 días que estremecieron a Honduras,* Editorial Universitaria, Tegucigalpa, 1996, and Marvin Barahona, *El silencio quedó atrás: testimonios de la huelga bananera de 1954,* Guaymuras, Tegucigalpa, 1994.

4 Community Culture and the Evolution of Hog Production: Eastern and Western Oklahoma

Chris Mayda

CONTENTS

Introduction 33
Background 34
Texas County: Guymon 36
Hughes County: Holdenville 38
Demographics 39
    Texas County 39
    Hughes County 40
Ethnicity and Employment 41
    Texas County 41
    Hughes County 42
Corporate Culture: Why Seaboard and Tyson Located Where They Did 42
    Seaboard 42
    Tyson 43
Social and Cultural Factors 45
    Texas County 45
        Schools 47
        Housing and other Indicators 48
    Hughes County 49
Conclusion 50

INTRODUCTION

The rise in Oklahoma's hog-based concentrated animal feeding operations (CAFOs) in the latter 1990s produced a furor of public protest and legislative reaction. Two companies invested heavily in hog production in Oklahoma: Seaboard Farms in Texas County in the Panhandle and Tyson's Pork Group, whose base of operations is in Hughes County, southeast of Oklahoma City. Each produces hogs using different

0-8493-0917-4/01/$0.00+$.50
© 2001 by CRC Press LLC

production methods. Seaboard Farms is a vertically integrated hog producer, corporately operating all phases of hog production from birth through processing, whereas the Pork Group operates sow and nursery units with contract farmers.

This chapter establishes the fact that the two approaches have not been by isolated chance. Changes in U.S. hog production are a complex result of society's millennial zeitgeist and intimately related to the particular cultures affected. A historical foundation of U.S. hog production, with definitions, is presented. Then a look is taken at how each county adopted a hog production method that best reflected its own geographic, agricultural, economic, and demographic culture, why hog production entered specific counties, and why the cultures of each county were or were not issues for the corporations. The social and cultural effects in each county resulting from these different hog production methods are explained. The chapter concludes by examining the reasons why hog production issues have been dealt with in a traditional empirical style of single observations, and why a system involving a complex interrelated agroecosystem is more beneficial for life and the land in the long run.

BACKGROUND

Oklahoma enacted an anticorporate farming law in 1971. Later, Farmland Industries and Tyson both began contract farming in Oklahoma and wanted clarification of the agribusiness law. This inquiry led to the 1991 change that endorsed corporate farming. Since that time, several lawmakers have rued the day this law was adopted, saying that the intentions have not met the reality.[1] Their intent was the nostalgically popular "save the family farm" by having farmers work as contractors with corporations. But the law has been interpreted to allow vertically integrated hog production facilities that are very different from contract farming.

Contract farmers raise company animals according to company specifications. The farmer provides land, buildings, and labor. Some farmers provide their own pigs, but increasingly, hog producers own and provide the pigs and hire the individual farmer to raise them. The company then provides the pigs, feed, medical services, and transportation. This is the way Tyson handled its hog production. In addition, the head of Tyson's Pork Group has an extensive background and education in the pig industry. Management for Tyson's Hughes County operations resides locally, which means that management is both local and from the Arkansas home office. Tyson's corporate cultural background is similar to that in Hughes County. For example, most corporate members were very active members of the dominant Baptist church in town.

A vertically integrated company brings together two or more successive steps of production or distribution under the ownership or control of a single company. With Seaboard Farms, the company owns everything from pigs through processing. Few individual farmers participate in Seaboard's hog production. The head of Seaboard Farms and his executives do not live in Texas County, but prefer the more urban areas

---

[1] *Daily Oklahoman,* April 28, December 22, 1997; January 8, March 12, 20, May 18, 1998.

TABLE 4.1
Number of Hogs and Pigs in Oklahoma 1992 (1997 Census of Agriculture)

	Number of Pigs 1992	Number Farms	Number of Pigs 1997	Number Farms	Inventory Ranking[a]	Sold Ranking[a]
Oklahoma	260,682	3,415	1,689,700	3,002	9	8
Texas County	13,513	39	907,046	30	1	1
Hughes County	481	31	125,474	42	3	2

[a] Oklahoma ranking among 50 states; county rankings among 77 Oklahoma counties.

of suburban Kansas City. The chief executives are not "pig people" as with Tyson, but rather certified public accountants (CPAs) who are more interested in economic opportunity than what the company does. Profit is the main concern for Seaboard officials, not pigs.

> We are not a regular type of company. . . . you look for the rhyme or reason. It's simply this. We had capital. We thought we could make money. We could produce chairs for conference rooms. That's Seaboard. It's an entrepreneurial organization. And it could just as much have been a power barge supplying electricity to the Dominican Republic which we do. Why'd we do that? Well, we thought we could make money doing that. (Mark Campbell, Vice President of Seaboard Farms.)[2]

In the past decade, hog production in Oklahoma has grown more than tenfold, from 187,351 pigs in 1987 to 260,682 in 1992 to 1,980,000 in 1998[3] to 2,190,000 in 2000, with more than 80% of the growth under CAFO corporate control (Table 4.1). Oklahoma went from 25th in hogs and pigs sold in 1994 to 8th in 1998. Texas County has seen the largest increase, from 13,513 pigs in 1992 to nearly 1 million pigs in 1997.[4] Hughes County also has seen a large increase, from 481 hogs in 1992 to 125,474 hogs in 1997.[5] Hughes County's approach, using local farmers as contractors, has been more accepted than Seaboard's corporate approach of using immigrant labor. In addition to labor differences, the geography of the two regions also has complicated environmental regulations. The two regions are entirely different geographically and environmentally, yet they are required to meet uniform statewide regulatory standards.

No state in the United States is culturally homogeneous. Historically, the state of Oklahoma has been divided between east and west along the Indian meridian, which today roughly equates with the north/south route of Highway 35. The Panhandle was a region tacked onto the territory just before statehood. This region is entirely differ-

[2] Interview, Mark Campbell, June 1997.
[3] 1992 Census of Agriculture, State Data: Oklahoma has the number of hogs and pigs at 1,689,000, which has since been estimated as increased to 1,980,000, Table 31, Hogs and pigs inventory; agricultural report, December 1998; Hogs and Pigs, National Agricultural Statistics Service, June 23, 2000.
[4] Oklahoma Agricultural Statistics Report, 1990–1997; 1997 Census of Agriculture.
[5] Oklahoma Agricultural Statistics Report, 1990–1997; 1997 Census of Agriculture.

ent from both the western and eastern parts of the state. At the most basic level, Oklahoma can be divided into these three regions, which can be viewed as distinct agroecosystems.

Although Texas and Hughes counties are both geographically and culturally miles apart, they suffered Oklahoma's common economic dilemma at the beginning of the 1990s and found similar "cures." Both Texas County and Hughes County emerged victorious against regional competitors and brought capital-intensive, factory-style hog production to their areas. The similarities ended there. The corporate, physical, and cultural geography of the corporations and the counties were reflected in the hog production methods they chose.

TEXAS COUNTY: GUYMON

> The Seaboard project will be as positive to our economy as the Dust Bowl was negative. (Guymon Mayor Jess Nelson.)[6]

The Panhandle of Oklanoma is part of the High Plains, a flat, treeless expanse where the wind blows from the southwest. Although flat, the Panhandle gradually rises from east to west, 2,500 feet at the eastern edge to 4,500 feet in the west, 168 miles away.[7] It is a semi-arid area, vegetated with short tufted grass and averaging 15 inches of rain a year, most falling during the critical summer growing months. The weather is mild, with 180 frost-free days and a mean temperature of 79°F in July and of 35°F in January. It is a place that few know except for its historical notoriety. This was the heart of the 1930s Dust Bowl.[8] Today, Texas County is again notorious, with Seaboard Farms building a hog processing plant and vertically integrated CAFO facilities. Ironically, the hog industry is interested in Texas County, not because it is dry, but because it has water. At the turn of the century, the Beaver River flowed through Texas County, which is the middle of three Panhandle counties.

> This Beaver River out here, you've crossed it and you don't think it ever had any water in it do you? My kids played in that river. I mean we lived on that river. . . . The old cotton wood trees? It used to be you could dig a posthole down that river and water would come up that posthole. . . . Those old trees never had to look down for any water. It just got sucked out from underneath them.[9]

Surface water in the county disappeared as irrigation escalated. Texas County might look as if it were "flat and plain,"[10] but it has two hidden blessings that make it one of the richest counties in Oklahoma. Deep below the surface lie the Ogallala Aquifer and the Guymon-Hugoton natural gas basin. One provides water for agricultural use and the

---

[6] *Guyman Daily Herald,* November 30, 1992.

[7] *Oklahoma Almanac 1995 1996,* p. 763.

[8] Worster, D., *Dust Bowl,* Oxford University Press, Oxford, 1979.

[9] Interview, Bill Newman, July 1997.

[10] A History of Hooker: A Diamond in the Rough, *The Hooker Advance,* Hooker, Oklahoma, 1983.

TABLE 4.2
Agriculture and Geography[a]

	Texas County 1990	Texas County 1998	Hughes County 1990	Hughes County 1998
Acres in wheat	300,000	260,000	10,000	3,000
Acres in corn	50,000	90,000	500	3,300
Acres in peanuts	0	0	7,800	3,400
Acres in sorghum	91,000	65,000	1,200	0
Cattle	344,000	270,000	49,000	51,000
		(1999)		(1999)

[a] Oklahoma Agricultural Statistics, 1990–1998.

other the energy to bring the water to the surface. Both are the largest of their kind in North America.

Although the presence of the vast aquifer was known during most of the 20th century, only after World War II did it become economically feasible to pump the water to the surface. In 1962 there were 270 wells in the Panhandle, but by 1995 the number of wells had reached 2,000.[11] Irrigated fields for crops use 95% of the water in Texas County, with municipal and livestock use sharing the remainder.[12] Irrigated corn feeds its other agricultural bounty: cattle feed lots. More than half a million cattle are "finished" today in Texas County alone, with several million more in the surrounding High Plains region.[13]

In 1997, Texas County had 415,600 harvested acres, with 785 farms averaging 1,384 acres per farm.[14] With its combination of Ogallala water, good soil, and level land, Texas County is Oklahoma's number one agricultural producer, with 14% of all Oklahoma agricultural receipts. In 1998, Texas County produced almost half of Oklahoma's corn, sorghum (milo), cattle, and hogs, ranking 4th in Oklahoma's wheat production and 23rd in the nation in total agriculture receipts[15] (Table 4.2).

Although the Ogallala has shaped Texas County's prowess agriculturally, its limitations are not known definitively. The large amounts of water being used for feed corn and milo and the incumbent animals eating it have raised environmental questions about groundwater usage and pollution. Scientific studies have not had time to analyze fully the effects of the massive water usage or the effects of manure applications on its fields. Preliminary reports show agricultural chemicals in the groundwater more than 200 feet below the surface. It is still too early to see the effect of hog production on groundwater quality, but the rapid pace of technology and population

[11] Wahl, K. and Tortorelli, R., Changes in flow in the Beaver-North Canadian River Basin upstream from Canton Lake, Western Oklahoma, USGS Water Resources Investigations Report 96–4304, Oklahoma City, 1997.
[12] United States Geographic Survey Mark Becker, 1998.
[13] *Cattle-Feeding Capital of the World: 1998 Fed Cattle Survey,* promotional piece by Southwestern Public Service Company: A New Century Energies Company, Plainville, TX.
[14] Oklahoma Cooperative Extension Service: Texas County Agriculture; 1997 Census of Agriculture.
[15] Oklahoma Cooperative Extension Service: Texas County Agriculture.

growth leaves little time for scientific analysis before application. Changing production methods and a swift rise in livestock production continue in the area.

HUGHES COUNTY: HOLDENVILLE

[Holdenville's] site is commanding and picturesque in the center of a beautiful rolling prairie, skirted by native forests of oak and hickory. The variety of fertility of the farms surrounding Holdenville, in the great valley between the North Fork and Canadian Rivers, is unsurpassed for richness and variety of products anywhere west of the great Mississippi.[16]

Whereas Texas County is dry and arid, Hughes County receives 40 inches of evenly distributed rain annually. It is hot and sticky in the summer, a result of the Gulf coast influence, and cold and crisp in the winter, with only the wind to remind the residents that, yes, they are in Oklahoma. The mean January temperature is 39°F, whereas July's mean is 82°F. In Hughes County every mile seems to have a stream or pond. Water is everywhere. The landscape also is entirely different, with trees such as cedars, Osage, and elms dispersed between the dominant scrub post and blackjack oaks. The county rests on the edge of the Cross Timbers region of Oklahoma in the Arkoma basin, a geologic province characterized by bedrock of shale and sandstone and overlying huge oilfields that flourished and eroded the area in the 1920s. The erosion from oil field drilling and salt water pumping made much of Hughes County farmland sterile.

Agriculture and livestock have not been dominant in Hughes County, but much of the region still is classified as agricultural, although little is actually grown. In 1997, 355,192 acres were farmed, with the average-size farm being 396 acres. However, a comparison of crops and acreage with those of Texas County (Table 4.2) shows a wide agricultural gap. Hughes County's main crops have been cotton, corn, and peanuts. It ranked fifth in peanut production in 1998, growing more than 5.6 million pounds, but corn production dropped considerably, and cotton was non-existent (Table 4-2).[17] As with Texas County, hogs were not part of Hughes County's traditional landscape before Tyson's arrival. In 1992 there were 481 hogs in Hughes County compared with 125,474 in 1997.[18]

Water issues also are very different in Hughes County. Instead of groundwater issues, there are surface water issues. Because the new agricultural laws are oriented more toward controlling groundwater pollution, contract farmers in Hughes County feel that they are not heard. All the attention goes to Seaboard.

We're a totally different animal than the other part of the state. . . . Everything that anybody asked, or anything that was said that was derogatory was pointed at Seaboard, not at us. . . . I think what we are doing, we aren't hurting anybody's ground water and it

---

[16] *Holdenville Daily News,* September 16, 1901.

[17] Oklahoma Department of Agriculture, Oklahoma livestock, 1998: In 1997 Hughes County had 4,000 acres of wheat and 3,800 acres of corn; Number 3 rating, *Oklahoma Agricultural Statistics,* 1996.

[18] 1997 Census of Agriculture.

looks to me like we're off in a way that's good for everybody concerned, and so why don't they change the state law to be the way we're doing it if that is better?[19]

Water issues are not the only problem. Economic and cultural differences also affect Tyson and Seaboard. These differences have not been addressed or recognized, but they are important in both analysis of information and how economy and culture work in relation to agriculture and geography.

DEMOGRAPHICS

TEXAS COUNTY

There had been a contact from a—what would you call this guy? He was just hunting an area to do this deal with Seaboard. We talked to him and then he kind of disappeared into the woodwork and nobody heard from him again. And then another person with Seaboard contacted us. I think they had their mind made up as to where they wanted to come—wherever they got the best deal. Oklahoma was friendly to confined animal feeding operations.[20]

Texas County population declined from 17,781 in the 1980s to 16,429 in the 1990s. There were several reasons for this decline, such as the farm crisis with its consolidation of farms and ranches, and the 1987 loss of Guymon's largest employer (200 employees) industry, Swift Meat Packing. Home values decreased from an average of $63,378 in 1980 to $50,850 in 1990.[21]

The Guymon-Hugoton gas field below Texas County's surface peaked during the 1960s. Many were employed at that time, but gas production has dwindled ever since.[22] In the 1980s the farm crisis took a toll on agriculture, Texas County's other major source of income. By the late 1980s, Guymon pursued economic development as an answer to the boarded up shops on Main Street and the decline in population.

In 1993 the population declined further to a decade low of 16,035 before bouncing back as Seaboard opened and began operations. By 1996 the population almost recovered its 1980 height (17,409), continuing to rise as Seaboard went to a double shift in 1998, reading 18,329 in 1999 (Table 4.3). Per capita income rose from $15,368 in 1990 to a pre-Seaboard $22,107, then dropped 14% to $19,204 as low-wage workers in the hog industry began to move into the county.[23] Although the average hourly wage in the 1980s for laborers in meat packing facilities was nearly $10, the beginning hourly wage at Seaboard in 1999 was $7, with few benefits.[24]

---

[19] Interview, Leroy Phillips, September 1998, p. 1.

[20] Interview, Bill Newman, July 1997, p. 1.

[21] NCRCRD, The Impact of Recruiting Vertically Integrated Hog Production in Agriculturally Based Counties of Oklahoma, North Central Regional Center for Rural Development, Ames, IA, 1999.

[22] Johnson, K. S., Minerals, mineral industries and reclamation, in *Geography of Oklahoma,* Moris, J. W., Ed., Oklahoma Historical Society, Oklahoma City, 1977, p. 98.

[23] U.S. Department of Commerce, 1997.

[24] Oklahoma Employment Security Commission, February 1999.

TABLE 4.3
Economic Profile[a]

	Texas County	Hughes County
Population		
1980	17,781	14,353
1990	16,429	12,975
1993	16,035	12,730
1996	17,409	13,052
1999	18,329	14,064
Per capita income		
1980	$9,831	$6,407
1990	$15,368	$11,098
1993	$22,107	$12,437
1996	$19,204	$13,549
Average wage per job		
1990	$13,854	$16,421
1993	$15,048	$17,791
1996	$21,128	$19,422
Farm income		
1990	$43,872,000	$1,362,000
1993	$116,161,000	$186,000
1996	$26,415,000	[$3,658,000]
Average home value		
1980	$63,378	$35,152
1990	$50,850	$26,717
1998	$54,675	$30,000

[a] Oklahoma Agricultural Statistics, 1990–1998.

HUGHES COUNTY

Pork was coming to Oklahoma no matter what. There were 11 communities competing for pork, right or wrong or indifferent. We, as a community went to that window faster than the other 10. We didn't think out the whole process but we knew the same problem would be here if Tyson was in Atoka, or Drumwright Oklahoma. Whatever. They would be somewhere right? Because they were coming to Oklahoma. That was a corporate decision.[25]

The Hughes County population dropped from 14,353 in 1980 to 12,975 in 1990 and continued its downward trend before beginning a modest rise in 1996 to 13,080 and 14,064 in 1999. Meanwhile, the per capita income has undergone a steady 29% rise from $11,098 (1990) to $13,549 (1996), although it is 74th among 77 counties in Oklahoma. Less than 1% of the income has been from farming.

---

[25] Interview, Jack Barrett, Mayor of Holdenville, Hughes County, Oklahoma, August 1998.

The average farm size is much smaller in Hughes County than current mechanized processes can accommodate, so farming has been largely abandoned. Most of the livestock production has been cattle.

The largest employers in Hughes County are Davis Correctional Center, a privately owned prison opened in 1996, and Tyson's Pork Group. Tyson employs 165 workers in addition to approximately 50 contract farmers, whereas the Correctional Center employs 210 workers. Both are relatively new employers in a county with a labor force of 5,270.

Hughes County's unemployment rate is consistently among the top five among the counties in the state. However, its average wage increased from $13, 854 in 1990 to $21,128 in 1996, a 52% increase. The average wage in "poor" Hughes County is currently higher than in relatively wealthy Texas County. The 1990 average wage in Texas County was $16,421. Although it has risen 18% to $19,422, it still is $1,706 below Hughes County's average wage. These figures do not tell the whole story because a great deal of Texas County's wealth is in unearned income from oil and gas rights. These per capita income numbers tell us that those who earned money in Texas County earned even less than the "average" figure. A few high unearned incomes can skew the averages. Texas County remains one of the lower wage-earning counties in the state. On the other hand, Hughes County's increase per capita shows a positive increase in income, largely because of its two new employers.

ETHNICITY AND EMPLOYMENT

TEXAS COUNTY

> Immigration, all of a sudden, becomes this huge issue. It's as much about Mexicans as it is about hogs.[26]

Historically, Texas County has been demographically homogeneous, with a white population of 88% in 1990 and 98% in 1999.[27] As Seaboard began to hire workers, there was a 49% increase in Hispanics, from 1,634 in 1990 to 2,690 in 1998. This change was particularly evident in the 1996–1997 fiscal year, when the county's population grew the fastest in the state at 3.86% from 17,400 to 18,100, largely from the influx of immigrant workers.

Texas County has historically had a low unemployment rate. In August, 2000 that rate was 2.2%. Economic development usually is pursued in a region to create jobs or else people leave the county. Texas County had only about two hundred unemployed people when Seaboard offered more than 1,000 jobs. As with many other meat-processing towns in America, immigrant labor filled the low-paying, dangerous jobs.

---

[26] Interview, Seaboard Farms Employee, July 1997.

[27] Although the number for the "white" population is 98%, it includes Hispanics who are tabulated twice—white, for race and Hispanic for ethnicity.

HUGHES COUNTY

Holdenville has always been known far and wide as a white man's town. No people of color live there. There has always been a very decided objection by a great many people of the town against people of color settling here, and this has always prevented them from living in Holdenville.[28]

The demographic makeup of Hughes County remains stable. The population of Hughes County peaked in 1930 with 30,334 residents, but declined in 1990 to less than half at 13,032.[29] In 1990, the census recorded 10,354 whites, 384 blacks, 2,232 American Indians, and 81 Hispanics. In 1998, 14,100 marked a very slow growth, with 10,420 whites, 790 blacks, 2,240 American Indians, and 270 Hispanics.[30]

Hughes County diverges from Texas County with its high unemployment rate of 15.6% in 1993, 9.8% in 1997 and 5.0% in August, 2000.[31] High unemployment and low per capita income fit into the model of low literacy rates that so many of Oklahoma's counties share. In Hughes County, 24% are at level 1 literacy, one of the worst records in the state.[32]

CORPORATE CULTURE: WHY SEABOARD AND TYSON LOCATED WHERE THEY DID

SEABOARD

In 1992 Seaboard, a Delaware Corporation with its home office in Shawnee Mission, Kansas, just outside Kansas City, opted to build a state-of-the-art pork processing plant in Texas County using its open space, feed, water, and location to create a new hog production market. Following the trail-blazing footsteps of Wendell Murphy's 9 million hogs in a hog- and corn-deficient North Carolina, Seaboard decided to leave traditional Midwest markets and establish pigs in the Panhandle.

Seaboard knew it wanted to be in the High Plains region. Earlier in the decade, Seaboard had purchased and experimented with a defunct meat processing facility in Minnesota. The experience taught them that they did not want to be in the already heavily competitive Midwest. Instead, they wanted to move to a region where there was access to corn, a sparse population, and better proximity to the Japanese export market. Seaboard went shopping among the many dying towns willing to give corporate tax incentives to "save their town." It appeared to be a win-win proposition. Seaboard needed the lucrative export market for its Oklahoma operation to be profitable. The state-of-the-art technology providing 5-week-old "counter fresh" meat to the Japanese market brought Seaboard to the High Plains. Guymon, in turn, got the economic boost for which it had lobbied.

---

[28] *Holdenville Daily News,* July 29, 1904.

[29] Oklahoma Almanac, The Oklahoma Department of Libraries, Oklahoma City, 1995/1996.

[30] U.S. Bureau of the Census, Oklahoma Department of Commerce, 1998.

[31] Oklahoma Bureau of Labor Statistics, 2000. http://www.oesc.state.ok.us/1mi/ Accessed 10/22/00.

[32] Oklahoma Department of Libraries, 1998. Adults at level 1 literacy have difficulty functioning in life because of lack of skills.

So this processing plant here in Guymon, Oklahoma is pretty ideally located to ship to California . . . and also export products to Japan and Mexico. The Japanese product goes out of southern California on boats as fresh product, and so delivery becomes a major issue. And because of our proximity to the coast, we have freight advantages as well as delivery advantages, having a fresher product than all those plants in the upper midwest. . . . If we cannot produce a high-quality product, and if we cannot enjoy the opportunity to ship to the Japanese marketplace, this processing plant can't make any profit.[33]

Texas County was not chosen for any of the aforementioned reasons, but rather because of economics played out through geography. Exporting meat furnishes the primary profit margin for Seaboard. By locating the plant in the Panhandle, it became the farthest Western processing plant other than Farmer John in Los Angeles, California. Seaboard planned to build its pig production in Texas County and surrounding areas so that the stock would be near the processing plant, unlike the California plant. Because of technological advances, meat can now be shipped to Japan, the most profitable export country, delivered as fresh, not frozen, meat. The Panhandle was as far west geographically as Seaboard could locate a large processing facility and have access to aquifer-watered corn, sparse population, and inexpensive land.

The usual economic motivation of job creation was not the incentive in Texas County. Seaboard employs more than 2,000 people, yet very few are Panhandle locals, and the remaining employees, in both management and labor, are highly transient and therefore have not assimilated into the community where the plant is located. There is little active community participation by Seaboard in Texas County, with high-level executives remaining in Kansas City.

Corporate America has little attachment to the land and its roots. Its "replaceable parts" attitude toward people and the landscape did not anticipate the vitriolic opposition that Texas County mounted against the hog farms. Seaboard was not prepared for people who did not want to move or sell their land. These were people who cared about where they lived and were attached to the land beyond economic "realities." Place meant something that few corporations could identify with.

When Seaboard chose the Panhandle, its managers objectively analyzed the location in relation to the Pacific market, the water availability, and the feed access, but they never included the subjective factor, people, in their analysis. They never synthesized all the information. They analyzed for what they were created to do: make a profit.

TYSON

In contrast, local community participation or connection is high in Hughes County in southeastern Oklahoma, where the Pork Group, a division of Tyson, employs 165 workers in addition to 50 contract farmers. It is difficult to find anyone who does not have some connection with the Tyson operation. Tyson's head of pork operations lives in the community, and most of the contract farmers live within an hour of Holdenville.

---

[33] Rick Hoffman, CEO of Seaboard Farms, at Task Force Meeting, August 1997.

Tyson established hog production operations in Hughes County in the 1990s. Contract farmers need to be reasonably close to the town because they receive their feed from the Tyson mill in Holdenville. Reasonable proximity also is needed for the transfer of hogs from one site to another. Most of Tyson's operations are contract sow units, with Tyson controlling the nursery units. Hogs leaving the nursery unit are shipped to the corn-rich Midwest for finishing.

Because Hughes County is a contract farming area with no processing plant, the need for immigrant labor is small as compared with Texas County. Although the hog culture has produced an opposition in Hughes County, it does not have the indirect racial element found in Texas County. Instead, the opposition in Hughes County is based overtly on such environmental problems as water pollution and odor control, although there are also indirect elements of resistance to the loss of family farmer control and the rise in corporate culture.

Hughes County's cultural heritage has Ozark Arkansas roots that made it "natural" for Arkansas-based Tyson to raise pigs in Holdenville. Previous experience in contract farming with chickens made the populace more accepting to the raising of contract pigs. Yet, many feel that contract farming is not saving the family farm, but rather making the family farmer a tenant on his or her own land.

> Well, as far as I'm concerned, getting into a contract relationship with one of these companies . . . if people look back, they will probably say it is the stupidest decision I ever made in my life . . . if they could go down the road 20 years and see the impact and then look back. People get into contract relationships largely out of desperation. When companies go into areas, rural communities, they make relationships with bankers, with the mayors, with the people in the know, and they find out who is in trouble financially. [34]

Others feel that contract farming has been a boon to the economy for southeastern Oklahoma, but not an issue for the Panhandle.

> Tyson and some of the other companies that have come in, they use more contract growers. . . . It gave them an opportunity to stay in the countryside. It gave them an economic base to keep their schools open. Southeastern Oklahoma was the least economic area. So this is really giving them a shot in the arm. . . . So it really has improved their economic base, as small family farms. [35]

Job creation, bringing jobs to places such as Hughes County, is often the emphasis in economic development. Corporations frequently do not look at community indicators, but rather at their own desires. In the case of meat processing, the location in relation to the market is most important. The proximity to Tyson's home state and the cultural similarities of the two areas were sufficient reason to bring Tyson to Holdenville, along with whatever economic incentives were mustered, such as property tax abatements. The cultural similarities, socioeconomic placement, small farms, and the requirements of the retired farmers were a good fit for Tyson's contract farmer

---

[34] Interview, Suzette Hatfield, Oklahoma Family Farm Alliance, August 1998.
[35] Interview, Hal Clark, Guymon businessman, July 1997.

needs. Tyson was able to maintain an image among many as the "good guy" saving the small family farm, rather than the cold impersonal corporation that Seaboard appeared to be. Saving the family farm was the image that legislators wanted to maintain, regardless of the reality. Even Tyson's representative, John Thomas, said, "This arrangement is not for everyone."

Why Texas County has so few contract farmers remains unclear. It could be a result of the county's affluence and Seaboard's lack of experience in hog production, or Seaboard may have purposely intended to integrate vertically, a wise decision in retrospect considering the 1999 pork prices.

The different corporate cultures had an effect on each respective county's social and cultural development. The effect of the isolated and imposed structure of Seaboard was and remains difficult for Texas County with regard to its cultural development, schools, and housing. The assimilation with Tyson, although not seamless, was more in tune with the place itself. Therefore, the adaptation by the county was easier, although environmental questions are still a open issues.

SOCIAL AND CULTURAL FACTORS

As the world becomes a global economy, it undergoes adjustments resulting from changing demographics in obscure places such as Worthington, Minnesota, Garden City, Kansas, or Guymon, Oklahoma, all new Meccas for non–English-speaking immigrant populations seeking jobs in which they can earn more than in their native country. These multicultures are celebrated on a superficial level: "Look at all the new Vietnamese and Mexican restaurants we have." However, celebration is not what these people experience in their new lives, nor is it the only multiculturalism in the states.[36] Multiculturalism exists within the borders of each state among the citizens assumed to be the same, but who are not.

The moniker for Oklahomans is "Okies," a stereotype that fits from time to time. But Oklahomans are no more all the same than the immigrants brought to Guymon to work in the pork processing industry. The state's citizens speak the same language, but they are not the same culturally or geographically. Although Texas County and Hughes County are both in Oklahoma, they are not the same culturally. They have little common heritage or history other than the happenstance jigsaw puzzle assemblage of the state just before statehood. Cultural and social differences, along with the geographic surroundings that helped to form these differences, are part of the reason why the hog industry operated differently in Hughes County and in Texas County.

TEXAS COUNTY

> The people were hospitable and friendly. Meals and beds were given when needed, doors were never locked, and nothing was stolen. Written contracts were not made, only verbal agreements and they were kept.[37]

---

[36] Stull, D., Broadway, M., and Griffith, D., Eds., *Any Way You Cut It: Meat Processing and Small-Town America,* University Press of Kansas, Lawrence, Lawrence, 1995; Mayda, C., *Passion on the Plains: Pigs in the Panhandle,* PhD thesis, University of Southern California, Los Angeles, CA, 1998.

TABLE 4.4
Demographics

	Texas County	Hughes County
Population		
1990	16,419	13,014
1998	18,600	14,100
Whites		
1990	14,525	10,420
1998	18,180	11,030
Blacks		
1990	81	325
1998	100	790
American Indian		
1990	286	2,231
1998	290	2,240
Hispanic		
1990	1,634	81
1998	2,690	270
Unemployment rate		
1990	4.3%	15.5%
1993	4.1%	15.1%
1999	2.5%	9.8%

I mean, morals and trust are getting to be where it's nonexistent. I went into business with a guy. We were running cattle through a feedlot, you know. I asked him whether it's going to be a partnership and I said do we need to set up a contract? And he said, "If you're the kind of guy I have to set up a contract with, I don't want to do business with you." And I said, "I feel exactly the same way." But you know, a handshake anymore isn't good enough and a few years ago that's all it took in this country.[38]

Texas County has a history of isolation, both from its status as no man's land at the end of the last century and from its separation from the "mainland" of Oklahoma. This has produced a willful and hearty people who have a strong individualist bent not conducive to accepting outsiders, whether they are corporations or state or national officials. Today, many Texas County families claim roots in the past, an anomaly in the locationally rootless corporate America of the 1990s (Table 4.4).

The High Plains is a difficult place in which to make a living, and those who have stayed are dedicated. With its external natural resources the county has become bifurcated in its economic structure, with a wealthy segment separated by the low-wage workers at the bottom of the state's wage scale. This bifurcation has been augmented by the importing of the processing plant and its incumbent low wages. Because of an

---

[37] Texhoma Genealogical and Historical Society, Ed., *Panhandle Pioneers,* 1969.
[38] Interview, Texas County Farmer, July 1997.

already low unemployment rate, labor had to be imported, often from beyond state and largely beyond national lines. This has produced a multicultural atmosphere in a largely homogeneous region.

In the 1960s a small contingent of Hispanics settled in Texas County and assimilated with little evident discriminatory stress. This is not occurring with the current imported labor, partly because of the inherently high turnover rate at processing plants. Many of the people who formerly worked in Guymon's Swift meatpacking plant also remained in the county, although few, if any, now work at Seaboard. Meat production wages have dropped precipitously in the past decade, and the turnover rate has increased.

> They have offered to buy some land. . . . [They] asked if . . . we would be willing to sell land. And of course [we] said, "No we wouldn't." Because . . . we wouldn't do to our neighbors what we don't want done to us. . . . Beside that we don't want to sell it. . . . Delmer inherited this land. . . . His mother came out here to farm her dad's land in 1929. And I don't remember how long her dad had owned land out here, but this is a family farm. And this is roots, and it's sentimental.[39]

Schools

> There's more economic development. I mean there's more dollars coming into the school. But there are more programs that have to be provided too. So, it takes a lot of dollars to provide education for those students we are concerned about. Will there be enough dollars to educate the influx of population? That's our main concern. And those programs are expensive.[40]

Before Seaboard, Texas County had a small Hispanic community that had settled, learned English, and assimilated with little discrimination. But the laborers at Seaboard were different. The meatpacking business is notorious for its Spanish-speaking immigrants, who are largely illegal, and its transient labor force, with an average 120% turnover.[41] This adds to the public service expenses and also impacts the schools, which now require additional money for English as a second language (ESL) and other special needs classes.

The school system grew, and Hispanic students increased 114% from 1990 to 1997. These students require increased educational expenses and attention, but the teacher–student ratio has risen, giving each student less attention.[42]

Texas County arranged a tax increment/financing package in which all property tax increases over the next 25 years are used to pay off infrastructure construction

---

[39] Interview, Vancy Elliott, Texas County, July 1997.

[40] Interview, Mel Yates, Guymon School Superintendent, July 1997.

[41] Stull, D., Broadway, M., and Griffith, D., Eds., *Any Way You Cut It: Meat Processing and Small-Town American,* University Press of Kansas, Lawrence, Lawrence, 1995; Mayda, C., *Passion on the Plains: Pigs in the Panhandle,* PhD thesis, University of Southern California, Los Angleos, CA, 1998.

[42] NCRCRD, "The Impact of Recruiting Vertically Integrated Hog Production in Agriculturally Based Counties of Oklahoma," North Central Regional Center for Rural Development, Ames, IA, 1999.

instead of being used to finance public schools. Oklahoma is one of the most challenged states in the United States in terms of literacy rates. In some areas an average of 20% are at the lowest literacy level. Part of the reason is the low property tax rates in Oklahoma. Texas County has been one of the better-educated counties, with only 13% considered as level 1 in literacy. However, the new population has less education than most in the state and also does not speak English. The county will require additional money, while tax increment financing erases the now needed property tax increases. Although Seaboard did arrange to pay an additional $175,000 per year over the 25-year period, its improvements, if taxed, would bring in nearly twice that amount in property taxes. This cultural shift in population demographics has affected the entire student population, lowering the standards just when the standards need to be augmented. Nothing has been done to address this situation directly, although there are many questions being raised about Texas County's educational problems.

Housing and other Indicators

There is a critical shortage of housing. There is no two ways about it. Especially the middle-income type housing. Affordable housing. We are working on that hard. There's been a lot of trailer parks going in and nobody wants a trailer park in our city. But we have it.[43]

The plant added a second shift in 1997, with a total employment of approximately 2,200. As housing was in short supply and prices for existing housing increased, the only place for the new workers to live was on the outskirts of town in *de facto* segregated units that were still unable to handle the large immigrant population. In late 1997, two thirds of all Seaboard workers lived out of the county, mostly in Liberal Kansas, 40 miles away. Those who lived in Guymon were isolated both linguistically and geographically on the edge of town, with no schools or shopping nearby. The police force still had no Spanish-speaking officers in 1999, although up to one half of all calls were directed to the area where the Hispanics lived. Other social indicators that raise questions about the changing quality of life in Texas County are the increased crime rate (up 75% since 1990), the rate of violent crime (up 378% since 1990), and the court caseload (up 165% since 1990).

These effects, along with Texas County's lack of a multicultural history, have raised many "racist" flags. People who have never had to understand the adjustments that must be made to live in a multicultural society are not prepared for the reality of multiculturalism beyond the simplistic "celebration" of demographic change. There is a great deal of emotion involved for many living in Texas County. Often, emotion is what comes to the forefront when issues such as multiculturalism and the changes taking place in America are not understood. Education regarding how America fits in the global economy and the changes needed to help alleviate the pain many feel toward what has happened in their town and county may help.

---

[43] Interview, Bill Newman, July 1997.

Most of these people . . . get all involved in their emotions and it's hard to deal with people's emotions. It's hard to change people's emotions. And that's what a lot of this is. A lot of this is strictly emotional.[44]

HUGHES COUNTY

We live in the Bible belt and this is the buckle. We all are registered Democrats but if we lived anywhere else in the world, we would be Republicans and probably to the extreme right. That is Oklahoma. That's us. Very conservative.[45]

Unlike Texas County, which is largely Republican, Hughes County has been almost entirely Democratic at the polls, although in the 1994 election "other" held the majority vote. The conservative cultural makeup of southeastern Oklahoma has its roots in its early settlers. They either were from Arkansas or had a strong Native American heritage that now is proudly proclaimed by many Hughes County citizens, a cultural factor almost completely lacking in Texas County. Hughes County is in the Creek Nation area of the old Indian Territory (Table 4.4).

The Arkansas heritage has a heavy tinge of Southern and Ozark cultural images, displayed in the various Confederate flags dispersed throughout the county, and also in a Blanche Dubois gentility that does not approach strangers in the same way as in Texas County. On arriving in Hughes County, the author was told by one local, "People here are shy, quiet, and scared to death." People were much more guarded and wary of the stranger than in Guymon. Not one interview occurred without an introduction, and few people in Hughes County travel without firepower assurance. Guns are kept casually in a pant pocket or handy beneath a chair or truck seat. The author was told more than once, "Never know when you will run into a rattler," which is true enough in a largely rural state with many poisonous snakes.

After becoming acquainted with Hughes County's citizens, the author found that they were still very different from the citizens of Texas County. They were the same—warm, sincere, and friendly—in what can only be called small town America, but with a Southern sensibility. Visiting in Hughes County is not momentary. You do not just say "Hi" and move on. You stay and pass the time in Hughes County in a slow, genteel, and neighborly fashion. But first, you must break through the ice.

As in Texas County, there is a split between the city and county in Hughes County, but it is the opposite of the Panhandle experience. The people of Holdenville often are long-time residents, born in the area and settled in the town, but people in the county, although often from the area, have not lived there throughout their lives. A constant high unemployment rate and a poor farm economy have led many residents to leave the county to live in Tulsa, Oklahoma City, or other neighboring towns where there are more opportunities. The clarion call from the family acreage has often drawn people home for retirement, although farming acreage in Hughes County is much smaller than in Texas County.

---

[44] Interview, Jess Nelson, Mayor of Guymon, July 1997.
[45] Interview, Jack Barrett, Mayor of Holdenville, Hughes County, Oklahoma, August 1998.

Oklahoma is full of beautiful streams, water, trees and mountains and the people were starved out and had to leave during the depression. Now these people are getting to reach their retirement days and they want to come back and live in the mountains and fish in good streams. They don't want hog barns nowhere around them. It's devaluating the price of our property.[46]

These retirees usually do not farm because there has not been money in that, but they do raise cattle. Lately, however, they find they cannot make it raising cattle alone because of the downturn in the market. When Tyson came to town with its spreadsheet analysis of hog production, many saw this as the way to save the family farm. It is not always the family farmer who has tried to make a go of it, but also the retired oil field worker, teacher, or wife who wants to bring in extra income for the family.

CONCLUSION

The focus of most Oklahoma hog production literature has been on Panhandle environmental problems, centering on Guymon, in Texas County. More than 1 million hogs are raised from farrowing to finish each year in Guymon, supplying a processing plant built and operated by Seaboard Farms since late 1995. Seaboard has built a vertically integrated hog production facility controlling the pigs from conception to consumption. In doing this, Seaboard has created an empire in Oklahoma, a state with differing laws, culture, and geography. On the other hand, Tyson's Pork Group has operated in a more amenable way in Hughes County where culture and experience better reflect the corporation's experience and culture. One would hope that Tyson would "win out" in the Oklahoma corporate hog struggle, but in 1999 Tyson Food reorganized and attempted to sell its operation and continues to seek a buyer as this volume goes to press. Seaboard continues to bully its way into the Panhandle culture and economy, making few friends along the way.

Although the issues around hog-based agroecosystem are complex, corporations, legislators, and communities often over simplify them. Cultural differentiations are seldom addressed in legislative debates. Instead, Oklahoma's 1997 and 1998 hog bills are directed to environmental regulations, rather than to the problems created by cultural identifications. No bills were introduced in 1999. This has been difficult for both the corporate farms in Texas County and the contract farmer in Hughes County. In addition, the vastly different geographic regions of the two counties make statewide regulations difficult to enforce, with one area being fed by surface water and the other groundwater oriented. The hog production dilemma of the family farm versus corporate farm CAFO production has spread across 28 states in recent years. The inability of states to cooperate with each other and deal with water pollution issues has prompted the Environmental Protection Agency to step in to regulate waters nationwide. No one is happy with lumping entire sections of the United States into one general region with general regulations.

---

[46] Interview, Hughes County farmer, August 1998.

The use of a uniform law is the result of dealing with only reductionists' measurable scientific problems rather than with the more complex problems of ecosystems and humanity. The landscape and the "lifescape" are expected to change to support economic profit, as if economy were separate from the land and the organisms that live on it. As long as issues are resolved only on reliance on seemingly objective "sound science" and ignore multicultural realities, and as long as we continue to ignore the reality that the earth and its beings are interrelated as a whole, we deserve the conundrums and resultant problems of our existence.

5 Forest Conservation and Degradation in a Sub-Subsistence Agricultural System: Community and Forestry in Mexico

Daniel James Klooster

CONTENTS

Introduction 53
Communities and Agroecosystems 54
Community, Agriculture, and Forestry: The Case of San Martin Ocotlán 55
Milpa Cultivation and Other Agricultural Activities 56
The Economics of Sub-Subsistence Agriculture 57
Complements of the *Milpa* 58
A Forest-Dependent Peasantry 58
The Community Dynamics of Timber Poaching 60
Social Structures, Values, and the Forestry–Agriculture Relationship 61
Distribution of Benefits 61
Social and Political Structures 62
The Forest as an Economic Subsidy 63
Forest Management 63
A Struggle for the Forest 64
Conclusion 65
Acknowledgments 65
References 66

INTRODUCTION

As agroecologists have noted for some time, forests are vital components of agroe-cosystems because forest fallows produce food and fiber, restore soil fertility, and contribute to weed control. They provide agriculture with a "subsidy from nature" (Alcorn, 1989; Hecht, Anderson, and May, 1988). Agroecology less often considers the role of forests as sources of economic subsidy to agricultural systems increasingly

0-8493-0917-4/01/$0.00+$.50
© 2001 by CRC Press LLC

unable to provide the sole basis for community reproduction in a context of global-ization. Understanding the sustainability implications of this connection, however, requires broader consideration of the relation between agroecosystems and the human communities that make them function.

COMMUNITIES AND AGROECOSYSTEMS

Communities and agroecosystems are composed of interpenetrating components that influence one another in a co-evolutionary process of change. The activities related to producing food, fiber, and cash crops provide the link between the biophysical components of an agroecosystem and the social system. Many communities in rural settings affirm and recreate themselves largely through productive activities nested in their agroecosystems. Through routines of work, such as planting and harvesting, related rituals, and the social relationships that accompany these activities, people create and maintain communities. Through production, they forge an identity as com-munity members, both to themselves and to their neighbors. Agricultural production is particularly important in historically agrarian settlements, where community-affirming rituals surround planting, weeding, and harvesting, and where cultivating land helps to define an individual's relationship to the community. Sources of change may be endogenous because of soil depletion or local population growth, for exam-ple, or exogenous because of factors such as climate change or the growth of off-farm employment opportunities. Figure 5.1 illustrates these relationships.

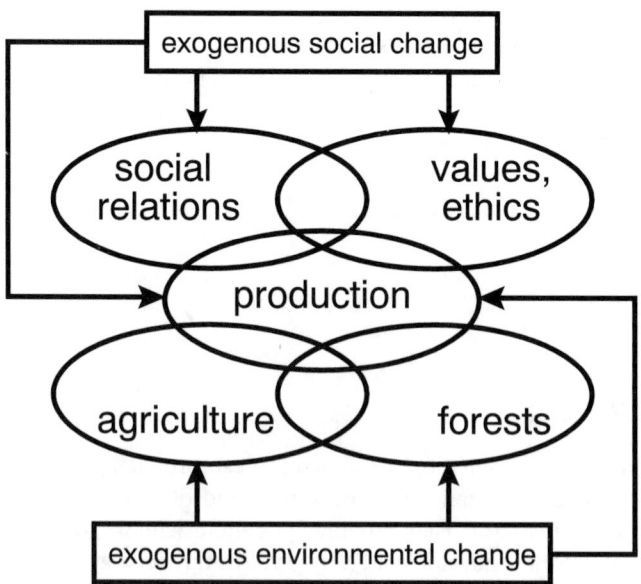

FIGURE 5.1   Social and Environmental Changes on Agroecosystem Production

It is often the case, however, that the products of traditional agroecosystems are of little value in the global and regional marketplace, and this threatens the economic viability of rural communities' productive bases. In this co-evolutionary model, people have agency. Their choices, although limited and framed by both internal and external factors, help to shape the evolution of the agroecosocial system. Because rural people often identify themselves with their communities and greatly value them, they often seek ways to maintain them, even when this is no longer economically beneficial. Many rural Mexicans, for example, refuse to abandon their rural communities completely or give up agriculture. Instead, they supplement it with temporary migration, craft production, and forest-based activities in order to survive as members of rural communities, despite persistent signals from the market, government policy, and national society to move elsewhere and produce other things.[1]

Forests frequently provide economic supplements to support the agroecosystem-based productive activities that underlie and maintain rural communities of great value to their members. Internal social stratification shapes the distribution of the economic subsidy from forests, however, and this has repercussions in agroecosystem change and the possibility of sustainability.

This chapter explores the relationship among community, forest, and agriculture in a comparative case study. First, it examines the role of the forest as a supplement to subsistence agriculture. Once this economic link between the forest and cropland components of the agroecosystem is established, it explores the way the community-level social processes affect the exploitation of the forest.[2] It describes the way these social processes influence values and rules in a case study that fails to constrain members completely from the timber poaching that might threaten the ability of the forest to continue subsidizing the rest of the agroecosystem. Finally, this article briefly considers the experience of other communities that illustrate a more successful relationship among forests, agriculture, and community.

COMMUNITY, AGRICULTURE, AND FORESTRY: THE CASE OF SAN MARTIN OCOTLÁN[3]

In Mexico, the vast majority of forests are community forests. Approximately 9,000 communities own 80% of remaining forests as common property territories, ranging in size from 100 to 100,000 hectares. These communities, known as *ejidos* and *comunidades agrarias,* are political entities whose members own the forests and rangelands surrounding their villages as common properties, combining forest activ-

---

[1] Given free trade agreements that pit Mexican rainfed maize against international producers and the higher wages in cities and areas of capitalist agriculture in the United States and Mexico, it is surprising that so many Mexicans stay in rural areas and choose to channel earnings back to rural areas when they do migrate. I am thankful to David Barkin for providing this insight during several extended conversations. See Ch. 14 for further discussion on this point.

[2] The role of exogenous social factors is addressed in more detail elsewhere (Klooster, 2000; Klooster, 1999).

[3] San Martin is a pseudonym. Data for this section comes from 16 months of participant observation and interviews, enriched with archival research in the Oaxaca Agrarian Reform archives, financial audits, and the 1995 payroll for the forestry business (Klooster, 1997).

ities of varying intensity with a production portfolio often centered on agriculture and livestock production. Formal gatherings of community members determine what will be done with common property forests. They elect a president and other communal authorities to represent them. Several thousand of these communities operate logging cooperatives, and they typically elect community members to oversee this activity. National policies and programs also affect forests, but over time, the federal government has exercised decreasing authority over forests, in favor of communities (Bray and Wexler, 1996; Klooster, 1996; Wexler and Bray, 1996; World Bank, 1995).

One of these agriculture/logging communities is San Martin Ocotlán, in the state of Oaxaca. San Martin is home to some 600 households and 3,300 inhabitants. Slightly less than half of the total population lives in the community's capital, San Martin Ocotlán Village. One fourth of the residents speak the indigenous Mixtec language in addition to Spanish. During 1995, logging activities were concentrated in the forests above the hamlet of Benito Juárez, population 500.

MILPA CULTIVATION AND OTHER AGRICULTURAL ACTIVITIES

Forest covers well over half of San Martin Ocotlán's territory, grasslands one third, and approximately 13% (1,998 ha) of the community is divided into family-held agricultural parcels. Nearly all households plant *milpa:* maize combined with beans, squash, potatoes, various leafy vegetables, and flowers. A number of small, arable plots are irrigated by diverting the three perennial creeks, and they comprise approximately 11% (213 ha) of the community's agricultural area. Farmers plant *milpa* in these plots and, occasionally, alfalfa or wheat for fodder. The major part of agricultural lands, however, approximately 89% (1,708 ha) is rainfed. Some rainfed lands are flat enough to be arable with draft animals, but most are *retoñeras:* hillside oak groves where farmers periodically clear, burn, and plant *milpa.*

A handful of other agricultural activities supplement the *milpa,* notably small orchards of peaches, apples, and pears, which occupy some 3% (54 ha) of agricultural land. Virtually all households (88%) have a few chickens, pigs, sheep, or goats, and some 24% of households (135) own cattle. Sheep and goats are the preferred grazing animals, however, with 2,000 goats and 1,500 sheep in the community (INEGI, 1991).

Agricultural activities are concentrated during the May to October rainy season, when the area gets 800 to 900 mm of rainfall from frequent storms. Agricultural activities during the winter months are limited by both lack of precipitation and frequent frosts[4].

The exact timing of agricultural activities varies both according to altitude, which ranges from 1,700 to 2,900 m above sea level and according to whether the field is rainfed or irrigated. In the highlands, where soils retain moisture, and in irrigated plots, planting takes place in late February or March. Harvests begin in June or July in irrigated plots, and extend to February in the highland rainfed fields. At all altitudes and in all types of field, farmers manually weed the *milpa* twice before harvest. Failure to weed on time results in yellow, stunted plants, earns the ridicule of

---

[4] INEGI, 1981; INEGI, 1984.

neighbors, and can cost the harvest. Planting, harvesting, and especially weeding require substantial labor concentrated in fairly brief periods.

THE ECONOMICS OF SUB-SUBSISTENCE AGRICULTURE

In San Martin Ocotlán, *milpa* production generates very little money. Only 5% of households sell part of what they grow (INEGI, 1991), usually small amounts of young squash, husk tomatoes, and fodder. Farmers rarely sell grain. But *milpa* production does require money. In most cases, chemical fertilizer is indispensable, and it must be purchased at prices sensitive to international exchange rates for the Mexican peso. Few families have a plow and oxen, so those with arable fields pay for plowing. Family labor is rarely sufficient at the bottleneck weeding periods when farmers hire additional labor at about 20 pesos a day. Finally, tools must be purchased.

Why do farmers spend time and money on something that costs money but makes none? Farmers in San Martin Ocotlán cultivate their *milpas* for a variety of reasons. They enjoy the variety of foods from their *milpas,* many of which mark seasonal cycles and play important roles in community rituals. The *milpa* and *milpa* fallows also provide fodder for domestic animals, flowers for the dead and for the saints, and, in the case of recently fallowed plots, firewood for the kitchen and for sale. A family of five with the land and labor to plant a hectare of *milpa* can hope for a year of food self-sufficiency, and a store of grain in the house represents a significant source of security in a volatile economy. In addition, farming uses family labor that has little outlet in local or regional economies.

The act of farming also helps define community membership. Engaging in agriculture confirms the right to usufruct of land, especially where titles are imprecise and based on community recognition. Furthermore, the usufruct of agricultural land ties farmers into an intergenerational family project of acquiring, maintaining, and passing on land. To summarize, *milpa* benefits include the following:

- Provides maize and other *milpa* foods for home consumption
- Uses family labor
- Provides fodder, firewood, auxiliary crops for occasional sale
- Provides a 400 pesos/ha cash subsidy from the federal government that is meant to ease maize farmers' inclusion in the North American Free Trade Agreement (ProCampo)
- Demonstrates plot ownership
- Demonstrates good character to neighbors
- Confirms community membership to others
- Confirms self-identity as a community member
- Participates in an intergenerational project of land ownership
- Provides flowers for the saints and other products for ritual activities

Nevertheless, farmers know that in money terms, what they are doing does not make sense. "A ton of fertilizer costs more than a ton of maize!" they say. "It's better to just buy the grain." Even so, very few give up farming completely (Klooster, 1997, p. 185).

COMPLEMENTS TO THE *MILPA*

In addition to required money investments in labor and fertilizer for *milpas,* families need money to purchase clothing, school supplies, medicines, and other sundries, including tobacco and liquor. They also need money to buy maize because harvests often fail, stored grain is perishable, and only a minority of peasant households manage enough land and labor for 12 months of food self-sufficiency. To make ends meet, therefore, farmers of San Martin Ocotlán must cobble together a livelihood portfolio that includes more than just agriculture.

Temporary migration is one important component of this strategy. As family size, *milpa* schedules, and opportunities permit, some men find work in Oaxaca City, commonly as construction workers. Single men, and occasionally entire families, migrate for extended periods to Oaxaca City, Mexico City, Tijuana, and rarely, to the United States seeking work. To maintain rights to agricultural lands and home sites during extended absences, community members must inform communal authorities that they are leaving. Community records show that 13 men left San Martin Ocotlán Village during an 18-month period, a rate of 2% a year. El Manzanito, the community's smallest hamlet, had the highest rate: eight men emigrated since 1994, rate of nearly 15% per year.[5]

Although many people leave the community, many more choose to stay. Furthermore, emigration often is cyclically tied into household production strategies and family life cycles. Emigrants are often young men who work outside the community until aging parents make land available to them. Others leave families at home in the community and break extended absences with sporadic visits. During the economic crisis of 1995, a number of emigrants returned to San Martin Ocotlán. Fed up with job loss, crime, and pollution in Mexico City, they took their accumulated funds, bought land and cattle, built houses, and returned to farming.

Most of the farmers who remain in the community must complement the *milpa* with additional, money-generating activities inside of the community. The *milpa,* however, constrains labor availability. Weeding and other labor bottlenecks in the *milpa* cycle require labor inputs at precise times, so farmers seek work that is flexible enough to allow them to tend to their *milpa* at these times, which might mean the difference between 8 months of food self-sufficiency or none at all (Klooster, 1997).

A FOREST-DEPENDENT PEASANTRY

One key complementary activity is tending livestock. Selling an animal provides money needed to buy fertilizer or hire laborers for weeding. The forest, however, provides the most accessible supplementary activities to the *milpa.* Working in the communal forestry business, cutting firewood and making charcoal for sale, and engaging

---

[5] These estimates certainly underestimate the actual rate of emigration because communal authorities do not keep track of the young men who lack the status of registered *comuneros* and thus are more likely to emigrate than are land-possessing, registered *comuneros*. Communal authorities do not have any way to estimate the number of women who leave San Martin Ocotlán either.

in timber poaching are vital sources of cash income and crucial components of livelihood.

Commercial forestry in the community goes back to 1958, mostly under a series of concessionaires. In 1980, the community broke free from the concession system and formed its own logging business. Currently, the community owns several logging trucks, a motorized winch, and a sawmill. Some 366 men, half of all working-age men in the community, received between 250 and 10,000 pesos in wages during the 1995 logging season. The average payment was 1,300 pesos for sporadic work, but 62% of the workers earned less than that (Klooster, 1997).

In addition to work in legal logging, however, the sale of oak firewood and charcoal offers additional means to earn the cash needed to supplement the *milpa*. The only equipment needed is an axe, machete, and a mule, although chainsaws are increasingly common. Woodcutters also cut oak building posts, which are highly valued in regional informal construction markets, and these fetch from 8 to 15 pesos each in Zaachila and Oaxaca markets. Woodcutters also find a small market for firewood in San Martin Ocotlán Village, where a mule load of wood fetches 5 pesos. Local truck drivers with permits purchase larger volumes and transport firewood for resale in the weekly markets at Zaachila and Oaxaca, where the price nearly doubles.

Charcoal makers add value to oak and reduce its weight for transport. Like firewood cutters, charcoal burners sometimes market small quantities themselves, carrying gunnysacks to the roadside on mule back and hitching a ride on a public bus. Larger quantities require a deal with a truck owner. Truckers charge 400 pesos to transport a load of 100 gunnysacks of charcoal to Oaxaca City, where wholesalers pay roughly 12 pesos per bag. Charcoal burners also can sell their product in the forest, for about 8 pesos a gunnysack.

A riskier but more lucrative complement to the *milpa* is timber poaching, the preparation and sale of rough-hewn boards, roundwood posts, beams, and rafters from pine.[6] After felling the selected tree, cutters carve it into 8-ft lengths with a chainsaw, square the logs, and mount them on a makeshift platform to provide a flat, dry, clean space on the forest floor. Taking strings impregnated with the black lead powder from the insides of batteries, expert cutters mark the lines along which to cut boards from the squared-off log, and then skillfully slice the log into boards. Truck owners coordinate the activity, carrying the wood to Oaxaca Valley markets along a network of old logging roads, dodging roadblocks and patrols along the way.

Poaching generates benefits much greater than other area labor opportunities, both to truck drivers and to cutters. One driver who fell into the hands of local authorities reported that his pickup load of wood took three people 4 hours to cut and would have sold for 1,000 pesos in Oaxaca. After discounting the costs of gasoline and oil for cutting and transport, that still nets each cutter more than 200 pesos[7] for 4 hours of work, in an economy wherein the going rate for an entire day weeding someone's

---

[6] Timber smuggling evolved from traditional forest usage in the context of the imposed restrictions of scientific forestry, which raised barriers to trafficking in wood at the same time it created infrastructure, skills, and opportunities to do so (Klooster, 2000a)

[7] In 1995, the exchange rate averaged 6.4 pesos per U.S. dollar, according to International Monetary Fund statistics.

milpa is worth only 15 to 25 pesos. In the forest, a board goes for 6 pesos, whereas in Oaxaca that board sells for 10 pesos, so a driver expects to earn a 400-peso return from a typical load of 100 boards purchased from other cutters (Klooster, 1997; Klooster, 2000a).

THE COMMUNITY DYNAMICS OF TIMBER POACHING

Community dynamics affect the intensity of timber poaching, which has the potential to degrade the forest. In 1995, communal authorities aggressively patrolled the forest to control unsanctioned cutting. They meted out fines, temporarily decommissioned vehicles and chainsaws, and reported repeat offenders to federal authorities. Their activities greatly reduced rates of timber poaching as compared with previous seasons.

Andrés, a young man who was a full-time poacher in the past, estimates that in 1995, only two trucks a week made it past local patrols and roadblocks. In 1994, however, when enforcement was lax, he smuggled five truckloads each week. He estimated there were approximately 20 trips per week between 10 smuggling trucks, which would translate into an illegal cut of 4,500 m^3 of standing timber, 10 times the estimated 1995 rate. Oaxaca lumber merchants blame cheap lumber from the community for driving several legal lumberyards out of business.[8]

In a total free-for-all, in which most of the community's 35 pickup trucks participated, poachers could conceivably cut more than 20,000 m^3 of pine per year. The annual sustainable cut, as calculated in the community's forestry study, is only 16,000 m^3, so timber poaching could potentially inflict drastic changes on forest cover and composition (Klooster, 2000a).

Barring a total breakdown of community control over poachers, however, the immediate threat from timber poaching is degradation. Timber poachers focus their work on mature trees with straight, branchless boles, close to roads. Even when poachers thin the forest by cutting young pines for posts and beams, they also choose the best trees available. The effects of such high grading in Mexican pine forests are well known.

> The superior provenances [trees best adapted to a particular site and set of environmental conditions] are frequently the first to be removed. The trees left to regenerate are often of poor form, stagheaded [or rogue], and as seed trees produce genetically inferior progeny (Styles, 1993, p. 415).

The community's prescribed silvicultural method, in contrast, calls for culling undesirable individual trees in thinning cuts, saving the best trees in the forest to serve as seed trees in small partial clearings. Poachers take precisely the trees forestry supposedly reserves for seed trees. The end effect of unrestrained poaching will be genetic impoverishment of pines, lack of pine regeneration, oak dominance, and drastic, long-term decreases in available commercial volume. It potentially threatens the sustainability of the agroecosystem's forest component.

---

[8] *El Imparcial* (Oaxaca City newspaper). June 4, 1995.

Despite the high return to labor from timber poaching, a free-for-all has not been unleashed on the forest. Timber poaching is not unrestrained. Checks come from two sources: enforcement through community structures of authority and local ethics and values regarding proper behavior in the forest commons. Many community members see timber poaching as a form of theft against the community because individuals cut community-owned trees, sell the lumber, and leave nothing of common benefit. These community members have concerns that the pressures to cut and clear could get out of hand and threaten the forest. Thus, there is still substantial support for restrictions and the enforcement efforts of communal authorities. Unfortunately, community social processes threaten these sources of restraint. Community social processes affect the way San Martin interacts with its agroecosystem.

SOCIAL STRUCTURES, VALUES, AND THE FORESTRY–AGRICULTURE RELATIONSHIP

To clarify the relationship between San Martin's social and political structures, values, and production strategies, this section compares and contrasts San Martin with seven other forestry communities that avoid problems with timber poaching. These seven communities illustrate a more successful relationship of community and agroecosystem.

Site visits, interviews, and literature reviews provide comparative information on the following communities: San Antonio, San Martin's immediate neighbor and a member of the Union of Forestry *Ejidos* and *Comunidades* of Oaxaca (UCEFO); several communities in the Union of Zapotec and Chinantec Communities in the Sierra Norte of Oaxaca (UZACHI); Ixtlán de Juárez, also in the Sierra Norte; and Nuevo San Juan Parangaricutiro, Michoacán. These communities are among the most successful forestry communities in Mexico because they log conservatively, reforest aggressively, leverage substantial rural development benefits from forestry, and successfully control timber poaching (Klooster, 1997; Klooster, 2000b)

With the exception of San Antonio, *milpa* and livestock occupy relatively less area and labor in the aforementioned communities than in San Martin, whereas fruit production and services play much greater roles. However, *milpa* remains an important activity in all of them, although it is less important than in the past. All of these communities possess highland pine and oak forests, and like San Martin, they have logging businesses integrated into community political structures.

DISTRIBUTION OF BENEFITS

The successful communities invest the proceeds from logging into public works, including schools, churches, and road improvements. San Antonio and Ixtlán also distribute a portion of forestry proceeds to community members. Each community member of San Antonio received $690 in 1994. The other communities dedicate these proceeds to economic diversification designed to create more local jobs.

In San Martin, distribution of the proceeds from community forestry is much less equitable. Although forestry proceeds do fund public works projects such as the

Catholic temple, a cobblestone street, a health clinic, government buildings, and the community-owned sawmill, these all are aggregated in the central village. The outlying settlements, meanwhile, consistently see their requests for electrification, schools, roadwork, and communal pickup trucks rejected. Although forestry generates sporadic earnings for nearly half of the working-age men in the community, a select few earn substantially more than the average, and this group is disproportionately from San Martin Ocotlán Village.

In 1995, after 6 years without any profit distribution from the forestry business, dissidents demanded audits that uncovered loans of money and wood to a group of wealthy men from the central village who owed the forestry coffers of nearly 208,000 pesos, equal to 40% of payroll. The overwhelming majority of recent debtors were from the central village. Many owed sums in excess of 10,000 pesos, and most refused to acknowledge their debts. A sawmill audit uncovered an additional problem with the misclassification of boards, which represented a bonanza for truck owners from the central village who could buy cheap and resell at a higher, more expensive classification. In San Martin, the proceeds from legal forestry concentrate in a few hands (Klooster, 2000b and Klooster, 1999).

SOCIAL AND POLITICAL STRUCTURES

The successful communities integrate production into participatory community political structures. Vigorous, regular, and well-attended community assemblies are standard features. In Nuevo San Juan, for example, most of the community's 1,000 members convene each month (Sanchez, 1995). In Ixtlán, failure to attend community assemblies results in fines deducted from forestry profit shares. Community assemblies determine how to distribute forestry revenues and elect leaders and forestry administrators.

The seven communities share accounting and reporting practices that provide community members with healthy flows of information. In UZACHI and San Antonio, special committees receive financial training and assistance from nongovernmental organizations (NGOs) to oversee forestry finances, whereas in Ixtlán and Nuevo San Juan, a traditional oversight committee, the *Consejo de Vigilancia,* takes on this function. Effective oversight enables accountability, and communal leaders who misappropriate funds have been quickly removed from office in UZACHI (Ramirez and Chapela, 1995) and Ixtlán.

This situation contrasts with the situation in San Martin, where a forestry elite in the central village dominates the logging business.[9] Most of the community's members neither participate in the decisions regarding collective use of the forest nor share in the jobs and economic benefits that commercial forestry produces. In effect, a forestry elite usurps the community's forest.

The formal political institutions of common property management should provide controls against mismanaging of the communal forestry enterprise. Like the

---

[9] This elite developed when concessionaires cultivated a clientele in a community already stratified along family lines.

seven other communities, the basis of local power in San Martin Ocotlán is supposed to be an assembly of household heads (mostly male) that elects community members to positions of authority. These include a president and executive committee to represent the community to outside authorities, administrators of the logging business, an oversight committee to monitor the other authorities, and a forest patrol committee to combat timber poaching. Similar positions exist in the other seven communities. They should provide a system of checks and balances to maintain the accountability of communal authorities and forestry administrators, but the forestry elite find ways to circumvent these checks on its power.

The elite dominate communal institutions through intimidation, manipulating elections and discouraging participation in community assemblies. In contrast to the appropriate accounting and reporting institutions of the other communities, San Martin authorities hire a professional accountant, and the opacity of his public reports does not facilitate accountability. Threats, violence, bribes, and the manipulation of reciprocal obligations are common tools of internal politics. The elite comprise the Council of Distinguished Men, a traditional body of authority parallel to the community assembly, and this traditional institution often circumvents the community assembly in decision making. The aforementioned concentration of forestry proceeds reflects the unequal distribution of power in the community.

These weapons of the not-so-weak perpetuate the forestry elite's power and privilege while undermining the democratic potential of community political institutions. The distribution of power in San Martin is reflected in the concentration of forestry proceeds already mentioned.

THE FOREST AS AN ECONOMIC SUBSIDY

In the seven successful communities, forestry employment and proceeds help to maintain community viability by providing alternatives to emigration. Interestingly, proceeds from organized forestry also explicitly subsidize agriculture and the establishment of new agricultural activities including floriculture, fruiticulture, and agricultural marketing initiatives. These communities decided to use forestry to support agriculture for two reasons. First, it was an important gesture of solidarity to members of the community who did not participate in logging, preferring to farm. Second, it was part of a strategy to increase employment without increasing dependence on the forest. For this reason, the successful communities also use funds from forestry, together with organizational structures strengthened by forestry management, as a fulcrum to leverage new economic activities, such as ecotourism (Bray, 1991; Chapela, 1997; Klooster, 1999; Klooster, 2000b; Lemus, 1995; Sanchez, 1995). San Martin, in contrast, shunts forestry proceeds to an elite minority instead of using them to subsidize agricultural production or promote productive diversification.

FOREST MANAGEMENT

The successful communities act to conserve and enhance their forests. Reforestation efforts exceed those required by law, with Nuevo San Juan digging down through several feet of volcanic ash from a 1943 eruption to reach soil. These communities also harmonize local forest management goals with legal logging plans, maintaining reserve areas and protecting the watersheds from which they obtain water for drinking and irrigation (Bray, 1991; Chapela and Lara, 1995). The forest managements of Nuevo San Juan and UZACHI have earned certification for good management from Smart Wood, an international certifying organization accredited by the Forest Stewardship Council. The successful communities invest in future forest productivity, despite the increased costs and decreased timber volumes implied by these efforts (Chapela and Lara, 1995).

In San Martin, on the other hand, logging practices associated with a corrupted and mismanaged communal forestry business favor short-term benefits over long-term forest care. In response to local complaints, inspectors from the environmental enforcement agency visited part of the community's logging area. They found cutting intensities higher than permitted. Instead of the thinning cuts called for in the official management plan, loggers targeted sawmill-diameter logs. The inspectors also noted that debris from logging had not been adequately cleaned, creating a fire hazard, and that reforestation efforts were below goals. Instead of conducting forestry for the long-term health of the forest, the community's hired forester directed logging on sawmill-quality trees, as directed by leaders of the forestry business.

Timber poaching is under control in the successful communities. Community members participate in decisions about forestry. They can monitor and hold their leaders accountable for the financial management of the forestry business, and they enjoy fairly distributed benefits from employment and investment in public works generated by their community-owned logging businesses. Not surprisingly, they perceive restrictions on cutting, burning, and grazing as legitimate and comply with them (Klooster, 2000a; Klooster, 2000b).

As a contrast, in San Martin, elite dominance of forestry undermines the community basis for restrictions on timber poaching because the same communal authorities who enforce restrictions on clearing and cutting are associated with corruption in the logging and milling business, and this compromises their legitimacy and provides timber poachers with a mantle of legitimacy. Pedro Diaz, an admitted timber poacher and one of the loudest critics of village authorities, clarifies this link: "When we are cutting one tree, they want to throw us in jail, so how can they cut 50,000 trees and not produce anything? You can't say I'm going to cut 50,000 trees, and you, two gunnysacks of charcoal. There are two kinds of timber poachers. Some have licenses but still leave nothing for the community" (Klooster, 1997, p. 263).

A STRUGGLE FOR THE FOREST

In San Martin, the association of community leaders with corruption and inequity does more than lubricate timber poaching. The same local ethic of propriety in the use of the forest commons that continues to put a brake on timber poaching generates opposition to the inequity in legal forestry. In 1995, the community assembly called for audits and then demanded accountability for the unpaid loans these audits uncovered. In one high point of internal conflict over natural resource management, members of Benito Juárez, an outlying village, put up a chain to stop logging trucks from reaching the forest.

The heated exchange between community leaders and the chain's defenders clarified the motivations. Together with perceptions that enforcement against timber poaching and agricultural clearings was unfair, the loan issue particularly grated against community values and expectations for forest management. The denunciation that "just a few are using our resources while the rest are in abject poverty!" became a call to action (Klooster 1997, p. 301, see also Klooster 2000a).

These events brought logging to a halt. In late 1997, dissident members of outlying settlements were elected to positions of communal authority. This is the first time in 20 years that these positions did not fall to members of the central village. Efforts to resolve conflicts, control timber poaching, and return to logging under different institutional arrangements were still under way in early 1999, however. Successful conflict resolution and control of timber poaching were by no means assured, but remain a possibility.

CONCLUSION

In all of the communities considered here, the forest represents a vital means for earning the cash income needed to maintain livelihoods centered on a *milpa* agriculture that neither generates money nor guarantees subsistence. Income from forest-based activities helps community members to continue living and producing as members of an agrarian community. The continuation of this subsidy and the sustainability of both the agroecosystem and the community, however, depend on the form of this relationship. In San Martin Ocotlán, individualistic production strategies rob funds from the community's coffers, sap established structures of community governance, and degrade the community's forest.

The other seven communities considered in this study successfully integrate forest production into traditional structures of community control. They execute forestry in a way that reinforces the community's social and political structures, and resonates with community values. They invest in both the forest and agriculture, increasing the long-term ability of the agroecosystem to help sustain the community. They demonstrate that alternatives are possible, and that forests, agriculture, and communities can sustain each other.

ACKNOWLEDGMENTS

Without the cooperation of villagers and village authorities from San Martin Ocotlán, the research on which this article is based would have been impossible. For support in the field, I thank the Inter-American Foundation, Fulbright-Hayes, and the University of California, Los Angeles. Support during writing came from the Princeton Environmental Institute, the Science Technology and Environmental Policy Program at Princeton University, and William and Jane Fortune. I thank them for their support. Thanks also to Peter Vandergeest, David Barkin, and Cornelia Flora for commenting on earlier versions of this manuscript.

REFERENCES

Alcorn, J. B., Process as resource: the traditional agricultural ideology of Bora and Huastec management and its implications for research, *Adv. Econ. Bot.,* 7, 63, 1989.

Alvarez-Icaza, P., Forestry as a social enterprise, *Cult. Surv. Q.,* 17, 45, 1993.

Bray, D. B., The struggle for the forest, *Grassroots Dev.,* 15, 13, 1991.

Bray, D. B. and Wexler, M. B., Forest policies in Mexico, in *Changing Structures of Mexico: Political, Social and Economic Prospects,* Randall, L., Ed., M.E. Sharpe Press, Armonk, 1996, 217.

Chapela, G., El cambio liberal del sector forestal en Mexico, in *Semillas Para El Campo En El Campo: Medio Ambiente, Mercados y Organizacion Campesina,* Pare, L., Bray, D. B., Burstein, J. and S. Martinez V., Eds., UNAM-IIS, Sansekan Tinemi, and Saldebas A.C., Mexico, 1997, 37.

Chapela, F. and Lara, Y., *El Papel De Las Comunidades En La Conservacion De Los Bosques,* Concejo Civil Mexicano para la Silvicultura Sostenible, Mexico City, 1995.

Hecht, S. B., Anderson, A. and May, P. The subsidy from nature: shifting cultivation, successional palm forests, and rural development, *Hum. Organ.,* 47, 25, 1988.

INEGI (Instituto Nacional de Estatística, Geografía e Informática), *Carta De Efectos Climaticos Regionales Mayo-Octubre 1:250,000,* Secretaria de Programacion y Presupuesto/Instituto Nacional de Estadistica e Informatica, Mexico City, 1984.

INEGI (Instituto Nacional de Estatística, Geografía e Informática), *Censo Agropecuario, Estado De Oaxaca,* INEGI, Mexico City, 1991.

Klooster, D., Community-based conservation in Mexico: can it reverse processes of degradation, *Land Degrad. Dev.,* 10, 363, 1999.

Klooster, D., Community forestry and tree theft in Mexico: resistance, or complicity in conservation, *Dev. Change,* 31, 281, 2000a.

Klooster, D., Como no conservar el bosque: la marginalizacion del campesino en la historia forestal mexicana, *Cuadernos Agrarios,* 14, 144, 1996.

Klooster, D., *Conflict in the Commons: Commercial Forestry and Conservation in Mexican Indigenous Communities,* Ph.D. thesis, University of California Los Angeles, Los Angeles, 1997.

Klooster, D., Institutional choice, community, and struggle: a case study of forest co-management in Mexico, *World Dev.,* 28, 1, 2000b.

Lemus, O., Plan de Manejo Integral de los recursos naturales de la comunidad indigena de nuevo san juan parangaricutiro, in *Experiencias Comunitarias En El Manejo De Recursos Naturales: UZACHI-UCFAS,* Martínez, L. J. A., Chávez, L., and Ramírez, G., Eds., mimeo, 1995, 22.

Ramirez, R. and Chapela, F., La union de comunidades Zapoteco-Chinanteca, in *Experiencias Comunitarias En El Manejo De Recursos Naturales: UZACHI-UCFAS,* Martínez L., J. A., Chávez, L., and Ramírez, G., Eds., mimeo, 1995, 29.

Sanchez, M. A., The forestry enterprise of the indigenous community of Nuevo San Juan Parangaricutiro, Michoacan, Mexico, in *Case Studies of Community-Based Forestry Enterprises in the Americas;* Presented at the Symposium on Forestry in the Americas: Community-Based Management and Sustainability, University of Wisconsin-Madison, February 3–4, 1995, Forster, N., Compiler, Land Tenure Center and Institute for Environmental Studies, University of Wisconsin Madison, Madison, 1995, p. 137.

Styles, B. T., Genus *Pinus:* a Mexican purview, in *Biological Diversity of Mexico: Origins and Distribution,* Ramamoorthy, T. P., Bye, R., Lot, A., and Fa, J., Eds., Oxford University Press, New York, 1993, 397.

Wexler, M. B. and Bray, D. B., Reforming forests: from community forests to corporate forestry in Mexico, in *Reforming Mexico's Agrarian Reform,* Randall, L., Ed., M.E. Sharpe, Armonk, 1996, 235.

World Bank, *Mexico Resource Conservation and Forest Sector Review,* Natural Resources and Rural Poverty Operations Division Country Department II Latin America and the Caribbean Regional Office, Washington, DC, 1995.

6 Community Fruits and Vegetables for Export: The Impact on Two Mexican Ecosystems and Rural Communities

Magdalena Barros Nock

CONTENTS

Introduction 69
La Loma, Jalisco 72
Tlaca, Morelos 79
Threats to the Agroecosystems and Rural Communities 82
References 83

INTRODUCTION

By examining two communities that have gone from traditional small farmer agriculture to irrigated agriculture and from local to international markets, I analyze the ecologic, economic, and social impacts and the effects of global dependency on place, the rural community, and the ecosystem. I also analyze the development of winter fruit (melons) and vegetable (onions and cherry tomatoes) production, how the enterprises began, and how each became one of the main economic activities in the community. I present the strategies that farmers and brokers implemented to increase their room for maneuvering in international markets. Farmers created local organizations to obtain credit, inputs, technical assistance, labor, and access to the international market. The problems and conflicts that led farmers to close their packing plants and stop producing for the international market are identified. Both communities developed strong farmer organizations to respond to the emerging agroecology opportunities provided by irrigation and market opportunities resulting from international trade regimes, including the North American Free Trade Agreement (NAFTA).

Farmers in these communities are integrated into commodity markets that have developed their production and reached the international market. Nevertheless, two bad seasons in a row can push the farmers back into poverty and into the lines of emigrants leaving their community to search for employment in the United States. They are vulnerable: fruit and vegetable production involves a very insecure market, high levels of financial resources, intensive labor, regular technical assistance, and day-to-day information. In the two reported cases, agriculture shifted away from natural biologic interactions toward intensive monocropping practices. This shift caused environmental problems including soil nutrient depletion, erosion, contamination of the soil, increased vulnerability to pests and crop diseases, and the lowering of water tables.

Mexico has increased fruit and vegetable production dramatically since 1940. Not only has internal consumption increased, but export also has increased dramatically, going from 629,240 tons between 1980 and 1984 to 2,104,295 tons between 1995 and 1997 (de Grammont et al., 1999, p. xiv).

The main recipient of Mexican fresh and processed fruits and vegetables is the United States. In 1990, 85% of the total exports went to the United States and 10% to Canada, and by the end of the 1990s, 98% went to the United States. All tomatoes, 94% of the strawberries, and all watermelons and cucumbers are exported to the U.S. markets.

Restructuring of fruit and vegetable production shifted, as did control of that production. The states of the north–Sinaloa, Sonora and Baja California–increased their production of nontraditional fruits and vegetables such as peas, eggplant, grapes, and a kind of squash (called *cabocha* and produced exclusively for the Japanese market) destined almost entirely for the United States. Other states have reduced their export volumes considerably. In 1988, 27 of 32 states exported fruits and vegetables to the U.S. markets. By 1991, 11 of these had dropped out of the international market completely or reduced their production for export significantly (Standford, 1996, p. 147).

Land has become increasingly concentrated in rural areas (de Grammont et al., 1999). During the past decades of neoliberal policies and free trade agreements, the Mexican agrarian structure has polarized. It is dominated by large exporting enterprises with more than 1,000 hectares that represent 0.28% (12,487) of the total number of enterprises but cover 44% (48,010,873 hectares) of the total amount of land. On the other hand, 59% (2,620,399) of units of production have fewer than 5 hectares each and cover only 5% (5,574,769 hectares) of the national land used for forestry and agriculture (de Grammont et al., 1999, p. 5). The enterprises capable of a steady production for export are located in the states of Baja California, Jalisco, Estado de Mexico, Michoacan, Nayarit, Sinaloa, Sonora, and Tamaulipas. One of the case study communities is in Jalisco, but in an area of small producers. The other is in Morelos, a state with more traditional agriculture.

High productivity in fruits and vegetables for export results from the use of modern technological packages. Modern technology is applied mostly in the northern states of the country, where production is geared mainly to the international market, and where North American capital finances and markets most of the export production (Gomez et al., 1993).

Neoliberal policies of privatization and a decreasing role for the state have resulted in a radical decrease of public sector research on fruits and vegetables. Most of the technological packages come from the United States and are implemented through production contracts, processing industries, or the Mexican associates of these firms. Family farmers, who cannot afford such modern and expensive technological packages and cannot attain the quality standards required in the international market, produce for the domestic market.

This discussion analyzes the interaction between agroecosystems and rural communities in the face of rapid changes in agricultural structure by looking at the experience of two communities that recently have penetrated international fruit and vegetable markets. The method of articulation with the international market has important implications for the agroecosystems and the communities.

Most farmers in the two case studies belong to the *ejido* sector. The *ejido* is a form of land tenure that grants rights of usufruct to agrarian reform communities. These include individual or collective plots and common land. In 1992 the government changed the constitution to allow the privatization, and thus the disappearance, of the *ejido*. However, the *ejido* continues to play an important role in the rural areas. Only approximately 3% (841) of the *ejidos* grow fruits and vegetables. Of these *ejidos*, 282 produce horticultural products, excluding sugar cane. Crops for the domestic market usually are grown by *ejidatarios* on *ejidos*, which are totally or partially irrigated. Some 30% of the *ejidos'* irrigated land is cultivated with fruits and vegetables.

The *ejidos* and small private landholders with irrigated land who export their production usually work under production contracts with transnational firms, although some try to export their produce directly. The case studies presented are drawn from the small group of *ejidos* that have irrigation, have been able to produce fruits and vegetables, and have penetrated the international market.

Transnational firms relate to farmers in different ways. Transnationals have developed strategies for production and market contracts to ensure a constant supply of raw material for their agro-industries. Generally, transnational firms do not rely on intermediaries for the acquisition of the necessary raw materials, but have their own brokers. Farmers who supply transnationals generally are capitalist entrepreneurs with medium to large landholdings, irrigation, and elaborate and expensive technological packages. These farmers use the most advanced techniques for agricultural production. Working with transnationals means good business for the capitalist producers, because transnationals strive for stability in both supply (quality and quantity) and price. A stable price does not always coincide with the lowest price.

Most *ejidatarios* and small private property owners do not meet large transnational firms' demands for a stable supply. They have to sell to small and medium firms that are searching for cheap produce and are interested in doing business with small and medium farmers. These small foreign firms have brokers or firm representatives in different regions searching for new land to use for producing and packing fruits and vegetables, especially during the winter, when U.S. production decreases. These firms supply larger firms, retailers, and wholesale markets and supermarket chains throughout the United States. They deal with farmers through either production contracts or consignment sales. These instruments further vertical integration, serving U.S. interests by securing steadier quality supplies with less risk to the brokerage houses. Many of these firms are owned by Mexican American or Mexican immigrants. Their personnel also are Mexican or of Mexican origin who use their family and friends in Mexico to build social networks in producing regions.

Mexican producers must respond to a rapidly changing market. Exports depend on the demand of the U.S. domestic market and on production within the United States and other countries. The flexibility of small and medium farmers producing fresh fruits and vegetables is limited in many ways. The perishability of the product and the farmers' dependence on foreign capital are two major constraints. Fresh fruit and vegetable production does not permit day-to-day manipulation of the market by farmers, as might be the case with nonperishable commodities, which can be withheld until prices rise. Land use must be planned months ahead, on the basis of earlier market trends, prices, supply, and input costs. Small farmers lack adequate up-to-date information about demand. For brokers and transnational firms, the market for perishables is difficult because of seasonal changes in supply, the need for fast transportation from the packing plant to the market, and hourly changes in prices at the border. All Mexican exports to the United States are inspected by the U.S. Department of Agriculture and must pass through certified packing stations. This frequently has increased costs for farmers as shipments are not allowed to pass or are delayed, resulting in spoilage.

The two reported case studies deal with farmers in small communities that were able to create strong local organizations, construct their own packing plant, and export their produce through small foreign firms. I analyze the problems and conflicts farmers faced throughout the decades with the different actors involved. It is my intent to show how the proliferation of these small firms (especially since NAFTA) is producing an environmental problem in the regions where they work. Furthermore, it is precisely the strategies that firms and producers require for competition in the international market that lead themselves to unsustainable production practices and eventually to failure.

LA LOMA, JALISCO

The agrarian town of La Loma, with 8,950 inhabitants, is located southwest of the state of Jalisco. The *ejido,* La Loma, is one of the biggest *ejidos* in the region, with 405 members and 5,850 hectares. Approximately 18% of *ejido* land is located in the Valley of Autlán. Each *ejidatario* has eight hectares of land on the average for agricultural production. The rest is used collectively for cattle grazing and wood collecting.

Since the 1940s, migration has been one of the main economic alternatives for the people of La Loma. Most migrants have gone to the United States to seek seasonal jobs, and more recently, to become permanent U.S. residents. Every family in La Loma has at least one family member in the United States. For many peasants, life has consisted of working on the land at the beginning of the rainy season and then migrating as soon as the land is sown with maize and beans. Other members of the household, mainly women and children, harvest the crops. Normally, those who migrate to the United States send money regularly to their families for their subsistence and, if possible, they use part of their earnings to improve their homes, buy agricultural implements, or invest in off-farm activities.

Irrigation changed the situation for those *ejidatarios* whose land was located in the Valley of Autlán. A central irrigation system, called El Operado, was constructed by the government in the late 1950s and early 1960s. El Operado irrigates both *ejido* land and private property, serving 300 smallholders with an average farm size of 12 hectares. In the *ejido* sector, 17 *ejidos* received water (59% of the land in El Operado is *ejido* land. These *ejidos* have land scattered throughout the valley. El Operado does not irrigate the entire valley. Therefore in many *ejidos,* only part of the land is irrigated. Approximately 1,084 of the 5,850 hectares of *ejido* land in La Loma receive irrigation.

It took several years for peasants to learn how to use the new infrastructure. During the first years, they grew maize and beans twice a year, rather than only once. They gradually introduced vegetables and fruits, including experimentation with watermelon, tomato *cascara* (a variety of tomato common in Mexico), onion, garlic, and chili, which they sold in the region and in the state capital, Guadalajara.

By the 1960s, irrigation had changed the agrarian structure of the region in many ways. Overnight almost half of the *ejidatarios* found themselves with resources that gave them new economic possibilities. That accented the social differentiation within the *ejido* and in the town. Farmers who had irrigation were able, with the help of money earned in the United States, and later on with the introduction of winter melons, to accumulate capital.

Attracted by the new irrigation, in the late 1960s American firms started renting *ejido* land in La Loma during the winter months to plant melons for export. These brokers had worked mainly with private landowners in the area. They first approached the local *cacique* (local power broker), Chucho, who later introduced them to other farmers.

Brokers made deals with farmers individually. Most farmers at this stage rented out their land. Farmers were introduced to the know-how of fruit and vegetable production. During the first years, farmers were in a relationship of subordination with the foreign firms. They did not participate in the decision-making process, and their earnings were not enough to sustain their families. Farmers had no mechanisms through which to oblige foreign firms to fulfill their agreements. By 1970, farmers of La Loma were linked to the international market.

Farmers knew that being a small community, far from packing plants and markets, was a disadvantage. Also, they understood that foreign firms did not like to deal with small farmers. They had to become attractive to brokers. Some farmers started to organize themselves into a producers' organization, producing winter melons for export and building their own packing plant. At first, only 10 farmers were interested in the idea. Two years later, 30 farmers were interested.

Farmers formed a Society of Social Solidarity (SSS), which enabled them to legalize their packing plant and to receive certain benefits such as credit and tax exemption. Special permission was needed to sow and export cash crops. Farmers had to be organized in a Local Agriculture Association (LAA) to get that permission.

Bureaucratic procedures were endless. A great number of federal and state government authorities were involved in the legal organization of the packing plant. Few *ejidatarios* knew of these authorities, and most of them were located in other towns and cities, meaning extra cost and time. After many trips to the state capital and negotiations with state employees, the SSS of La Loma was founded in 1975. Chucho became president of the organization. Most members were *ejidatarios* whose land had benefited from the irrigation system, El Operado, but the SSS included private landholders and some landless people able to rent land with irrigation.

Currently, 264 Mexican LAAs are combined into 17 Regional Agrarian Unions (RAUs), which form the National Confederation of Horticulture Producers (NCHP). The NCHP was created in the 1960s to organize and regulate Mexican exports through the distribution of official obligatory certificates of origin and shipment permits. With these tools, the NCHP was supposed to regulate the national market to prevent overproduction, defend farmers' interests, and help farmers' associations penetrate the international market. They provided information on potentially interested American firms and offered information on border prices for different products.

Guided by the information that NCHP provided, Chucho contacted several U.S. firms, trying to interest them in buying the melons produced in La Loma. He traveled, accompanied by a *compadre* (godfather of one's child) to McAllen, Texas, to contact firms that might be interested. Chucho did not succeed: most of the firms recommended by the NCHP were not interested in negotiating production contracts with small and medium *ejidatarios*.

In the nearby town of El Grullo, production started to decrease because of crop diseases and soil erosion. A Texas firm called El Manto that was working with the association from El Grullo became interested in La Loma's production.

The agreement reached between Chucho and the firm consisted of the following: El Manto would finance the construction of the packing plant, buy the machinery, and manage it during the harvest season. With their production, farmers would pay half of the sum lent for this, with interest. After paying for their half, the farmers would receive 50% of the profits made by the packing plant and become partners with El Manto. The firm gave farmers $200 to start production and supplied them with seed brought from the United States. Farmers, in turn, committed themselves to packing their melons only in this packing plant, no matter what the price might be in other packing plants. The first fruit, harvested in April of 1976, was packed by young women of La Loma and exported to the United States.

This way of organizing the packing plant meant that the foreign firm also benefited from the subsidies that the government gave the SSS. Mexican resources in the form of subsidies were transferred to the international sphere, benefiting foreign firms and subsidizing foreign consumers. Many of these small foreign firms work behind *ejidos* or *prestanombres* (name lenders), avoiding all responsibility toward the country in which they are working, including taxes.

The introduction of winter melon production brought many changes. It introduced a new production process to the ecosystem as well as new social relations of production in agriculture to the community. Local farmers had to work with a new product and new technological packages, acquire the know-how, build infrastructure, mechanize production, and contract large numbers of wage laborers. The packing plant meant new organization of postharvest production and new activities, such as accounting and administration, that had to be learned and managed. Farmers had to deal with a country completely different from their own, through foreign firms and brokers who had different discourses, rules, and ways of doing things. A process of internationalizing agriculture was well on its way. The economic differentiation among farmers was accented. The distinction between the farmers who had irrigation and those who did not became sharper.

The SSS took care of constructing and running the packing plant and negotiating with El Manto, and the LAA took care of aspects related to agricultural production and export permits. Several issues were negotiated with the Ministry of Agriculture (SARH, now SAGAR) through the LAA. The most important were water and technical assistance, which had always been deficient and now are even scarcer because of recent cuts in personnel. Farmers also obtained from SARH the documents necessary for export, such as the phytosanitary permits that certify produce quality. This procedure was adhered to closely. If a permit was not issued truthfully and melons had any crop disease or did not have the necessary quality, the load would be sent back from the border, increasing farmers' losses.

El Manto negotiated with the SSS the terms of the contract for all farmers. Through Carlos, the firm's representative, the firm provided each farmer with seeds and $200 every year, which was not enough to cover the costs of production. El Manto delegated the distribution of seeds and credit to the leaders of the organization.

The brokerage firm deducted their loans from the first fruit delivered to the packing plant, no matter which farmer produced the fruit. Afterward, farmers had to sort out who had not delivered. If a farmer did not repay the loan, the rest of the farmers paid for him or her. The firm did not risk anything.

Some farmers in La Loma had acquired experience in growing melons while working in U.S. melon fields. They had the basic know-how, which was supplemented by the relatively limited technical assistance given by the brokerage firm and the SARH. The firm told the farmers what inputs to use according to the requirements of the U.S. Department of Agriculture, even selling farmers the necessary inputs.

The average annual investment required to produce 1 hectare of melons ranges from U.S. $500 to $700.[1] A lack of financial resources forces many farmers to negotiate disadvantageous production contracts with brokers and to have dealings with brokers who have dubious reputations. Lack of financial resources has forced farmers to cut back in fertilizer and insecticide application and the number of day laborers used for harvest.

State policies and the ability of farmers to negotiate credit collectively have influenced the farmers' degree of financial vulnerability. Farmers have had access to four sources of financial resources besides production contracts: banks (the state bank, Banrural, and private banks), moneylenders, self-financing, and sharecropping.

---

[1] Interviews with farmers and SARH employees.

The availability of credit for fruits and vegetables from Banrural is irregular. Banrural has given preference through the decades to basic grains to increase the national food supply. Since the 1970s, the Mexican government has used the *ejido* to produce basic foods. Maize and bean production for national consumption was supported by, if not imposed on, the *ejido* sector. Credit allocation favored maize producers. In La Loma, most farmers grow maize and sorghum during the summer in the rainfed land. These products have received constant support from the state. Some farmers save part of the credit from maize and sorghum to supplement fruit and vegetable production costs.

Farmers able to accumulate some capital have rented land from other *ejidatarios,* especially women. They have also sharecropped with *ejidatarios* who did not have enough capital, but did have land and labor. A few farmers have rented out part of their land to obtain cash to start melon production in the remaining hectares.

A common source of credit in the rural areas are moneylenders, who usually are the well off persons in a community. They usually demand high interest rates. Moneylending also can be a risk for the lenders: if the crop does not succeed, then repayment of the loan might be delayed or altogether impossible. Moneylenders lend only to people they know well, with whom they have a certain kind of tie or social link. For many farmers, production contracts have been preferable to moneylenders.

The benefits of producing for the international market were soon evident. After 3 years of producing melons for the international market, 80% of the LAA members were able to buy a truck. Melon production increased steadily. In 1976, of the 156 *ejidatarios* with irrigation, only 28 ventured into this business, sowing an average of 2 to 4 hectares each to reduce risk. Amazing profits were made, and soon many more farmers wanted to join the organization. For 3 years, LAA membership increased. By 1979, 148 farmers were producing melons in more than 900 irrigated hectares, a ninefold increase over the 3 years. Production contracts negotiated through the SSS had an homogenizing effect: credit and technical assistance were made available to many farmers who otherwise would not have been able to produce fruits for export.

In 1979 melon production exceeded what the packing plant could pack and market. The market became saturated and prices fell. Farmers waited for more than 15 hours to have their melons packed, standing in long queues under the sun, which decreased the quality of the fruit. At the height of the season, the fruit had to be dumped along the road, and many farmers suffered great losses. The next year, the amount of land sown with melons decreased to 249 hectares. Many farmers lost all their capital and even their property.

To fight market fluctuations, 12 farmers decided to try their luck with cherry tomatoes, buying a tomato selection band for the packing shed. In 1980 they had very good returns. The next year, they did even better as a result of frost in the United States that destroyed American crops and caused prices to rise dramatically. That year they built a greenhouse. Motivated by the good price, 30 farmers sowed cherry tomato in 1982. That increase in production at La Loma was multiplied all around the winter vegetable regions of the country. Prices dropped, and many farmers lost what they had gained in the previous years.

Production costs started to increase considerably with the second oil shock in 1979 and the withdrawal of state subsidies. The intensive monoculture caused diseases to spread, which further increased production costs. More pesticides and better fertilizers were required. The amount of money received through production contracts and credit remained constant and did not cover the increase in production costs.

The El Manto representative's main complaint was that farmers did not use the necessary inputs, weakening the melon plant and making it more susceptible to crop disease. The firm manager complained that farmers were always trying to save money, or to retain the money for their own purposes, at the expense of the firm. However, the farmers said that the firm had never lent them enough money to buy the necessary inputs or to pay for the labor necessary to carry out the required work in the fields. The firm's average credit of about $200 per hectare did not meet the U.S. $500 to $700 required for production.

Lack of technical assistance has been a constant problem, despite the firm's contractual obligation to provide that assistance. Also, the technical assistance tended to come after the field was eroded and infected with crop diseases. In as little time as a week a whole field can become totally infested with pests. Well-off La Loma farmers started renting or sharecropping land elsewhere to continue producing melons.

The partnership with El Manto ended in the season of 1983–1984. After the harvest, when most of the melons had been packed and shipped to the United States, the brokerage firm informed the executive committee that the prices had dropped dramatically, and that the quality of the shipment had been poor. El Manto therefore was not able to pay any more than what had already been advanced to the farmers. In fact, the farmers owed El Manto money! The news was received with astonishment, provoking angry reactions from many farmers who saw all their work and money disappear. According to them, the firm still owes the farmers approximately U.S. $75,000.

All the farmers lost a considerable amount of money. They were in debt, and their land was unable to produce horticultural crops for export. The packing plant closed and remained closed for 4 years.

The deterioration of their land and the disappearance of the broker had a negative impact on the local organization. Conflict arose within the organizations to the point that it was not possible to continue working together. Some farmers suffered losses so large that they had to sell their trucks and other assets to pay back part of their loans.

Associations such as the LAA, SSS, RAU, and NCHP, whether local, regional, or national, have no real means to control what happens to their produce once they cross the border, and they have no means to force foreign brokers to comply with the terms of their contracts.

A new irrigation infrastructure was built at the end of the 1980s and the early 1990s by the government. This project was planned to irrigate approximately 2,000 hectares, allowing more fruit and vegetable production. Seeing the future advantages of keeping the packing plant open in 1987, several farmers and state employees started the necessary negotiations to open the packing plant once more. The only way to produce melons for export was to plant them in new land that had not been used before for this purpose, because the old land had not yet recuperated. New leaders, such as Manuel, were key in reopening the packing plant.

Manuel migrated to the United States in the late 1960s because his land had not received irrigation from the Operado dam. He lived and worked in California and Texas on farms that produced fruits and vegetables. He became a tractor driver and made many contacts with owners of different brokerage firms. His wife sowed maize on his land using the remittances he sent home. In 1973, Manuel started working for a Japanese man who had an import firm in the United States. Manuel worked for him in Mexicali for 7 years and learned the business. His family told him about the success of the packing plant in La Loma, so in 1980 he decided to come back and try his luck in his hometown. He rented land and started producing melons. He became a member of the SSS and LAA. When his land got infested with pest and the packing plant closed, he left again for the United States.

With the money earned in the United States, Manuel managed to buy a tractor and a truck. He built his house with these U.S. savings and with the profits from melon production. Because his land would be irrigated when the new dam was finished, he decided to return to La Loma and try his luck again. He participated in promoting the reopening of the packing plant and was elected president of the SSS and LAA executive committees. He worked for 3 years as president of the committee. In the season of 1987–1988, he sowed 20 hectares with melons. During the next season, he sowed 2 hectares in the *ejido* of La Loma and, in partnership with a broker, rented land in other *ejidos*, even opening a packing plant in a neighbouring *ejido*.

The SSS met and agreed that Manuel and another member should make a trip to the U.S. border (McAllen) in search of a buyer, replicating the contract with the brokerage firm running the packing plant and purchasing the fruit "at the gate" or taking it to the border on commission. The provision of credit by the bank was conditional on his finding a buyer.

Manuel, with the help of a *compadre* who lived in the United States, looked for possible buyers, especially Zuki, who had become godparent of his second daughter at her first communion. Zuki had a fruit brokerage firm. He bought fruit in the valley of Autlán and in Tonaya as well as in Michoacán. Zuki and Manuel agreed that the packing plant would be rented to Zuki's brokerage firm, and that the firm would supply the farmers with seed, in addition to $200 dollars each. Farmers would be obliged to sell all their fruit at gate prices to the firm. By activating the social network he had built when he migrated to United States, Manuel signed the contract on behalf of the SSS. The SSS lost control of the packing plant, and the farmers were charged for packing of their fruit. The returns would go to the firm. The melon price was set according to the price in Apatzingan, which usually was lower than the prices paid in the United States. The firm personnel took complete charge of the packing plant, made all the technical repairs needed, hired specialized workers from Michoacán, and bought the necessary inputs.

Farmers had to obtain financial resources to supplement those advanced by the firm. To do so, the farmers relied on sharecropping and moneylenders. The lack of subsidies and credit made production for export even more difficult for La Loma farmers than it had been in the 1970s.

At harvest, the packing plant was opened. After the third week, the price started to drop, until it equaled the price paid in the national market. Some farmers, mainly those who had not been satisfied with the contract from the beginning, began to sell their melons to intermediaries from the national market. This provoked a response: the prices offered by the firm started to rise again. The fact that some farmers sold their melons in the national market and not to the packing plant led to conflicts between farmers and the firm, and the firm declared that it would not work with this SSS again.

In 1991, the management and maintenance of the irrigation system was given to the regional farmers' organization. The La Loma farmers had to build new networks to negotiate with new actors. During the next seasons, farmers from La Loma were not able to obtain water for melon production and did not open their packing plant. Farmers tried with tenacity to keep their packing plant open and their production running, but they had been affected by several structural constraints: scarce financial resources, increased production costs, a rise in input price because of inflation and the withdrawal of subsidies on input, lack of water, crop disease, uncertainty in the international market, and a general lack of information.

In small communities such as La Loma, a few actors commonly monopolize social networks with outside social actors who provide support, financial means, and information, making them individual, not collective, networks. Small communities can be isolated and the communication infrastructure scarce and deficient: this makes it easier for powerful actors to monopolize knowledge and information. Hence, in small communities farmers are forced to rely on their leader and have no possibilities to check their actions. Most farmers in small communities find it difficult to be linked to multiple channels to obtain resources, information, and technical assistance. Small farmers do not have the financial means, the social networks, or the necessary information to enter the fruit and vegetable business individually. It is through local organizations and groupings that they can negotiate credit and production contracts with foreign firms and penetrate the international market. However, technical assistance remains absent.

Manuel and another farmer continue to market with a different broker. Each year they sow from 15 to 25 hectares of melons in different ejidos, searching for new land as used land becomes unproductive as a result of crop diseases and pests. However, most of the other farmers have planted their irrigated lands to sugarcane, an increasingly less lucrative crop (Otero and Flora, 1995).

Winter melon production for export is not sustainable in La Loma. Farmers interested in continuing melon production for export rent newly irrigated land by the Trigomil Dam or land located in other ejidos as disease and land erosion makes melon production impossible on La Loma lands.

TLACA, MORELOS

Tlaca is a small onion-producing community of approximately 2,000 inhabitants. The *ejido* of Tlaca has 132 members and 1,056 hectares. There also are private property owners. Between 1971 and 1975, the government constructed an irrigation unit with deep wells that gave water to 725 hectares, benefiting 294 farmers: 396

ejido hectares and 328 private property hectares. Irrigation brought many changes inside the community. The traditional crops produced in the community were maize and sorghum. By 1973, national intermediaries, mainly from Mexico City, were renting the newly irrigated land for the production of onions and tomatoes.

There were several layers of intermediaries between the onion farmers and the consumers. The intermediaries worked either directly with a foreign firm or for international brokers, who themselves were working for a foreign firm. The intermediaries either rented land themselves or contracted with a larger number of farmers either renting their land or sharecropping. Difficulties in acquiring financial resources accentuated farmers' dependency on intermediaries.

Federal support of ejidos increased in the 1970s. In 1974 the ejido of Tlaca was a founding member of the Union of Ejidos Emiliano Zapata (UEEZ). It was formed by 13 *ejidos* and 2,997 *ejidatarios* with 3,700 hectares. The main objectives of UEEZ were to support sorghum production, obtain credit, and find marketing channels in an attempt to fight intermediaries.

Neoliberal policies replaced policies supporting ejidos in the decade of the 1980s. The government of Morelos created the Regional Union of Horticulture Producers of the State of Morelos (Regional Union) controlled by large landowners. The Regional Union, formed by bringing together the Municipal Agricultural Association (MAA), undermined the UEEZ activities in the region (Barros, 1998).

The main objectives of the Regional Union were to increase onion farmers' collective marketing to the national and international market, obtain better prices, and regulate production in an attempt to avoid overproduction in the region. Limiting its activities to onions, technical assistance and research were not among their objectives.

The impact of the Regional Union was felt immediately in the region. To begin with, it increased the harvest period for export through negotiation with the federal SARH. The farmers were able now to export most of their production. A bad season in the United States caused the price of Mexican onions to rise. The next season, as happened with cherry tomatoes in La Loma, there was overproduction all around the country, and prices fell dramatically. The Regional Union tried to reduce the volume produced and control the number of hectares sown, mainly by limiting the market access of small farmers such as those in Tlaca.

Farmers from Tlaca attempted to pack and export their onions through the UEEZ and then with the Regional Union. Neither attempt was successful due to lack of information and experience, and corruption. In the 1980s, the community discussed building a packing plant that would give farmers direct control in the decision-making process. Several attempts to organize a packing plant started and stopped. Then the Montes family, a father and two brothers who all owned land with irrigation and had been able to invest in off-farm activities, organized a local cooperative. Better-off and younger farmers made up the small cooperative membership, which constructed and ran a community packing plant. In 1983, the farmers organized a Municipal Agricultural Association (MMA), and by 1984 it had 72 members with 326 hectares. The members of the cooperative became its leaders. The executive committee was the same in the two organizations. The cooperative negotiated production contracts with foreign firms that benefited all members in the two organizations. The MAA grew at a fast rate, with many farmers packing and selling their onions through the cooperative, which paid more than intermediaries in the onion fields. The layers of intermediaries were reduced. The cooperative charged only a fee for the packing, and the prices offered by brokers were the same for all the farmers. Members of the MAA benefited from the production contracts negotiated by the cooperative with foreign firms. Farmers who were not able to obtain credit at least had seeds and initial capital to start production.

The cooperative's main objectives were to work together in the production, selection, and marketing of onions; to construct basic infrastructure (packing plant, storage room, and offices); to pack, transport, and market onions collectively; and to open a savings account and negotiate credit for all members with the state, private lenders, foreign firms, and national banks. The first year, the cooperative obtained credit through government programs for all those interested in producing onions.

Most farmers sowed from 1 to 4 hectares with onions, with the exception of the cooperative members, who sowed an average of 8 hectares each. The amount of land sown with onions fluctuated from 214 to 326 hectares. This was a modest increase in the number of hectares sown with onions considering that there were approximately 725 irrigated hectares. Farmers were cautious and produced onions only for export on small extensions of land. Onions have high production costs and require high amounts of labor, which is hard to obtain.

In the first years, the cooperative worked very well. The fact that there were only eight members made organization easier. As in the case of La Loma, they tried to diversify their production and bought a selection band for peanuts. This increased their revenues considerably. They built an office and bought a tractor and a truck for the cooperative.

After 6 years of producing onions, crop diseases and pests intensified. The lack of sufficient financial means prevented the farmers from using adequate amounts of fertilizer and insecticide, and they could not pay for the required labor needed to do the necessary work in the fields.

The only possibility small farmers had to produce and export their produce was to negotiate production contracts with brokers and small foreign firms. In the 1989–1990 season, Tacho, a broker, approached the cooperative. He wanted to buy the cooperative's onions and peanuts. By now, production costs were increasing because of crop diseases, and Tlaca farmers were having difficulties meeting U.S. quality standards. Despite the poor prognosis, the cooperative managed to obtain a credit from a private bank. Tacho's wife took care of the cooperative's finances and the negotiations with the bank. The members of the cooperative gave the installments to repay the credit to her, and she was supposed to make the deposits in the bank. At the end of the harvest, Tacho and his wife disappeared. No payments had been made to the bank. The broker took all their production and never paid the farmers the full amount. He had paid the farmers only enough to repay the bank, but his wife had not made the deposit. There were many other complaints against him from other farmers and packers. However, once such brokers leave the country and enter the

United States, there is little that can be done. The cooperative appealed for help to the Regional Union without success. The Peasant National Confederation and the Mexican government declared that there was nothing they could do.

This left the cooperative and the MAA with yet another problem: division among themselves. Mistrust consumed the organizations. Some members blamed the executive committee directly for their problems, believing they had participated in the fraud scheme and gained money from it. Lack of trust decreased their strength in negotiations with the bank, the Regional Union, and government. Seeing the ineffectiveness of the organization, members stopped participating. The cooperative ceased to be an arena for discussion and collective decision. The internal community problems, together with the infestation of pests and diseases in the fields that produced onions for export, led them to close the packing plant.

THREATS TO THE AGROECOSYSTEMS AND RURAL COMMUNITIES

Irrigation is crucial for the production of fruits and vegetables, especially for winter production for export. The Mexican government provided some farmers in the ejidos of La Loma and Tlaca with irrigation, but not the means to use it. These farmers irrigated by flooding, which not only wasted water, but also spread diseases and pests through the fields. There are better irrigation methods that waste less water, but they are too expensive for general use. The increase in winter fruit and vegetable production has meant an increase in water use, decreasing ground water levels.

In an attempt to gain control over their resources, the farmers in both La Loma and Tlaca created local organizations for the production and packing of fruits and vegetables. It was through organization that the farmers were able to negotiate better production contracts with foreign firms and credit from state programs and private banks. Local organization enabled many farmers with scarce resources to obtain resources and access to markets. They would not have been able to obtain these resources on their own. The economic advantages of participating in these organizations were clear.

Through production contracts, new products, high-yield seeds, insecticides, and fertilizers were introduced into La Loma and Tlaca. Farmers in the two communities negotiated production contracts with small foreign firms and brokers interested in working with small and medium-size ejidatarios. Small farmers and small brokerage firms take greater risks. Production contracts provide advantages and risks for all the actors concerned.

Production contracts provide financing and function as mechanisms for technology transfer in the rural areas. There is considerable room for negotiation, as well as for exploitation and conflicting interests, between foreign firms, brokers, and farmers. Farmers depend on brokers and foreign firms for financing, so they have to produce what they want to buy. Usually, foreign firms and brokers specialize in only one or two products, leading farmers to monocropping, which eventually leads to land erosion and soil deterioration. Lack of alternative financial opportunities, adequate information about market possibilities, and technological assistance further limited the options of farmers on the ejidos of La Loma and Tlaca.

International market standards of quality cause farmers to introduce nontraditional inputs in their fields, which often have an adverse effect on the land. The U.S. seeds provided by the production contracts are sometimes highly susceptible to local diseases and pests. In their attempt to work with as many farmers as possible, brokers sometimes also cut corners in the quality of seeds and inputs sold to farmers.

The fruit and vegetable business has had important environmental repercussions. Land erosion and uncontrollable crop disease are found in many communities that venture into this business. In La Loma, nearly 1,000 hectares of melons were replanted with sugarcane because farmers were not able to halt the pest infestations. In the case of Tlaca, farmers were forced to change to a more resistant crop, as the poor quality of their produce excluded it from the international market.

With structural adjustment policies, research and technical assistance to small communities has virtually disappeared. Even during the years before budget cuts, small farmers received almost no technological assistance and thus were not able to keep the land healthy nor control diseases.

Most foreign firms and brokers are interested in short-term profits at minimal cost, leading to intensive land use. After a foreign firm has been active in a region for some years, natural resources suffer from exploitation, including such environmental degradation as soil erosion, pests increases and high levels of crop diseases, and depleted water resources. Three or 4 years after crop disease starts to intensify, the brokerage firms disappear with whatever produce they can get from the farmers (Barros Nock, 2000). As Raynolds (1994) argued in her analysis of oriental vegetable production in the Dominican Republic: "What is at stake here is not only short-term foreign exchange earnings and domestic employment, but the long-term productivity of critical natural resources."

The analysis of the two reported cases in this article shows the negative effects that the introduction of fruit and vegetable production has had on small communities and their ecosystems, rendering it an unsustainable business. With NAFTA, small foreign firms in search of new irrigated land are increasing (Barros Nock, 2000). This could benefit small farmers if research and technical assistance as well as credit and information become accessible to small farmers. As it is, only large landowners and firms are benefiting from the fruit and vegetable business. The government repeatedly mentions the comparative advantages that Mexico has over the United States and Canada in fruit and vegetable production, and how small and medium-size farmers should change from grain production to fruits and vegetables for export. However, the cases show the difficulties that farmers face under the actual conditions described in the article and the important consequences that this business has had on the ecosystem.

REFERENCES

Barros Nock, M., *From Maize to Melons. Struggles and Strategies of Small Mexican Farmers*, CEDLA, Amsterdam, 2000.

Barros Nock, M., *Small Farmers in the Onion Business: A Case Study in Morelos, México*, working paper, Department of Social Studies, Colegio de la Frontera Norte, Tijuana, Mexico, 1998.

de Grammont, H., Gomez Cruz, M. A., Gonzalez, H., and Schwentesius Rindermann R., *Agricultura de Exportación en Tiempos de Globalización,* CIESAS (Centro de Investigaciones y Estudios Superiores en Antropología Social), Juan Pablos, Ed., UNAM (Universidad Nacional Autónoma de Mexico), Mexico City, D.F., Mexico, 1999.

Gomez Cruz, M. A., Schwentesius, R., and Merina Sepulveda, A., La producción de hortalizas de México y el Tratado de Libre Comercio con EUA y Canada, in *Reporte de Investigación 6.* Chapingo, CIESTAAM (Centrode Investigaciones Económicas, Sociales y Tecnologícas de la Agroindustria y la Agricultura Mundial), Universidad Autonoma de Chapingo, Chapingo, Mexico, 1993.

Otero, G. and Flora, C., Sweet neighbors? the state and the sugar industries in the United States and Mexico under NAFTA, in *Mexican Sugarcane Growers: Economic Restructuring and Political Options and Transformation of Rural Mexico,* Singelmann, P., Ed., Center for U.S.-Mexican Studies, University of San Diego, San Diego, No. 7, 1995, 63.

Plan Municipal de Desarrollo Urbano (PMDU) de El Limón, *Gobierno del Estado de Jalisco,* Jalisco, Mexico, 1986.

Raynolds, L. T., The restructuring of third-world agro-exports: changing production relations in the Dominican Republic, *The Global Restructuring of Agro-food Systems,* McMichael, P., Ed., Cornell University Press, Ithaca, NY, 1994, 214.

Stanford, L., Ante la globalización del Tratado de Libre Comercio: el caso de los meloneros de Michoacán, in *La Inserción de la Agricultura Mexicana en la Economía Mundial,* Vol. 1, Lara Flores, S. M. and Chauvet, M., Coord., INAH, UAM, UNAM and PyV, Mexico City, Mexico, 1996, 141.

7 Communities of Interest and Agroecosystem Restoration: Streuobst in Europe

Felix Herzog and Anja Oetmann

CONTENTS

Introduction 85
History of Streuobst 86
Streuobst and Nature Protection: Ecologic Functions .89
Streuobst and Society: Sociocultural Functions 92
Profitability of Streuobst: Economic Functions 94
Nongovernmental Organizations and Farmers for the Protection
and Use of Streuobst: A Success Story of Cooperation 95
Synthesis and Outlook 98
References 99

INTRODUCTION

The German term *streuobst* stands for standard fruit trees scattered ("gestreut") in the open agricultural landscape. "Arboles en diseminado" (Spanish), "fruit tree meadows" (English), and "près vergers" (French) basically designate the same traditional fruit production system, but are limited to fruit trees on grassland. However, standard fruit trees can also be underplanted with arable crops. The most common fruit tree types in Germany and Switzerland are apple (*Malus domestica* Borkh.), pear (*Pyrus communis* L.), plum (*Prunus domestica* L.), mazard (*Prunus avium* L.), and sour cherry, (*Prunus cerasus* L.). Tree densities vary from 20 to 100 or more per hectare. The trees' logs measure 1.60 to 1.80 m or more, but half-standard trees occur as well (1.00 to 1.20 m)

Fruit production is the primary objective of streuobst. Still, it is an agroforestry system, which combines multipurpose trees (fruit, wood, various services) with undercropping or livestock raising (Figure 7.1) (Herzog, 1998). Intensively managed orchards, on the other hand, consist of dwarf trees, pyramids, spindle bushes, and trellis managed solely for fruit production. They are not considered a form of streuobst, nor are fruit trees in homegardens.

FIGURE 7.1   Streuobst on grassland near Leipzig in Saxony, Germany.

Streuobst is a particularly interesting example for the study of interactions between a community of place and various communities of interest. In contrast to other agroecosystems such as the major annual crops, for which there is comparatively little interest from the population in general, streuobst has attracted various interest groups throughout its history, including intellectuals, artists, religious groups and, more recently, ecologists and proponents of nature protection. National and local governments have consistently shown particular interest in maintaining streuobst, intervening with laws and regulations. After a historical review covering the evolution of streuobst, which illustrates the impact of these different groups of interest, the ecologic, economic and sociocultural functions of the system as perceived today are analyzed. Then the discussion concentrates on the interaction between agriculture and nature protection, which determines the current situation of streuobst and its future development.

The focus of this chapter is on Germany and Switzerland, two countries for which there is abundant historical and modern literature on streuobst. However, a belt of streuobst stretches through northern France, southern Germany, and Switzerland to Poland. Approximately 1 million hectares of streuobst exist in 11 European countries (Herzog, 1998). Mixed crop and fruit tree forms of streuobst have disappeared almost completely, and today the trees are grown almost exclusively on grassland.

HISTORY OF STREUOBST

The beginnings of fruiticulture date back to prehistoric times. Temperate fruit trees originate from the Caucasus, Asia Minor, and the Americas, and fruits have always been collected by hunters and gatherers. The domestication of fruit trees started with the development of permanent settlements. Excavations of Neolithic pile villages show that fruit trees were planted on the territory of today's Germany as early as 3500 to 2000 BC. The "pile village apple" was probably one of the first varieties to be cultivated. During the reign of the Roman emperors at the beginning of the Christian era, new types of fruit trees and improvements in their cultivation spread in western and central Europe. From the sixth century onward, such laws as *lex salica* and *lex bajuvariorum* punished people who inflicted damage on fruit trees and stole fruit. At the end of the eighth century, the decree *capitulare de villis* of Charlemagne ordered the planting of all types of fruit trees throughout his empire. About 800, the Carolingians, a religious order, were the first to elaborate recommendations for particular varieties (Friedrich, 1956; Poenicke and Schmidt, 1950).

With the spread of Christianity, the clergy promoted fruiticulture through the installation of fruit gardens in monasteries and writing down instructions on their installation and care. They also did extension work with farmers in regions particularly suited for fruit production. In return, the monasteries required fruit from their congregation and contributors. Distant monasteries exchanged varieties and introduced new varieties from the orient that were acquired during the crusades. Following the example of the clergy, secular authorities and the noble landlords advanced the planting of fruit trees around settlements. The forestry regulations in Wurttemberg, for example, permitted the removal of wild fruit trees from forests if they were to be grafted and transferred to the open land (Lucke et al., 1992).

Until the 15th and 16th centuries, most fruit trees were raised for subsistence in home gardens. As market production started to gain importance, fruit was increasingly processed into products that could be stored over longer periods, such as must, cider, dried fruits, fruit purée, preserved fruit in syrup and walnut oil. Fruit tree varieties selected for these purposes were increasingly planted in the open landscape. This development was strongly supported by government authorities. In 1564, for example, the Saxonian elector, August, wrote the first fruit tree manual in German: (*Das künstliche Obstbüchlein*). Reinforcing his recommendations with laws and regulations, he issued a law obliging each land owner to plant two fruit trees at the occasion of his marriage. This law remained valid until the 18th century and was copied by many other rulers, who linked permissions for citizenship or marriage to the obligation of planting fruit trees (Anonymous, 1941). Stealing fruit and damaging fruit trees continued to be severely punished (Jordan, 1939).

Wars and epidemics of those times strongly affected fruiticulture. For example, fruit and vine production were drastically reduced during the Thirty Years' War (1618–1648). Many plantings were deliberately destroyed or degraded due to neglect. After that war, the authorities created fruit tree nurseries to furnish the planting material for streuobst. There also was a particular focus on planting along streets in order to use this unproductive land for food production. However, fruit trees also were planted on arable (mainly community) land and in vineyards where wine production was abandoned because of unfavorable climatic conditions and pest attacks (grape phylloxera). Thus, many vineyards were transformed into streuobst, with crops and vegetables interplanted (Weller et al., 1986).

Extension activities intensified during the Enlightenment. Priests and school-teachers were requested to spread information on fruiticulture. In 1752, Frederic II (the Great) issued the following decree:

> In each village, a cooperative, well-furnished tree nursery must be installed and run by a man trained in the handling and nursing of trees and capable to educate the villagers. In these tree nurseries, an adequate stock of fruit trees must always be available such that, once all gardens have been planted, the planting can be extended to streets in and near the village. If a surplus of fruit is produced, it is to be sold to the cities. (Lucke et al., 1992, p. 22.)

In the early 19th century, literate groups of priests, medical doctors, chemists, and teachers gathered in pomologic societies to collect, describe, and evaluate the numerous existing fruit varieties to select those of particular value (Kittel, 1895). However, mainly relying on locally well-adapted and robust varieties, their impact on farmers' practices was very limited. The need for standardized quality production led to the replacement of local varieties only when the development of railways made it possible to sell fresh fruit on the urban markets on a larger scale. Over decades, the process of standardization was driven by extension services (Lott, 1993).

In the 20th century, streuobst further increased, and the government's regular and detailed censuses of fruit trees illustrate the importance of this sector of agriculture. In 1900, the average density of fruit trees on agricultural land in the German Empire was 4.8 trees per hectare. On the average, there were 3 trees per inhabitant, ranging from 0.5 (Hamburg) to 7.7 (Lower Franconia) (Kaiserliches Statistisches Amt) [KSA], 1902). In 1913, the number of trees of the four dominating species (apples, pears, plums, cherries) had increased by 13.2% (KSA, 1915). Southern Germany was and still is a center of fruit production in Germany.

Figure 7.2 shows the development of streuobst fruit trees over the course of this century in Baden-Wurttemberg. Between the individual censuses, the collecting guidelines changed somewhat, so the figures cannot be compared directly. Still, the number of streuobst fruit trees remained almost constant until the early 1950s. After World War II, there was a last effort to replant streuobst, which can be explained by the necessity for subsistence fruit production (Weller et al., 1986). Later, fruit trees were planted increasingly in an orchard system, particularly for market-oriented fruit production. This technological development, together with regulation measures taken by national governments and the European Economic Community (EEC), led to a subsequent decline of streuobst.

In Switzerland, the evolution of the streuobst stands was somewhat similar, and currently there are approximately 42,000 hectares of streuobst left (of an estimated 140,000 hectares in 1951) (Bundesamt für Statistik [BFS], 1992, 1993). The share of the agricultural area planted with streuobst is particularly high in hilly areas where the topography limits the potential of more intensive forms of land use.

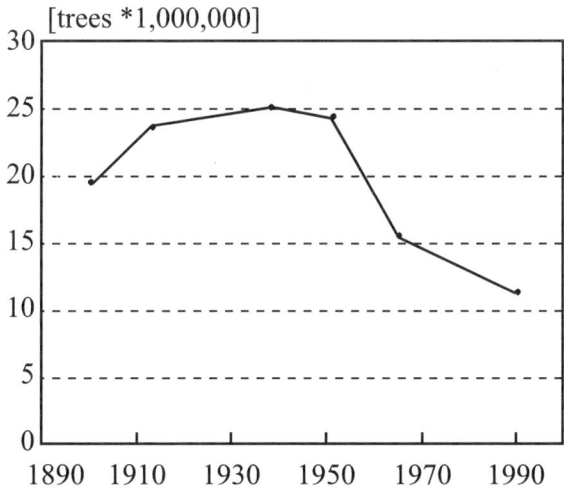

FIGURE 7.2   Number of fruit trees in streuobst systems in Baden-Wurttemberg. In 1900 and 1912, trees in home gardens are included. (Sources: KSA, 1902, 1915; SRA, 1940; Kemmer, 1950; SBA, 1954, 1966; Stadler, 1983, Maag, 1992.)

STREUOBST AND NATURE PROTECTION: ECOLOGIC FUNCTIONS

In a densely populated country such as Germany, there is a strong competition for land, with 48% of the total area devoted to agriculture and 30% covered by forest Bundesministerium für Ernährung, Landwirtschaft, und Forsten ([BELF], 1997). Only a few areas, such as National Parks, are relatively untouched by people. Thus, nature protection must take place in managed areas. The main goals are the maintenance or even the increase of biological diversity and the protection of endangered species. Biologic diversity in streuobst systems is significantly higher than in intensively managed orchards, and numerous articles describe the potential of streuobst as a habitat for animals and plants. In a bibliography of the German Federal Nature Protection Agency, 174 publications on the flora and fauna of this system are listed (Bünger & Kölbach, 1995). In an inventory of streuobst sites in Rhenish Palatinate, 2,391 plant and animal species were counted, 408 of which were rare or threatened with extinction (Simon, 1992). This richness of species can be explained by the ecologic variation in space (i.e., closely intertwined gradients: dry/moist, shaded/sunny, mown/not mown, exposed to/protected from wind), as well as by the ecologic variation in time resulting from the influence of season and management. This leads to a variety of ecologic niches that offer a range of habitats for plants and animals with different environmental requirements. Subsequent horizontal layers (soil, moss, herbal, and several tree layers) host different species. There is abundance of easily decomposing biomass. The overall biomass of arthropods can be higher in streuobst than in forest

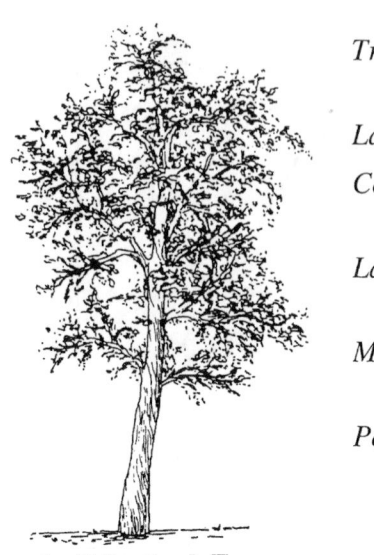

Treetop buzzard, crow, magpie

Lateral treetop shrike

Central trunk spotted woodpecker

Lateral branch fieldfare, chaffinch

Main trunk green and gray-headed woodpecker

Periphery goldfinch

FIGURE 7.3   Nesting opportunities for birds in an old pear tree (adapted from Ullrich, 1987).

ecosystems (Funke et al., 1986). New species of insects are still being discovered (Rudzinski, 1992; Rudzinski and Drissner, 1992).

Streuobst is an important refuge not only for arthropods, but also for small mammals (including bats), reptiles, and amphibians. It provides habitats for numerous bird species whose populations are declining, endangered, or even threatened with extinction (Rösler, 1998). Woodpeckers, nuthatches and tree creepers feed on insects in the trees' wood and bark. Holes or crevices in older trees provide nesting opportunities for birds that nest in caves (Figure 7.3).

Farmers encourage species diversity by providing nesting opportunities for titmice, which feed on the codling moth, a fruit pest (Wiesinger and Otte, 1991). In addition, they often mix streuobst with apiculture, with benefits for both fruit and honey production. The mixture of tree species and varieties, which flower at different periods, increases the availability of nectar and pollen over time, and there are no pesticides to affect the bees, as is often the case in intensively managed orchards.

Streuobst links agriculture to nature protection because it provides habitats which have become scarce in the course of this century. Industrialization and construction, land improvement, modern agriculture, and farmland re-allocations have caused a rapid decline in landscape diversity. As an example, the evolution of land use in a typical rural area of Saxonia is shown in Figure 7.4. From 1912 to 1989, ecologic infrastructure such as hedgerows, tree rows, and fruit tree alleys declined by 53%. Streuobst has resisted change better than most other elements of the landscape mosaic. In the rural Torgau district (western Saxonia), streuobst sites and fruit tree alleys, together with the remaining hedges, tree rows, and the like form an ecologic

FIGURE 7.4 Landscape change in a rural area in Saxonia, eastern Germany.

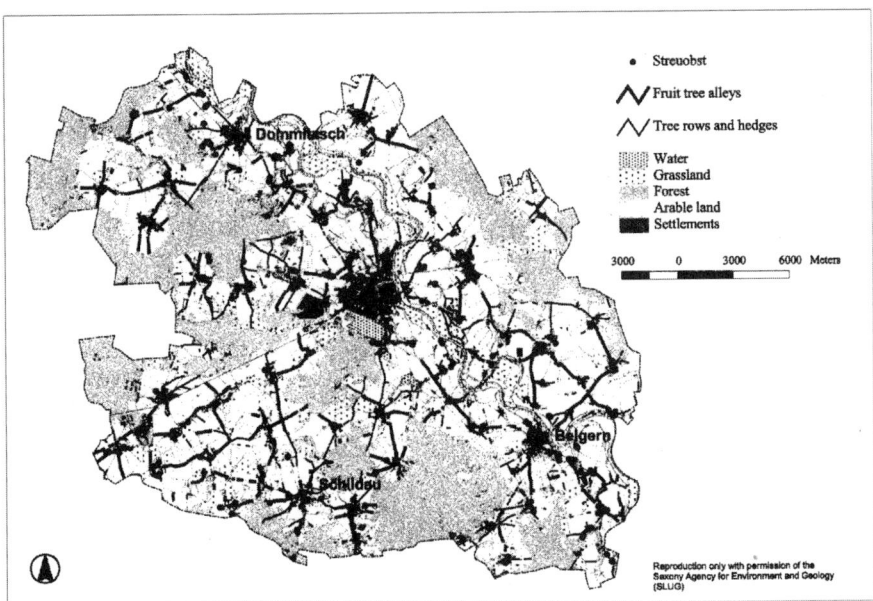

FIGURE 7.5 Ecologic network consisting of streuobst sites and fruit tree alleys in the Torgau District, eastern Germany. (Source: Digital biotope map of the Saxony Agency for Environment and Geology [SLUG] in Dresden.)

network (Figure 7.5). Although streuobst makes up only 2.1% of the total area, it contributes significantly to biodiversity.

In addition to the interspecies variability, there is an enormous amount of intraspecies variability in streuobst systems. Regionally differing fruit varieties evolved because selection took place locally by farmers, and each variety is adapted to particular site conditions. Also, each streuobst site normally consists of several varieties and types. Farmers choose the fruits according to their differing abilities to use and store the fruit, to pollinate, their different times of maturity and to spread the harvest throughout the season. Although many varieties were lost in recent decades, there are about 1,400 varieties of apples and 1,500 varieties of pears, cherries, walnuts, and plums in Germany (Rösler, 1995). To preserve this genetic material, national and international institutions as well as nongovernmental organizations (NGOs) conserve local varieties in gene banks, and more recently, on the farm *in situ*. Fruit tree varieties preserved in streuobst systems create important synergies between the preservation of genetic material and nature conservation (Herzog and Oetmann, 1997).

STREUOBST AND SOCIETY: SOCIOCULTURAL FUNCTIONS

Nature had and still has for many of us a mythologic dimension. Landscape aesthetics and recreation, essential services of cultural landscapes, are closely linked together. Both are prerequisites for human well-being.

Trees have always played an important role in myths and customs. In ancient societies, trees were symbols of fertility and well-being. In the German myth of the World Tree, the ash-tree "Yggdrasil" roots down to the underworld. Its branches embrace the earth, and its crown reaches up to the kingdom of the gods and supports the sky. Among the fruit trees, cherry and apple trees are mentioned particularly often with myths, beliefs, and customs. Up to the present day, on Barbara's Day (December 4), branches of cherry trees are cut and taken into the house. Their flowering on Christmas Day was a good omen, indicating happiness, fertile domestic animals, a good harvest, and favorable weather conditions in the year to come.

According to the beliefs of the Germanic peoples, apple trees were protected by the gods. Even Donar was not allowed to destroy them with his lightnings. Therefore, apple trees often were planted near houses to protect them. For several millennia, the apple were attributed to the goddesses of love and fertility in various ancient cultures: Babylonian, Egyptian, Indian, Greek, Roman, and Nordic. Because of their round shape, apples have always been a symbol of perfection. The imperial orb, therefore, was part of the insignia at the coronation of the German emperors (Bänninger, 1998; Haerkötter and Haerkötter, 1989; Schmeil, 1951). These traditional beliefs reinforce the support that streuobst is receiving among large parts of the population. Even in modern, industrialized societies, the traditional ways of life remain important for identification with the native place. Fruit trees and streuobst landscapes serve as a source of identity among a broad range of people (Herzog, 1994).

Cross-cultural experiments demonstrate that "the landscapes most preferred internationally are characterized by moderate to high depth or openness, relatively smooth or uniform-length grassy ground surfaces, and scattered trees or tree-clumps" (Porteous, 1996 p. 27). This is a description of an East African savannah, where *Homo sapiens* are thought to have originated, and this may explain preference for savannah type landscapes. The "prospect and refuge theory" (Appleton, 1975, 1996; Herrwagen and Orians, 1993) suggests that human beings experience pleasure in landscapes that satisfy their biologic needs. Both hunters and huntees (animals as well as early humans) have appreciated landscapes that provide the possibility to see (prospect) without being seen (refuge). Savannah-type environments such as streuobst have a good balance between prospect and refuge (Herzog, 1998).

In a more humanist approach to the aesthetic properties of streuobst, fruit trees are perceived as part of an ensemble consisting of harmoniously contrasting landscape elements of darker forest, structured arable fields, and green meadows interspersed with small woods, hedgerows, and the like. Fruit trees often are grouped around settlements, connecting them to the open agricultural land. As patches, rows, scattered individuals, and even single trees, streuobst can enhance or counterbalance the local topography. With varying shapes and sizes as well as differing colors of blossoms, leaves, and fruits, fruit trees enrich the scenery's variety and diversity in both space, and time. Even in the winter, the bizarre forms of the leafless branches and the individual shape of each tree catch the eye, and in many regions, the blossoms of fruit trees are a symbol of spring (Lucke et al., 1992; Weller et al., 1986). These properties, strongly appreciated by the public, are a major reason why the nonfarmer population often protests against the replacement of streuobst by intensively managed orchards (Jacob et al., 1986).

The aesthetic qualities of streuobst enhance the attractiveness of landscapes for recreation (Herzog, 1998). On hot summer days, the advantageous microclimatic features (i.e., comparatively lower temperature and higher humidity of the air) attract people seeking shade, relaxation, and moderate physical exercise. The farm animals often associated with streuobst on grassland are an additional asset. Especially in southern Germany, hobby fruit production during leisure time is a popular type of recreation. Numerous parcels of streuobst are owned or rented by city dwellers who seek recreation by moderate physical work in the open. Many families want their children to establish a closer relation with nature and to experience the production of

FIGURE 7.6 Streuobst trees near Basel, Switzerland, wrapped in polyester by Christo.

a natural and healthy food product from the planting and nursing of the tree, to harvesting and processing fruit in the household or in local cider mills. These families make it a point not to mix batches, but to provide everyone with the juice from his or her own fruit.

Recently, an artistic transformation underlined the particular aesthetic role that trees in general and streuobst in particular play in landscapes. In an event called "Wrapped Trees," the famous artist Christo wrapped 178 trees with polyester in northern Switzerland (Figure 7.6). The event was highly publicized, even in the international press, and attracted tens of thousands of visitors. It was linked to an arts exposition on "The Magic of Trees," and to the presentation of nature protection activities of several NGOs. Aesthetics, myths, and environmental protection crystallized around trees.

PROFITABILITY OF STREUOBST: ECONOMIC FUNCTIONS

Streuobst generally is not seen as a profitable branch of farming activities (Herzog, 1998). Conventional evaluations come up with annual losses of 0 to 55 European Currency Units (U.S. $0–$55) per tree (Berger and Roth, 1994; Hitz and Locher, 1996; Rösler, 1996a; Schnieders, 1997). This is explained mainly by the relatively low labor productivity, as compared with that of intensively managed orchards. However, there are some contradictions and ambiguities, which suggest that the global judgment of streuobst as unprofitable might be premature:

- Apples from streuobst have a considerable impact on the European market, and the yield of apples is negatively correlated with the market price for cider and dessert apples (Lobitz, 1997; Rösler, 1996b).
- The overall monetary value of the apple harvest from streuobst exceeds the value of apples from plantations (Weller, 1996).
- Most economic evaluations deal only with the profitability of fruit production, excluding the other products provided by the system (e.g., livestock, wood).

There are specific conditions under which streuobst has comparative advantages over other land-use types. It is still an important form of land use in the hilly parts of temperate Europe, namely in northern France, northern Spain, Switzerland, southern Germany. There, it integrates well into medium-size family farms with cattle and crop production, if the fruit-picking season is between the labor-intensive planting and harvesting seasons and if family members are available to help with the harvest (Herzog, 1998). Streuobst can be profitable for family farms which need not pay for hired labor.

Private economic efficiency is not sufficient for assessing the sustainability of production systems. Ecologic and social services must be taken into account to judge their "social efficiency" (Barbier, 1990). Streuobst has a high capacity to yield such services, namely in the fields of biodiversity and landscape aesthetics. These are public goods, and as such, extremely difficult to price (Funtowicz and Ravetz, 1994). Therefore, they are not included in conventional assessments of the economic performance of streuobst.

The economic dimension, however, is central for the on-farm conservation and management of agroecosystems. In Germany, this is widely claimed, especially since 1993, when the Convention of Biological Diversity was signed and ratified by the government. The decrease of streuobst was caused mainly by a failing economic efficiency. Therefore, only a sufficient profitability through sustainable use will guarantee the long-term protection of the agroecosystem, with its connected ecologic and social benefits. The following section describes and analyzes the cooperation between nature protection and farmers that has led to the preservation of streuobst.

NONGOVERNMENTAL ORGANIZATIONS AND FARMERS FOR THE PROTECTION AND USE OF STREUOBST: A SUCCESS STORY OF COOPERATION

One main strategy of nature protection is to maintain and/or to revitalize historic management forms, which are more diverse and less intensive than current agricultural practices, and thus allow for higher biodiversity. This often leads to conflicts between agriculture (community of place) and nature protection (communities of interest) (Marschall, 1993). According to the German Nature Protection Law, the damage done by ecologically relevant and area-consuming impacts, such as road construction, must be compensated. For this purpose, "valueless" (from the nature protection point of view) agricultural land is transformed into "valuable" biotope types. This often is done regardless of the farmers' economic and social needs.

One exception to this rule is the "protection by use" approach to streuobst preservation. It can be seen as a kind of success story for cooperation between nature protection and agriculture. After a short chronology of the main events concerning streuobst in Germany, the reasons for this success are analyzed.

Several European countries have supported the clearing of fruit trees by national programs since 1954. At the end of the 1960s, the production of fruit in the European Economic Community (EEC) was still higher than consumption. Therefore, the council regulation for the "redevelopment of the common fruiticulture" came into force in 1969. Each member state had to prime the clearing of apple, pear, and peach trees. Farmers who applied for that bonus were obliged not to replant fruit trees in the following 5 years. At the same time, the EEC prohibited any financial support of fruit tree plantings by the member states (Opitz, 1970).

The main objective of the regulation was to reduce fruiticulture in less productive and less profitable systems in order to support intensively managed orchards, which were competitive on the world market. Between 1970 and 1973, approximately one third of the German fruit tree stands were cleared (Lobitz, 1997; Petzold and Hahn, 1973; Stadler, 1983). The regulation thus proved to be efficient, and this probably was the reason why it ended in 1974 (Rösler, 1997a).

During the same period, however, resistance against the clearing of streuobst began. In 1971, the Report on the Environment of the Ministry of the Interior of Baden-Wurttemberg, one of the federal *Laender*, stated that the reduction of streuobst had negative effects on both the social and the environmental functions of the landscape (Rösler, 1997a). In the middle of the 1970s, several scientific publications called for the maintenance of streuobst systems, mainly from an ornithologic point of view. The first request for finances to support the conservation of streuobst, and consequently the maintenance of its ecologic and social functions, was made in Bavaria in 1978. However, it took another 9 years for Bavaria to pick up this demand (Rösler, 1993).

Increasing awareness of the important services streuobst provided motivated numerous NGOs to engage in the protection of this agroecosystem. In addition to numerous local initiatives, the two large organizations active at the federal level, German Society for Nature Protection (NABU) and Friends of the Earth Germany (BUND), made streuobst one of their priorities. In 1983, the NABU introduced the slogan "Mosttrinker sind Naturschützer," which can be translated as "Consumers of juice or cider from streuobst protect nature" (Rösler, 1997a). This concept integrated the conservation of an agroecosystem, its sustainable management, and the economic dimension. In 1987, the BUND began to implement an alternative marketing system based on a bonus price for streuobst products. Farmers obtain a financial bonus in addition to the market price if the fruit are produced in streuobst. This makes it attractive for the farmers not only to harvest the fruit, but also to maintain the streuobst trees. Of course, this results in higher prices for streuobst products. Therefore, in 1988 the NABU developed a quality label for streuobst products (Rösler, 1997a). They are labeled as particularly environmentally friendly and healthy, and an increasing number of consumers are ready to spend extra money for this kind of food.

Parallel to this engagement of NGOs, some of the Laender implemented specific programs to support the management of streuobst systems: Rhineland-Palatia (1986), Bavaria (1987), Northrhine-Westfalia (1988), Saar (1990), and Saxony-Anhalt (1991) (Rösler, 1997a). Between 1992 and 1994, 5 of the 16 federal Laender included streuobst in their nature protection laws as threatened biotope types under special protection.

Along with the reform of the common agricultural policy of the European Community, the council regulation 2078/92 "on agricultural production methods compatible with the requirements of the protection of the environment and the maintenance of the countryside" came into force in 1992. On the basis of this regulation, several Laender created programs supporting farmers' management, and in some cases, replanting of streuobst systems: Baden-Wurttemberg (1992), Brandenburg and Thuringia (1993), and Hesse and Saxony (1994). Already existing programs in other Laender were reorganized on the basis of regulation 2078/92. A major problem of this regulation is its restriction on the management and replanting of streuobst. The development of marketable products and new markets cannot be subsidized.

Baden-Wurttemberg improved its program in 1994 by supporting the control and marketing of separately harvested fruits from streuobst (Rösler, 1997a).

In 1997, the Germans consumed approximately 5 million liters of juice produced from fruits harvested in streuobst systems (Rösler, 1998). Approximately 90 initiatives in Germany currently organize the production and marketing of streuobst products, increasing sales to approximately 4 million European Currency Units (Lobitz, 1997). Streuobst marketing has been so successful that it currently is subject to misuse. Fruit products from intensively managed orchards are labeled as streuobst products by those seeking a higher price. The efforts of NABU to reach a political and juridical definition of the term streuobst in order to prosecute its misuse are a prerequisite for mutual trust between producers and consumers (Rösler, 1997b).

The NGOs also call for a modification of the European Union marketing regulation for dessert fruits, which prohibits the marketing of apples, pears, apricots, and peaches as dessert fruit if they are smaller in diameter than defined by the regulation. This excludes part of the streuobst harvest from marketing, reducing economic efficiency. Recently, the regulation was slightly modified as a result of the combined effort of a local government and several NGOs. To allow the marketing of apricots from the northernmost production region in Saxony-Anhalt (30,000 trees), the minimum diameter was lowered from 30 to 25 mm for apricots produced and consumed in that region (Rösler, 1998).

In Switzerland, streuobst products also are labeled, and nature protection is a marketing argument. Organically (biologically) produced apple juice from streuobst is available in grocery markets, whereas it is sold mostly by alternative distributors (direct marketing by farmers or nature protection agencies) in Germany. Streuobst trees are part of the Swiss agri-environmental scheme, which requires that each farmer allocate 7% of his or her farmland to seminatural biotopes to qualify for government subsidies.

What are the reasons for this successful cooperation between nature protection (community of interest) and agriculture (community of place)? First, the remaining streuobst systems often are extensively managed, and thus, per se, valuable for nature protection. Nature protection did not require significant management changes, avoiding one of the main potential conflicts between these two communities. Second, the typical approach of nature protection, which aims at reducing human influence to a minimum, is not appropriate for the conservation of streuobst. In fact, streuobst is much more labor intensive than other agricultural areas valuable for nature protection, such as chalk grassland and litter meadows. Because of these particular streuobst properties, in combination with the actual and potential area covered by this agroecosystem, farmers needed to be involved. Communication between the two social groups, farmers and nature protection activists, was indispensable.

Another unique aspect of streuobst is that standard fruit trees are perennial plants that resist change longer than annual plants and survive periods of abandonment. For this reason, many old local varieties can still be found in streuobst systems. These old varieties guarantee the special quality of streuobst products and their potentially higher prices as compared with products of intensively managed orchards. In contrast to chalk grassland and litter meadows, streuobst systems deliver a periodic fruit harvest. This was the basis for the development of market niches that could restore the economic sustainability of streuobst. A major advantage of streuobst is that the long-term conservation of valuable agroecosystems by communities of place is sustainable only if the management is profitable from an economic point of view.

Thanks to the activities of numerous NGOs, people are becoming convinced concerning the high quality of streuobst products. Many are forming a "sustainable consumer group" that is securing the economic attractiveness of streuobst for farmers and for the regional fruit processing industry. Aspects of ecologic sustainability are gaining importance for marketing. A growing number of consumers appreciate the low energy input in the production of streuobst products, as well as the short transport distances. They are aware of the interdependence between sustainable use and conservation of typical open landscapes such as streuobst, and therefore accept higher prices. Furthermore, consumers more and more identify the streuobst products with their region and appreciate the streuobst system as a typical landscape element they want to be maintained. This aspect is also relevant to farmers, as some regions benefit from increasing tourism.

SYNTHESIS AND OUTLOOK

Figure 7.7 outlines conceptually society's perception of streuobst's importance as it has evolved over time. Originally, the combined production of fruit and arable crops was the major purpose of streuobst. Undercropping declined in the first half of the 20th century and was replaced by grassland, which is easier to farm. With the development of intensively managed orchards, the perceived importance of streuobst fruit production decreased dramatically. However, this perception is somewhat distorted. In reality, streuobst fruit production still contributes significantly to the market supply, especially cider fruit. The main factor contributing to reversal of the trend was the spreading of bonus price systems.

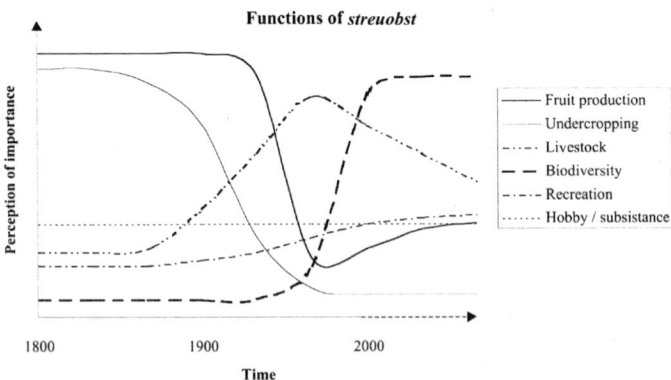

FIGURE 7.7   Temporal evolution of the importance of major functions of streuobst systems, as perceived by the population.

Although the production function of streuobst has decreased in importance, awareness of its other functions has been augmented since the 1970s. Biodiversity has become the major argument for preserving streuobst. Recreation has always had significance. In the Romantic Era, hillsides of river valleys planted with fruit trees and vineyards were famous for their beautiful scenery (Lucke et al., 1992). Hobby and subsistence were and still are important. In times of war and scarcity of food products, subsistence production contributes to the population supply.

The flexibility of the streuobst system is illustrated by its recent history. In the former German Democratic Republic, cooperative farmers were allowed to farm a small portion of land (generally, 0.5 hectares) for subsistence and personal profit. Often, this area was planted with fruit trees, and some farmers undercropped it with potatoes, turnips, oats, and alfalfa. They opted for an increased intensity of production in western Germany. Where there was overproduction of fruit, streuobst systems were extensified, and hobby fruit production gained importance.

The future and degree of sustainability of streuobst strongly depend on the external factors of policy and society. If there is a trend toward sustainable agriculture within a more sustainable society, in which ecologic and socioeconomic damages and benefits make a real economic difference, then streuobst will gain importance again because it has a higher potential to yield public goods than intensively managed orchards. However, if the current wasting of fossil energy, encouraged by low prices continues, streuobst, together with other aspects of sustainable agriculture, will be maintained only "artificially" through particular groups of interest that engage in environmental protection and form an alliance with farmers who are ready to cooperate. Streuobst can teach us how to progress toward more sustainability in agriculture and its related sectors if we are willing to learn from it.

REFERENCES

Anonymous, Lebensgeschichten berühmter Pomologen: August, Churfürst von Sachsen als praktizierender Pomolog, *Deutscher Obstbau.*, 3, 3, 1941.

Appleton, J., *The Experience of Landscape,* John Wiley and Sons, Chichester, 1996.

Appleton, J., *The Experience of Landscape,* John Wiley and Sons, New York, 1975.

Bänninger, A., In einem Apfel steckt die ganze Welt, *Schweizerische Z. Obst. Weinbau,* 1, 6, 1998.

Barbier, E. B., Economics for sustainable production, in *Agroforestry for Sustainable Production,* Prinsley, R. T., Ed., Commonwealth Science Council, London, 1990, 389.

Berger, W. and Roth, D., *Kosten und Preiskatalog für ökologische und landeskulturelle Leistungen im Agrarraum,* Jena, Sonderheft der Thüringer Landesanstalt für Landwirtschaft, 1994.

Bundesamt für Statistik (BFS), *Die Bodennutzung der Schweiz—Arealstatistik 1979/85: Kategorienkatalog,* Bundesamt für Statistik, Bern, 1992, 191.

Bundesamt für Statistik (BFS), Eidgenössische Alkoholverwaltung. Schweizerische Obstbaumzählung: der Obstbau in der Schweiz (Feldobstbau/Obstkulturen), Bern, *Land. Forstwirtschaft,* 7, 200, 1993.

Bundesministerium für Ernährung, Landwirtschaft, und Forsten (BELF), *Statistisches Jahrbuch für Ernährung, Landwirtschaft und Forsten der Bundesrepublik Deutschland,* Landwirtschaftsverlag Münster-Hiltrup, 1997.

Bünger, L. and Kölbach, D., Streuobst: *Bindeglied zwischen Naturschutz und Landschaft,* Dokumentation Natur und Landschaft, Sonderheft 23, Bibliographie Nr. 69, Bundesamt für Naturschutz, Bonn, 1995.

Friedrich, G., *Der Obstbau,* Neumann Verlag, Radebeul, 1956.

Funke, W., Heinle, R., Kuptz, S., Majzian, O., and Reich, M., Arthropodengesellschaften im Ökosystem "Obstgarten," *Verhandlungen Ges. Ökologie,* 14, 131, 1986.

Funtowicz, S. O. and Ravetz, J. R., The worth of a songbird: ecological economics as a post-normal science, *Ecol. Econ.,* 10, 197, 1994.

Haerkötter, G. and Haerkötter, M., *Macht und Magie der Bäume: Sagen-Geschichten, Beschreibungen,* Frankfurt, 1989.

Herrwagen, J. H., and Orians, G. H., Humans, habitats and aesthetics, in *The biophilia hypothesis,* Kellert, S. and Wilson, E. O., Eds., Washington: Island Press, Shearwater Book, 1995, 138.

Herzog, F., Multipurpose trees in traditional farming systems, two case studies (Africa, Europe), in *Agroforestry and Land Use Change in Industrialized Nations: Proceedings of the 7th CIEC Symposium at Humboldt University Berlin,* May 30–June 2, 1994, Welte, E., Szabolcs, I., and Huettl, R. F., Eds., 1994, p. 219.

Herzog, F., Streuobst: a traditional agroforestry system as a model for agroforestry development in temperate Europe, *Agroforestry Syst.,* 42, 61, 1998.

Herzog, F. and Oetmann, A., In-situ-Erhaltung von Streuobst: Synergien zwischen Naturschutz und der Bewahrung Genetischer Ressourcen, *Natur. Landschaft,* 72(7/8), 339, 1997.

Hitz, T. and Locher, M., *Wirtschaftlichkeitsberechnungen in der Mostobstproduktion,* Diplomarbeit, Höhere Wirtschafts-und Verwaltungsschule, Aargau, unpublished, 1996.

Jacob, H., Breunig, T., König, A., von Königsmarck, K., von Lossau, A., Pelz, G., and Stähr, E., *Erfassung und Massnahmen zur Erhaltung des Streuobstanbaues in Hessen,* Forschungsanstalt für Weinbau, Gartenbau, Getränketechnologie und Landespflege, Geisenheim am Rhein, unpublished, 1986.

Jordan, E., Der Obstbau im Rechtsleben des Deutschen Volkes, in *Der Forschungsdienst, Organ der deutschen Landbauwissenschaft, Band 7,* Meyer, K. and Schönberg, M., Eds., Verlag J. Neumann, Berlin, 1939.

Kaiserliches Statistisches Amt. (KSA), Ergebnisse der Obstbaumzählung im Deutschen Reiche im Jahre 1900, *Vierteljahreshefte zur Statistik des Deutschen Reichs,* 11(2), 224, 1902.

Kaiserliches Statistisches Amt. (KSA), Ergebnisse der Obstbaumzählung im Deutschen Reiche im Jahre 1913, *Vierteljahreshefte zur Statistik des Deutschen Reichs,* 24(2), 49, 1915.

Kemmer, E., *Die deutschen Obstbauverhältnisse im Lichte der Vor- und Nachkriegsstatistik,* Merkblätter des Institutes für Obstbau 15, Wiesbaden, Limes Verlag, 1950.

Kittel, G., *Die Wertvollsten Obstsorten Deutschlands,* Düsseldorf, Ferd, Richter Verlag, 1895.

Lobitz, R., Streuobst in Deutschland: ökonomisch und ökologisch betrachtet, *Verbraucherdienst,* 42(9), 223, 1997.

Lott, K., Der historische Obstbau in Deutschland zwischen 1850 und 1910, *Geschichte, Dokumentation, Aussagen für den aktuellen Streuobstbau,* Dissertation, Band 1, Berlin, Humboldt Universität, Fachbereich Agrar- und Gartenbauwissenschaften, 1993.

Lucke, R., Silbereisen, R., and Herzberger, E., *Obstbäume in der Landschaft,* Eugen Ulmer, Stuttgart, 1992.

Maag, G., Zur situation im obstanbau, *Baden-Württemberg in Wort und Zahl,* 9, 445, 1992.

Marschall, I., Gebrauchte Landschaft: Möglichkeiten eines Bündnisses zwischen Landwirtschaft und Naturschutz, in *Landwirtschaft 1993: Der kritische Agrarbericht,* Agrarbündnis, Ed., Bonn, 1993, 271.

Opitz, W., Zuschüsse für Obstbaum-Rodung, *Deutsch Gartenbauwirtschaft,* 3, 20, 1970.

Petzold, R. and Hahn, O., Ergebnisse der Rodungsaktion in der EWG und in der Bundesrepublik Deutschland, *Der Erwerbsobstbau,* 15(1),5, 1973.

Poenicke, W. and Schmidt, M., *Deutscher Obstbau,* Deutscher Bauernverlag, Berlin, 1950.

Porteous, J. D., *Environmental Aesthetics,* Routledge, New York, 1996.

Rösler, M., *Arbeitsplätze durch Naturschutz: dargestellt am Beispiel der Biosphärenparke und der Modellregion mittlere schwäbische Alb,* Dissertation an der TU Berlin, FB Umwelt und Gesellschaft, 1997a.

Rösler, M., *Streuobst-Rundbrief,* Editorial, 3/98, 1, 1998.

Rösler, M., Einführung zum Thema Streuobst, in *Streuobst: Bindeglied zwischen Naturschutz und Landschaft,* Bünger, L. and Kölbach, D., Eds., Dokumentation Natur und Landschaft, Sonderheft 23, Bibliographie Nr. 69, Bundesamt für Naturschutz, Bonn, 1995.

Rösler, M., *Erhaltung und Förderung von Streuobstwiesen: Modellstudie dargestellt am Beispiel der Gemeinde Bad Boll,* Gemeinde Boll, Boll, 1996a.

Rösler, M., Gegen den Mißbrauch des Begriffes "Streuobst": Einführung eines geschützten Begriffes "Streuobst" in die EU-Gesetzgebung, in *Landwirtschaft 1997: Der kritische Agrarbericht,* Agrar büdnis, Ed., Bonn, 1997b, 309.

Rösler, M., *Marktwirtschaftliche Bedeutung des Streuobstbaus,* Auswertung des EU-Streuobstforums im Rahmen der 2, Rhöner Apfelmesse, Rhöner Apfelbüro, Zella, 1996b, 3.

Rösler, M., Streuobst, in *Landwirtschaft 1993: Der kritische Agrarbericht,* Agrarbündnis, Ed., Bonn, 1993, 169.

Rudzinski, H. G., Beiträge zur Kenntnis der Trauermückenfauna Nordwestdeutschlands (Diptera, Nematocera: Sciaridae), *Drosera,* 92(1), 35, 1992.

Rudzinski, H. G. and Drissner, J., Neue Sciariden aus Deutschland (Diptera: Nematocera), *Entomol. Z.* 102(12), 223, 1992.

Schmeil, O., *Leitfaden der Pflanzenkunde,* Quelle and Meyer, Heidelberg, 1951.

Schnieders, K., Aufpreis-Vermarktungskonzept für Streuobstprodukte als Beitrag zur Erhaltung und Nutzung von Hochstamm-Obstwiesen im Landkreis Göttingen, *Landschaftspflegeverband Landkreis Göttingen e.V.,* 1997.

Simon, L., Entwurf, Ergebnisse und Konsequenzen der wissenschaftlichen Begleituntersuchungen zum Biotopsicherungsprogramm "Streuobstwiesen" des Landes Rheinland-Pfalz, Landesamt für Umweltschutz und Gewerbeaufsicht Rheinland-Pfalz, Begleituntersuchungen zum Biotopsicherungsprogramm "Streuobstwiesen," *Oppenheim, Beitr Landespfl Rheinland-Pfalz* 15, 5, 1992.

Stadler, R., Der landschaftsprägende Streuobstbau und sein Einfluss auf den Erwerbsobstbau, Baden-Württemberg in Wort und Zahl, *Statistische Monatshefte,* 31,173, 1983.

Statistisches Bundesamt Wiesbaden (SBA), Ed., *Die Obstbaumbestände 1951,* Statistik der Bundesrepublik Deutschland. Stuttgart, Kohlhammer, 1954.

Statistisches Bundesamt Wiesbaden (SBA), Ed., *Obstbaumzählung 1965,* Fachserie B, Landund Forstwirtschaft, Fischerei, Reihe 2, Gartenbau und Weinwirtschaft. Stuttgart, Kohlhammer, 1966.

Statistisches Reichsamt (SRA), ed., *Obstbaumzählung 1938 und Obsternte 1938,* Berlin, Verlag für Sozialpolitik, Wirtschaft und Statistik, Statistik des Deutschen Reiches, Band 541, 1940.

Ullrich, B., Streuobstwiesen, in *Die Vögel Baden-Württembergs: Band 1: Gefährdung und Schutz,* Hölzinger, J., Ed., Karlsruhe, Ulmer, 1987, 551.

Weller, F., Streuobstwiesen, in *Naturlandschaft: Kulturlandschaft,* Konold, W., Ed., Ecomed, Landsberg, 1996, 137.

Weller, F., Eberhard, K., Flinspach, H. M., and Hoyler, W., *Untersuchungen Über die Möglichkeit zur Erhaltung des Landschaftsprägenden Streuobstbaues in Baden-Württemberg,* Stuttgart, Ministerium für Ernährung, Landwirtschaft, Umwelt und Forsten Baden-Württemberg, 1986.

Wiesinger, K. and Otte, A., Extensiv genutzte Obstanlage in der Gemeinde Neubeuern/Inn: Baumbestand, Vegetation und Fauna einer traditionellen, bäuerlichen Nutzung, *Berichte Bayrische Akademie Naturschutz Landschaftspflege,* 15, 69, 1991.

8 Transhumant Communities and Agroecosystems in Patagonia[1]

Monica Bendini

CONTENTS

Background 103
The Rural Community of the Crianceros 104
Development of Transhumance Practices 104
The Community 105
Transhumant Agroecosystem 107
Postoralism and Landscape: Images and Perceptions .108
Perceptions of the Landscape 108
State Policies 110
Peasants' Poverty and Survival 111
Conclusions 113
    Control and Resistance 113
    On Problems and Potentiality 113
References 113

BACKGROUND

In the northwest of the Argentine Patagonia, peasants, who call themselves criaceros, move sheep and goats in seasonal patterns on the southern Andes Mountains and valleys. That transhumant style of production is very fragile because it depends on (1) the rural community to establish norms of resource use, (2) annual and seasonal variations in the pasture and forage available, and (3) national and international fiber markets.

---

[1] A version of this article was presented as a paper in the 4th European Symposium on European Farming and Rural Systems Research and Extension into the Next Millennium, Enviromental, Agricultural and Socioeconomic Issues held in Volos, Greece, April 3–7, 2000.

THE RURAL COMMUNITY OF THE CRIANCEROS

The crianceros move their sheep and goats through valleys of the southern Andes mountains, mountain slope, woodland, and arid steppes in the Patagonian plateau (Bendini and Tsakoumagkos, 1994). The territory extends from the south of the province of Mendoza to the center of the province of Chubut. Most of the crianceros population is concentrated in the provinces of Neuquén and Río Negro.

Approximately 7,500 crianceros and their livestock—mainly sheep and goats, with some horses and cattle—live on this landscape. Their herd size, measured in sheep units, averages 250 to 500. Sheep shearing and the sale of sheep wool, mohair, and goat meat provide income to these households. Some meat and wool are used by the family, but the majority goes to the market. Local, extraprovincial and international brokers buy the wool and mohair. They seek specific quantities and characteristics they seldom reveal to the crianceros, so there is little feedback regarding selection and pricing on which the crianceros can base future production practices. The meat produced is primarily for local markets. The Neuquén provincial government implemented marketing programs and a new sales approach for added value, but support for these programs declined after 1985.

Between the arid and semi-arid plateau and the *cordillera* (Andes mountain chain), the region under study, crianceros and *puesteros* (livestock companions) are 90% of the rural population. The majority are *fiscaleros* (occupants of government land) and are different from the typical Argentine agricultural producer in other parts of the country.

There are three basic types of crianceros: (1) the *transhumance* crianceros, or nomad sheep and goat raisers, who move their animals from the lower arid fields in the winter to the high valleys for the Andean summer; (2) the sedentary crianceros from the arid fields in the plateau; and (3) the agricultural crianceros or farmers operating around small creeks and brooks where livestock raising is supplemented with some precarious produce cultivation: pasture, cereal, vegetable.

For all three types of crianceros, tending the flock determines the social organization of the local communities. Ethnically, the communities are indigenous (with or without legal recognition and different degrees of recognition of indigenous rights), Creole, or a combination of the two ethnic groups. In all of the communities, animal raising and land management are based on behaviors and conventions built on traditional social bonds.

The largest of the three types as regards numbers of people and animals, the transhumance crianceros or nomad breeders is the focus of this chapter. They have the strongest sense of community identity, which has been strengthened over the years as they have resisted outside pressures in order to maintain their way of life.

DEVELOPMENT OF TRANSHUMANCE PRACTICES

Seasonal migration of livestock, or transhumance, as a way of life and work goes back to the early settlers, who brought exotic small ruminants to lands previously grazed by an indigenous small ruminant, the llama. Because of the harsh environmental conditions involving extreme temperatures and limited precipitation, the early settlers adopted the transhumant practices of the region's indigenous people, organizing households and communities around livestock activity. Accessability to Chilean markets added to the attraction of land availability, drawing immigrants from other parts of Argentina and abroad to settle in large extensions of state lands.

The rich resources from the cordillera for summer grazing and the accessibility to markets resulting from free commerce with Chile favored the advancement of transhumance livestock production. Livestock fattened in the summer were sold in Chile, providing capital for development in the region. For centuries, the settlers of the region (Mapuche Indians, Chileans, and Creoles) responded to the ecosystem and the market through extensive nomadic livestock raising by family units.

At the end of the 19th century, settlement increased pressure on lands south of the Colorado River as a consequence of new land distribution policies at the national level. The best lands to the south of the Pampas region were taken away from the original inhabitants through military expeditions and the "Conquest of the Desert" (Gasteyer and Flora, 2000). Land with access to water and natural pasture most of the year is located in the longitudinal strip at the foothills of the cordillera. That north–south strip, suitable for raising cattle year round, is located between the arid fields of winter grazing and the summer grazing camps in the cordillera. Low-density Indian populations occupied the territory before the military invasion, but were exterminated or removed from the more productive lands through federal decrees and legal machinations of large landowners interested in expanding into the area. The best land for sedentary livestock raising was purchased by large farmers. That land is between the winter and summer grazing lands of the crianceros. Furthermore, much of the grazing takes place on government land, which is in the process of privatization. These policy decisions have greatly hampered access to the natural resources necessary for livestock production.

By the middle of the 20th century, tariff barriers and greater border control, inspired by the Argentinean and Chilean industrial development model, destroyed the regional wool and hair market, transferring its control to the national agroindustrial capital. At the same time, national and provincial government instituted a spectrum of policies intended to control livestock raising in transhumance. These policy-induced limitations on production and marketing increased financial insecurity. Grazing permits to access governmental pastures in the cordillera during the key period of animal growth are temporary and nontransferable. Without a permit, livestock cannot be moved to government lands.

THE COMMUNITY

Family transhumance animal production persists because of its internal logic, family labor intensification to maximize income (Cucullu and Murmis, 1980), and the logic of an economic system that provides little access to capital for local commercial and agricultural development (Tsakoumagkos, 1993).

These communities of loosely knit household production units are extremely attached to the land and the animals. They work hard to maintain their way of life, as defined by nomadic animal production. There is substantial continuity between generations. The son or sons, whom the father generally picks at 18 years of age or upon marriage, solicit the signal ticket and the pasture permit, indicating inclusion in the production cycle.

The criancero communities, as a result of unequal access to resources and lack of market negotiating power, have developed pluriactivity, a combination of on-farm and off-farm income generation as a survival strategy. During crisis periods, the family unit acts as a refuge. The family survival strategy depends on those who migrate and those who permanently reside in the family home place. Since the middle of the 20th century, an increasing dependence on wage labor to generate household income has converted the crianceros into peons, workers, or employees.

The perseverance of the crianceros during the processes of social differentiation (Bendini and Tsakoumagkos, 1994; Cucullu and Murmis, 1980) depends on large extensions of government land and the fact that the Argentina economic system is unable to absorb them in alternate activities. Currently, government plans to privatize the grazing lands through a new land-titling program threatens the production system and further undermines community stability. That titling has privileged large landowners.

Criancero communities have two modes of access to the land and several degrees of legal formalized property rights. Indigenous communal properties are recognized in legislation for indigenous reservations, with varying degrees of ownership title formalization. Users of government lands (fiscaleros) gain access through acqusition of a grazing permit from the provincial government. At the center of these communities are a community-determined number of shepherds that use specific grazing lands.

"The Tragedy of the Commons" (Hardin, 1968) draws on a similar situation: the herding on common land. Hardin describes a situation in which utility maximization includes a positive component, the economic gains obtained through the increase in livestock, and a negative component, the environmental costs generated by the same increase via overgrazing. According to the author, the tragic characteristics of the unavoidable ruin from the unlimited increase of the existing livestock could be explained by the individual freedom to use the public goods. Hardin's solution includes an ethical aspect, the "reciprocal coercion" agreed on by the majority affected, which includes the suffering from a common agreement on coercion, and a legal aspect, private individual property attached to a legal inheritance.

Public pastureland is cited regularly as a particular case of "public property resources" or "free access resources." Its basic characteristic is "no exclusion." The consequences are abuse of resources and the associated social inefficiency, lack of incentives to invest in production improvement, and the improbability, difficulty, or violability of agreement on use reduction. The solution to environmental degradation, according to this logic, is land privatization.

Common property is different from government property. Common property means that the community holds some property rights, with the state holding residual property rights. Public (government) lands belong to all and are the responsibility of no one. Thus, they are assumed to be open to free access, with destructive implications according to Hardin. In fact, governments seek to control the usage of public lands by controlling a variety of usufruct rights. With public lands, common property lands, and private lands, the users and the owners must come to an agreement on how the resources may be used.

The World Bank differentiates "common property resources" from the "free access resources" (BIRF, 1992). In both cases, resource organizations are crucial. The Bank identifies communities as the resource organizations with authority to determine who has access and the conditions of access. Problems arise when the "communal properties" become "free access resources" as a consequence of their nationalization. The World Bank urges use of the local organizations' (village or shepherds' associations) potential for improved land management.

The local communities (crianceros in indigenous communities or crianceros fiscaleros) constitute traditional organizations with strong social bonds that enforce the behavior and conventions linked to the common shepherding. These communities are closely connected to the spatial allocation of land and have developed reciprocal mutual agreements at the local level on livestock management.

These local communities attempt to access a determined territory and bar the access to all "noncrianceros" from the local area or other areas. Shepherding, as practiced by the Patagonia crianceros, differs from free access to resources.

TRANSHUMANT AGROECOSYSTEM

Livestock movement is regulated by the cyclical rhythm of the seasons, and household activities adjust to these cycles. Each year, a temporary change of settlement followed by a return marks the beginning of a new cycle. The transhumant system is attached to the landscape and weather, which determine grazing possibilities. Whereas the crianceros previously moved their livestock four times a year in response to forage availability, the privatization of the river valley land has reduced the cycle to moves in the summer (to the mountains) and the winter (to the desert).

The circuit consists of the summer pasture, the winter forage, and the livestock trail that connects them. In the summer season, the livestock are moved to high mountain valleys, 1,200 m above sea level. This environment provides pasture and water to the flock during the summer days. The length of stay varies from 3 to 5 months, based on the distance from the winter fields and the altitude of the summer fields. In the winter season, the livestock move to the plateau and lower valleys. Water and pasture availability are very limited by the end of the winter. Transhumance efficiency declined drastically with the formation of large farms in the best precordillera fields.

The distances involved in the circuit vary considerably, from a few kilometers to more than 200 km. Livestock feeding, livestock composition, number of livestock, and landscape characteristics determine the cycle length. The crianceros accompany the migration on horseback, with cargo animals (mules saddled with cargo baskets), or with old-model, deteriorated pickup trucks (Bendini and Tsakoumagkos, 1994).

Community norms and traditional transhumant practice were insufficient to counteract the deterioration of pasturelands and depopulation. The crianceros' standard of living has deteriorated. Many have taken work on the large ranches, which occupy the best lands, because they can no longer pasture their own herds there. But because international demand for animal fiber is low and ranches have been transformed into hunting reserves, that alternative too has declined.

PASTORALISM AND LANDSCAPE: IMAGES AND PERCEPTIONS

The major institutions making decisions affecting land use in the 1970s and 1980s viewed the crianceros as poor rural residents engaged in extractive subsistence activities, which, because of open access, degraded the environment. Criancero sheep and goat grazing was viewed as particularly environmentally destructive. Crianceros were treated as a social problem. Concern centered around their apparently exploitive, environmentally destructive activities.

The identification of the desertification risk in vast parts of the state of Neuquén at the end of the 1980s produced strong pressure on the transhumant herders. The territory to which they had winter access was steadily decreased and limited to the most arid corridors. Each summer they were obliged to go higher and higher into the mountains. The wire fences built around the fields previously used by the crianceros and government reoccupation of land that had been abandoned for many years are indicators of a strong environmentalism from various government agencies.

Sectoral policies influenced by this discourse promoted practices that appeared to be less environmentally destructive. Silviculture was highly promoted. Because of the need for land title and the long period between planting and harvest, credit and technical assistance promoting it favored the larger producers with a strong financial base.

As government lands became titled, the crianceros fiscaleros, whose livestock grazed on public lands, were further displaced, as the lands were privatized to those who had the money and connections to access lawyers and the land tenure bureaucracy. The environmentalist discourse, which began with a legitimate concern about natural resource use, became an ideological support for a new process of resource appropriation, completing that begun at the beginning of the 20th century.

The settlement of the crianceros in the more fragile lands of the arid and semi-arid plateau and areas of the cordillera made land management even more important than it had been previously.

PERCEPTIONS OF THE LANDSCAPE

Crianceros and the large cattle ranchers perceive the process of desertification and the transhumant system very differently. The testimonies of both groups are relevant because they manifest different logics of production and reproduction. The large cattle ranchers, who represent less than 10% of the population in the area, attribute pasture deterioration to the crianceros. They use this causal model to legitimize themselves as enforcers of exclusion, keeping the crianceros from grazing on any lands. This causal model makes the crianceros even more vulnerable.

The transhumant producer relates to resources by deciding on the number of animals to graze in a particular area and on the management of financial resources. The crianceros do not recognize the term "desertification." They do not refer to the area where they carry out their livestock raising activities as a desert, nor to the process of degradation and erosion as desertification. They do recognize that there is a "problem," attributing the process to natural conditions: loss of the pastures' fertility or decreased forage availability. Their own actions do not, in their view, have an impact on either of these natural conditions.

The large ranchers, in contrast, refer explicitly to the process of desertification, but see it as a situation unrelated to themselves. They view it as a process associated with the crianceros, originating in the concentration of population and overstocking. They particularly pinpoint overgrazing and the predominance of goats in the herds of the poor peasants as the source of desertification.

Both the crianceros and the large ranchers consider the problem of desertification in the context of their respective global production orientations (Nogués et al., 1993). The crianceros do not have a universal perception about the causes of desertification. Most of them refer to prolonged cycles of drought, and only a few refer to overgrazing or continuous grazing.

> The thing is that the fields don't have a chance to rest. (criancero in an indigenous community)

> There has been so much drought. There isn't any water, and this is the cause . . . the lack of water. . . . The summer pastures of this place, on these dry years which have come, have been going downhill. They get worse each year, and this is because of the drought in these pastures. . . . Before, it wasn't like this; the pastures did not ever fail. . . . Yes, they are recupriable [sic]. If it rains for us, the pasture will recuperate a lot. (criancero in a Creole community.)

Their identification of climate as the only cause for the decline of the pastures and their recuperation is associated with a concept of recurrent cycles and a fatalist response to the actions of nature.

Cutting and selling firewood has been very important in the mountain area. But, with the increased use of natural gas in the urban centers, the demand has decreased. Firewood consumption is now restricted to rural households, rural service centers, and poor urban areas. For the large ranchers, desertification is a result of overgrazing and cutting firewood. It is important to note that the ranchers' herds are primarily cattle, although they also have a few sheep and goats.

> The cause of desertification is the use of goats and overgrazing, . . . and firewood is another factor. (large rancher in the central valley at the foot of the Andes)

Although goats comprise 10% of his animals, the rancher attributes environmental decline only to the goats of the crianceros.

The large ranchers recognize the problem of pasture availability and stocking rates, stating that this is the principal problem of the indigenous reserves, where the poorest crianceros live. They perceive the importance of soil conservation practices as a way to achieve greater efficiency and profitability in their enterprise. In contrast, the crianceros' soil conservation practices are strategies of social reproduction. Although they are pressured to adopt technologies for the prevention and control of desertification, they do not adopt the recommended practices. In the criancero communities, differential access to resources (land and capital) conditions the perceptions and orientations of the households and the communities.

STATE POLICIES

In the 1970s two policy directives made an impact on criancero communities: wool commercialization and agricultural extension to increase technology adoption. Beginning in 1974, a new commercialization system was implemented by aggregating crianceros' wool and hair production through a livestock growers' association. This further integrated the marketing chain.

The government was not only an agent for technical training (for shearing machines), but also absorbed some expenses and financed others. These programs helped to increase the relatively weak negotiating power of the crianceros. The program's production and marketing objectives were to improve the quality of each criancero's product for its later aggregation and sale, and to increase the producers' income, obtaining a better price by eliminating the intermediaries. These policies were incorporated into a global development strategy for income redistribution. Peasant development was not part of that strategy. The agrarian policy in the 1970s stemmed from the notion of the state government as benefactor, a role not questioned by the traditional agrarian commercial bourgeoisie (Pescio et al., 1993). This general improvement in the standard of living and in community income widened the income possibilites of the propertied classes in Neuquén, as new consumers entered the formal market economy through increased income and rapid demographic growth.

These programs of extension and marketing expanded until 1983 and 1984. At that point, they stagnated. Despite the reinstitution of a democratic government in those years, programs aimed at the crianceros did not receive renewed support. The provincial government cut subsidies, producing social and spatial redistribution of income to the urban middle-class areas of Neuquén. Because the provincial government depended on local commercial capitalists for its support, the marketing program did not expand to new producers. Instead, the program became more complex, involving everything bought, sold, and consumed by those in the program, including basic services. There was only one marketing chain that integrated marketing with inputs and consumption. Paternalism, welfarism, and voluntarism in the execution of the program restricted its general expansion.

By the end of the 1980s, there was ongoing debate on the viability of these producers becoming self-sufficient. Unresolved structural issues around land access and control resulted in an increase in private appropriation of government lands occupied by crianceros, who were unable to take advantage of existing laws to legalize their title. The debate around resource deterioration attributed to transhumant practices diverted the attention away from social conflicts, which forced the crianceros either to reduce their herds to nonprofitable levels or to overgraze.

The implementation of national structural adjustment policies resulted in a profound crisis during the 1990s in the province of Neuquén, which lost its main source of revenue: direct national government transfers and federal coparticipation. The provincial government was no longer a major investor in public work. That activity in the 1980s had created employment and economic growth. Social policies to redistribute economic growth weakened in the face of a new accumulation model (Barsky, 1992). In the agrarian sector, institutional and political transformations emphasized control of livestock activities and the use of resources, mainly the land, within a frame of general deregulation. These controls revealed a creeping policy of excluding the fiscaleros from government lands. The expansion of the local landowners was predominantly speculative. Privatization of land is framed as care for the environment. With this approach, expensive technologies were required for erosion control. The high cost of the "environmentally friendly" land meant it was not available to most crianceros.

At the beginning of the 1990s, groups with ready access to capital gradually gained control of land use, although there was some resistance from crianceros, especially from the indigenous organizations. Neuquén government policy oscillated during those years as it responded first to landed vested interests, then to peasant and indigenous organizations, and then to environmentalists.

The general situation in the province worsened with the general increase in unemployment and poverty. Social programs were channeled to urban sectors where major emergencies and social eruption concentrate (Murmis, 1997). In such a context, there is less inclination to implement social policies in rural areas.

Concomitantly, a new player appeared: the transnational corporation. These corporations had an industrial base and integrated backward to raw material production. Land previously occupied by the peasant crianceros was made available to the market. More than 8½ million hectares of provincial land in the north of the Patagonia became available for purchase on speculation, despite a rhetoric stressing that title provision must be "mindful of the occupants' legitimate rights." The real estate market was privileged over peasant rural development. Provincial policies, which had limited crianceros' options since the 1950s, further limited these options with the land titling of the 1990s. The privatization of land releases the government from guaranteeing the permanence of the crianceros as viable producers. Criancero communities became increasingly poor and excluded (Murmis, 1994).

PEASANTS' POVERTY AND SURVIVAL

Table 8.1 compares Creole and Mapuche criancero communities at the beginning of the 1980s and middle of the 1990s on the basis of primary data gathered in the beginning of the 1980s and middle of the 1990s.

These households maintained their peasant character despite indicators of aging (reduced household size and a higher percentage of income from pensions) and

TABLE 8.1
Conditions in Criancero Communities, 1981 and 1995

	1981	1995
Household size	7	4.8
Family labor/total labor employed in the family unit	0.80 to 0.98	0.78 to 0.90
Gross value for market/gross production value (%)	79	47
Basic salaries generated by production	2.2	1.8
Most important source of income	Agriculture and livestock	Retirement income and pensions

decreased market participation. The crianceros' goal is to balance, while possible, what they buy with what they sell. The criancero tries to obtain the maximum income through the use of all family labor available with three ends in mind: production for the market, production for personal use, and off-farm labor. In this manner, the household achieves maximum satisfaction of needs compatible with sparse resources (Wettstein, 1982).

Traditionally, as prices fell, the crianceros increased their stock to ensure minimum income. In many cases the stock population reached the saturation threshold of land and pasture resources. Efforts of government and nongovernment agricultural extension to improve livestock management has partially reversed that response.

The change in the relationship between those who sell and those who buy is marked. In the 1980s, 76% of the total production went to market, whereas only 47% was sold in the 1990s. The increased household consumption of the total production expresses the impoverishment process defined as downward social decomposition. However, the peasants do not disappear because they have no employment alternatives. Yet young people are expelled from the community, as seen in the household size decreasing from 7 to 4.8 individuals. The strategy of combining family labor inside and outside the household production unit explains the capacity for perserverance of these producers.

The crianceros participate in different markets. In the product market, they participate as sellers. In the retail market, they participate as buyers. In the labor market, they participate as permanent sellers or as a temporary work force. Their participation in the bank credit and real estate markets has been almost nonexistent, because they receive only small amounts of subsidized credits. Although some form of payment in kind exists, these crianceros are immersed strictly in a mercantile economy. The scarce circulation of money has a greater relationship with the general poverty that goes beyond the rural areas than with the supposed nonmonetary character of such an economy.

Total income for peasant households, as indicated by the number of basic salaries, declined only slightly. Decrease in income resulting from the fall in the salary's buying power is much more significant, as the growing rate of personal consumption over total production indicates.

CONCLUSIONS

CONTROL AND RESISTANCE

Historically, the market participation of crianceros was not marginal, especially for those originally tending goat flocks which produced fine mohair. However, the dynamics of the globalization and concentration, environmental policies, and land privatization in the hands of large landowners makes criancero participation in the product market more risky and restricts their possibilities for inclusion into the regional labor market. Current policies and programs threaten to worsen the survival crisis for criancero communities.

Until the 1990s, there was reciprocity between the best lands owned by the large ranchers and the government lands occupied by the crianceros, both Mapuches and Creoles. This arrangement is ending, indicating capitalist expansion in a marginal area. The perspective of incorporating grazing land into the market intensifies the differentiation between viable producers and nonviable producers, increasingly excluding the crianceros.

Impoverishment and desertification have not destroyed the social bonds at the local community level. Community resistance to structural conditions and clientelistic policies reveal strong social networks organized around production practices.

ON PROBLEMS AND POTENTIALITY

The main problems that the crianceros face are the following:

- restrictions placed on access to soil, pasture land, and water
- institutional policies of land access that do not trust the behavior and conventions typical of the local communities
- the impoverishment process in Argentina of the past two decades
- lack of economic alternatives for the producers and their families

The base for alternative development of the crianceros' local communities in Patagonia lies in the cooperation present in livestock activities, the social networks maintained under conditions of poverty and resource scarcity, and resistance to external control.

REFERENCES

Barsky, O., Políticas agrícolas y reformas institucionales en la Argentina en el contexto del ajuste, *Ruralia* 3 Buenos Aires, 1992 p. 7.

Bendini, M. and Tsakoumagkos, P., *Campesinado y Ganadería Trashumante,* Editorial La Colmena-GESA, Buenos Aires, 1994.

BIRF, *Informe Anual sobre el Desarrollo Mundial 1992, Desarrollo y Medio Ambiente,* Washington D.C. Cap. 7, Política ambiental en las zonas rurales, 1992, p. 141.

Cuccullu, G. and Murmis, M., *Tipología de Pequeños Productores Campesinos en América Latina,* PROTAAL, IICA-OEA, San José de Costa Rica, No. 5, 1980.

Gasteyer, S. and C. B. Flora (2000) Modernizing the Savage: Colonization and Perceptions of Landscape and Lifescape. *Sociologia Ruralis* (40): 128–149.

Hardin, G. "The tragedy of the Commons. *Science* 162: 1243, 1968.

Murmis, M., Algunos temas para la discusión en la sociología rural latinoamericana: reestructuración, desestructuración y problemas de excluidos e incluidos *Revista Latinoamericana de Sociología Rural,* Valdivia, Chile, No. 2, 1994.

Murmis, M., Pobreza y exclusión social; sobre algunos problemas teóricos y de medición y la situación argentina, V Congreso Argentino de Antropología Social. Panel de Antropología Urbana, Buenos Aires, Mimeo, 1997.

Nogués, C., Bendini, M. and Pescio C., Medio ambiente y sujetos sociales; el caso de los cabreros trashumante, in *Debate Agrario,* Lima, No. 17, 1993.

Pescio, C., and Bendini, M., El desarrollo rural alternativo desde la integración binacional, in *Latinoamérica Agraria Hacia el Siglo XXI,* (Centro de Planificación y Estudios Sociales) *CEPLAES,* Quito, 1993.

Tsakoumagkos, P., Acerca de la descomposición del campesinado en la Argentina, in *Sociología Rural Argentina: Estudios en Torno al Campesinado, Posada,* M., Comp., Centro Editor de América Latina, Buenos Aires, 1993.

Wettstein, G., Cambios agrarios en los Andes de Venezuela, in *Comercio Exterior,* Mexico, Volumen 32, No 6, 1982.

9 Farm–Community Entrepreneurial Partnerships in the Midwest

*Cornelia Butler Flora, Gregory McIsaac,
Stephen Gasteyer, and Margaret Kroma*

CONTENTS

Introduction 115
Rural Communities... 117
Production Skills.. 119
Mechanical and Technical Skills 119
Financial Management Skills 120
Relational Skills 121
Risky Shifts 121
Making Risky Shifts: Understanding the Current Resource Flows 121
Farm–Community Entrepreneurial Partnership and the Risky Shift from Bulk
Commodity Production in Piatt County 122
Focus on Relationships....................................... 128
Conclusions 128
References 128

INTRODUCTION

When aboriginal peoples settled the tall grass prairie, it provided pasture for ruminants, rivers for fish and amphibians, and a vast diversity of plants for gathering. The savanna was a mosaic of bluestem, oak, and hickory, with occasional closed canopy forests, usually found near water bodies that inhibited fires (Lauenroth, et al., 1999). Native American nations managed the ecosystem through fire (Sauer, 1950) and species-specific hunting (Hames, 1987). The tall grass prairie supported a large, dynamic human and plant population (Schleister, 1994).

The first Europeans practiced *transhumance,* following animals to trap them and sell their skins. Animal availability depended in part on the riparian areas and wetlands that interspersed the tall grass prairie.

0-8493-0917-4/01/$0.00+$.50
© 2001 by CRC Press LLC

115

The ranchers and farmers who settled central Illinois replaced the original vegetation with crops, often draining wetlands and straightening streams to complete the
conversion (Vileisis,1997). Grains, hogs, and ruminants were raised to supply the
growing cities on the eastern and western coasts, with exports to Europe. Part of that
production was aimed at providing cheap food for the new migrants: basic calories
provided by small grains to keep people fed and wages low. Grains, particularly
wheat, were critical in that system. The tall grass prairie changed to corn and wheat.

Piatt County, Illinois, was formed in 1841. Its location between Decatur (Macon
County) and Urbana-Champaign (Champaign County) provided nearby markets for
the diverse agriculture commodities and shipping points for grain. With the building
of the railroads and feeder lines in the 1850s, shipping was made easier, and more
land was put into grain crops. Businesses in the county reflected the farming patterns,
with agricultural banks, grain elevators, and drainage system contractors.

At various times, the markets for grain have become saturated by high levels of
production. Food prices, in general, are "inelastic": no matter how cheap a particular
food item becomes, we will not eat more of it. Increased incomes changed food consumption patterns, with protein substituted for some of the calories from carbohydrates. More grain went to livestock, which were fattened and slaughtered to fill the
plates of a growing middle class and a more affluent working class. Commodity
meats, particularly beef and pork, produced on farms in Piatt County that once were
tall grass prairie, aided this transition.

By the 1940s, Piatt County's farms were mixed farming systems of corn, cattle,
and hogs. The introduction of soybeans in the 1950s and early 1960s (the Illinois
Soybean Association was founded in 1964) allowed many farmers to grow only
crops, and to ship bulk corn and soybean down the rivers and on trucks and railroads
to ports for export. By 1997, only 17 farms in Piatt County raised hogs; 47 had beef
cattle and 61 had cows and calves, whereas there was only one dairy farm. The country had twice as many farms ten years earlier; the number declined from 604 to 448
in that period.

Row crops, grown in monoculture (corn) or rotation (corn and soybeans) on an
increasing number of acres, used increasing amounts of nutrients and pesticides (U.S.
Department of Agriculture [USDA], 1997). The drained fields, tiled to move water
rapidly from the fields to channeled streams, carried dissolved chemicals to the rivers
and ultimately to the Gulf of Mexico, where the surplus of nutrients contributed to a
process of eutrophication at the mouth of the Mississippi River, which reduced the
availability of oxygen during the summer months. The resulting hypoxia (low oxygen) or anoxia (no oxygen) shifted the types and quantities of organisms that can be
supported in affected areas in the Northern Gulf of Mexico. The size of that zone
varies according to the amount of water conveyed off the fields; it is higher in years
of heavy rains and smaller in years of drought (Committee on Environment and
Natural Resources, 2000; Council on Agricultural Science and Technology, 1999).

By the beginning of the 21st century, the market for food had shifted again
(Jones, 1997). More people now seek more differentiated diets. The widely varied
ethnic groups in North America have always demanded specialty products as ingredients for national or regional dishes. That dietary diversity is increasingly valued by

other segments of the population. These foods, composed of specific, identifiable ingredients, combine qualities valued by particular end users, which include genetics that determine texture, taste, and process yields; how the ingredient is raised; how it is processed; and who raised it where. Furthermore, these choices are not just a passing fancy of maturing *yuppies* (young upwardly mobile professionals). When given a choice, people are willing to pay more for meat that has been raised humanely or vegetables that have been raised in ways friendly to the environment (AIR-CAT, 1996; Chaudri and Timmer, 1986; Harris, 1988; Malone, 1990; Ohlendorf and Jenkins, 1995; Ritson, and Hutchins, 1995).

The increasing markets for identity-preserved food, particularly certified organic and nontransgenic grains, suggest the rapid pace at which this shift to the third level of consumption is taking place. These changes in dietary patterns require different relationships between farmers and agroecosystems, with more attention to ecosystem health. Can these relationships develop, however, without changes in the rural community? In particular, if the institutions that provide the inputs (seeds, information, stock, financing, labor, and management) and handle the outputs (processing, marketing, and transportation) remain static, it is very difficult for an individual farmer to make isolated innovative changes to increase system sustainability.

RURAL COMMUNITIES

Rural communities in Illinois, as in other areas of the Midwest, were established in the era of low-value, high-volume crops, when grain was moved by rail and water to the rapidly growing urban centers. The local grain elevator, often a cooperative, provided storage, processing (grain drying), and marketing. Chicago provided a major market for Illinois farm products (Cronon, 1991).

As the demand for meat grew and Chicago became "hog butcher to the world" (Sandburg, 1916), local sale barns were created in rural communities to help farmers bring together and market hogs. Buyers who gathered at the weekly sales included other farmers looking for animals to fatten, brokers, and meat processors. These weekly sales, like the grain elevators, became community gathering places for the farm community.

In 1973, increasing oil prices triggered a worldwide change in the terms of trade, which increased world demand first for basic calories, then for meat. Farmers in Illinois and elsewhere, encouraged by the U.S. Department of Agriculture, planted from fence row to fence row, destroying many of the previous soil conservation measures put into place as the result of New Deal farm policies. Consequently, production increased rapidly, and the local elevators were unable to handle the volume. Low real interest rates because of high inflation, government credit programs, and rapid tax depreciation allowance encouraged many farmers to build on-farm grain storage and drying units. That enabled farmers to deliver grain to the elevator on a "just-in-time" basis for transportation to the next grain handler in the supply chain. Federal programs that subsidized grain shipments and provided soft credit (low interest and forgivable loans) for countries that purchased American grain further encouraged exports. Local elevators sold grains to a broker, who sold to a grain company, which sold it to a processor or a foreign government. Grain companies provided the transportation from the elevator to the destination. Little comparative advantage remained for the traditional elevator involved in undifferentiated bulk commodities.

The shift in volume and capital availability gave farmers the ability to store and dry their grain and beans, with the grain companies providing the shipping. The only function left for the country elevators was marketing, which most continued to do through the regional cooperatives or the grain companies. As contract cattle and hog growing increased, and as buyers from brokerage firms went straight to the farm to assess animals for purchase, the sale barn also became less central to livestock production.

The local community institutions in Piatt County were based primarily around the food systems of the 19th and mid-20th centuries. Dealing with bulk commodities and undifferentiated products that moved through many hands from producer to consumer, they were not adept at detecting market signals or at feeding those market signals back to farmers

- from producing commodities to producing products
- from many intermediaries to integrated supply chains linking producers more directly with the end user
- from specialization to flexibility in response to constantly changing markets in a setting of increased environmental awareness and concern for quality of life

Changes in agriculture and rural communities reflect broader changes in the business environment around the world. Corporations are becoming leaner and more capable of adapting to market demands (Hamel and Prahalad, 1996; Mintzberg, 1996; Nevis et al., 1995; Porter, 1996). Differentiating markets require constant change in product mix, affording new economic opportunities and new risks. Production of a reliable, high-quality, diverse food supply in ways that are profitable, competitive, and environmentally sound requires not only innovative alternative management skills but new strategic on-farm and off-farm relationships (Coaldrake et al., 1995; Lejeune and Cloutier, 1996; Sonka et al., 1995). The on-farm challenges relate to how well farmers are able to make choices amid myriad alternative options that will enable them to develop management systems that ensure continued economic vitality, maintain or enhance environmental integrity, and meet their quality of life goals. The challenge of rural communities and their institutions—market, state, and civil society—is to provide the institutional supports that facilitate farmers implementing farm-level choices that contribute to ecosystem health, economic viability, and social equity.

Granovetter (1985) provided a useful perspective for understanding how most economic behavior is embedded in social relationships. Flora and Flora (1993) developed the concept of entrepreneurial social infrastructure for a better understanding of why certain patterns of interacting and collectively approaching problems can contribute to a locality's ability to respond to challenges in a rapidly changing context. Economic development in rural America is highly related to a community's capacity to consider alternatives, form internal and external networks, and mobilize local resources (Flora, 1995; Flora et al., 1997). Moreover, the emerging economics of transaction costs suggests the importance of these relationships for farming success. These intangible assets that embed and connect farms and firms sociospatially are increasingly recognized as important in agrofood studies (Pritchard, 2000). The increasing importance of intangible assets means a shift in the skill sets of farmers.

PRODUCTION SKILLS

Success in agriculture has been based on production skills for at least 10,000 years. Producers have known their crops and animals and understood seasonal cycles as well as the need to adapt to climate and pest unpredictability. Knowledge was passed from parent to child and neighbor to neighbor. That knowledge tended to be what Kloppenberg (1991) called "mutable immobiles," knowledge specific to place that had to be modified when producers moved to a new area. Many of the early farm failures during the colonization and frontier periods in the United States and Canada came from failure to adjust agricultural practices well-adapted to soils, climate, and an array of pests in the home country to the very different conditions of the frontier. The first period of settlement in Central Illinois when the tall grass prairie changed to cropland was based primarily on production skills, although relational skills also were needed for communities to grow.

For agriculturists in the industrialized countries of the world, there is a progressive reduction in the different skills needed in the production process with more industrialization (Havens, 1986). For example, successful ranching once depended on production skills, which included an understanding of the local ecosystem and an ability to adapt to changes in nature, particularly weather and available forage (Gefu and Gilles, 1990). Maintenance of biodiversity was an important risk-reduction strategy (Baskin, 1997; Berkes, et al., 1994). Local knowledge, such as criteria for bull selection and an understanding of the microclimate aided both rancher and community survival. The production of those who did not develop and apply that local knowledge, such as cotton farmers in the South and small grain farmers on the Great Plains (Hurt, 1994), ultimately was not sustainable. The evidence is seen in "natural disasters" such as cottoned-out land, as well as the dying towns and ghost towns left in the wake of the Dust Bowl (McDean, 1986; Hurt, 1994).

MECHANICAL AND TECHNICAL SKILLS

Major changes in the skills needed to be successful occurred with mechanization. Productivity per hour worked was greatly enhanced, if farmers knew how to run and repair machinery. These skills were conveyed from parent to child (generally from father to son), and also in the formal education system through vocational education. The knowledge needed for working with machines involves primarily "mutable immobiles," knowledge that does not depend on context. A spark plug's relation to the distributor is similar no matter who made the engine. Although farms in the United States got cars before they got tractors (Jellison, 1993), work on any internal combustion engine prepared an individual to work on farm equipment and appreciate preventive maintenance. International development attempts to introduce machinery without the farm-based maintenance skills that characterized North American farms in the 1940s through the 1970s often were dismal failures. Although dealer and independent farm repair garages were an important part of each rural community, they generally dealt only with major problems. For most farm men, part of the winter was spent maintaining machinery. Those who did not like machines or had no aptitude for them found it much more difficult to make a living in farming. Because machinery allowed more land to be farmed by an individual or household, farm consolidation favored the mechanically minded. Those who left farming tended to be older farmers not raised in a mechanical tradition.

Whereas the end of the World War I brought mechanical skills to agriculture, the end of the World War II brought technical skills, particularly those related to the use of fertilizers and pesticides (Perkins, 1978). As it became possible to farm more and more land with the same crop, the need increased to add soil amendments and to apply pesticides required by increased pest concentrations accompanying monoculture. It was important to know enough science to talk to dealers, if not to apply it safely on the farm.

These two related additional skills acquired by farmers resulted in Fordist agriculture. Fordist agricultural production prevailed in the basic commodities supported by the federal farm programs. By mastering mechanical and technical skills, farmers were able to increase their output greatly. Over time, however, they received a smaller and smaller percentage of the surplus generated. They did not control the value chain, and their undifferentiated product required protected markets and subsidies to survive the vicissitudes of increasingly global markets.

Grain is a good example of Fordist agriculture. Volume was the single price signal. Thus, all efforts were on increasing production, which mechanical and technical skills as well as government support programs greatly enhanced. Grain was sold at a local elevator, then went through numerous hands before reaching an unknown end user. As Smith (1992) documented graphically, over time farmers received an ever-smaller proportion of the food and fiber dollar, with more going to the inputs and to the processors.

FINANCIAL MANAGEMENT SKILLS

In the farm crisis of the 1980s, a different set of skills became critical for agricultural success: financial management skills. Knowing how much each enterprise cost and rendered, when to buy and sell, and how to hedge and forward a contract allowed some farmers to get control of a little more of the value chain. Knowing tax laws became as important as knowing the farm programs. The temperament required to work carefully over a computer spreadsheet at times was diametrically opposed to the temperament that took joy in newborn animals or a freshly plowed field. Many midcareer farmers left farming as financial management, and changes from the time when they bought land, made it impossible to farm profitably.

The increasing layers of skills necessary for successful farming has resulted, on the one hand, in exits from farming (Knutson, et al., 1998). On the other hand, they resulted in an increased division of labor within agriculture. A different individual may provide each set of skills. Also, each set of skills retains a different proportion of the value it adds.

RELATIONAL SKILLS

Under the current situation of rapid globalization and industrialization, it is much more difficult for farmers to maintain a constant share of the value chain. These chains tend to be driven by relationships with input suppliers, particularly suppliers of knowledge; with markets, particularly in reaching emerging markets; and with fellow producers in new models of "cooperatition." Relational skills reduce the transaction costs in carrying out the increasing number of tasks critical to farm success. New generation cooperatives, flexible marketing networks, development and marketing of specialty products all demand broad networks. Granoveter (1985) referred to the market benefits from these as the "strength of weak ties." These networks are an important part of social capital (Hassanein, 1999). Rural areas, and farmers in particular, often are embedded in strong ties with relatively limited networks. New relationships are needed as new market mechanisms replace old governmental mechanisms (Folke and Berkes, 1998). The challenge for rural development will be to increase these ties in ways that allow rural communities and their citizens to retain a larger portion of food and fiber value chains (ACEnet 2000).

RISKY SHIFTS

Farmers and communities constantly make decisions. Most of the decisions, however, involve relatively minor change: how much nitrogen to put on a field, when to plant, and when to repair a street or building. These decisions allow the systems involved to continue functioning in the same way, only hopefully more efficiently. Risky shifts mean leaving the comfort zone and not returning to it. The shift can mean taking away or adding system components.

Knowledge about alternatives that yield more desirable results does not alone lead to changed behavior (Moscovici, 1985; Sabatier and Jenkins-Smith, 1993; Sapp et al., 1994). When decisions result in polarization or involve risky shifts, as in a shift from commodity production to more flexible, diverse systems, groups (social capital) are critical to that process. People change behavior because "we" do things that way.

MAKING RISKY SHIFTS: UNDERSTANDING THE CURRENT RESOURCE FLOWS

In the case of farms and communities, the current system is best understood by looking at current resource flows. This can be done from the perspective of a single field, a whole farm, a watershed, or a community. The resources include people, dollars, organization, knowledge, inputs, and natural resources such as soil, water, and biodiversity.

Environmental qualities result from site-specific interactions among living organisms and their abiotic environments. These interactions may provide benefits to human communities in the forms of regulation and purification of water, aesthetically pleasing landscapes, and recreational possibilities. The biophysical environment also may impose costs on human communities by harboring crop, human, and livestock pests. Farming systems directly change local environmental systems by manipulating soil and managing biota. Farming practices can also have an impact on distant environments by introducing chemicals and sediment into watercourses, and by removing habitat for migratory species. Evaluating the impact of agricultural systems on environmental qualities is, therefore, a multidimensional and multiscalar problem. In the case of farms in Piatt County, there is increasing evidence that chemicals from farms there are carried by water to the streams that feed the Sangamon River flowing into Lake Decatur, which is the municipal water supply for Decatur, 1971, where nitrate concentrations periodically exceed the drinking water standard (Keefer and Demissie, 1999; Kohl et al., 1971). From there, the water, along with the dissolved chemicals and nutrients it contains, flows into the Illinois River, then to the Mississippi River and the Gulf of Mexico (Goolsby et al., 1999).

Although there is much promise in community-based approaches for addressing rural development and natural resource management, there are significant problems to overcome as well. Communities may have a short-term focus and a single-problem orientation that restrict the range of alternatives or linkages that are considered in planning. Furthermore, communities may not recognize the value of the natural capital on which they depend, and they may not recognize or care about environmental problems they export to other communities (Rhoads and Herricks, 1996). Thus, there is a need to foster awareness of local environmental conditions, and to promote continuous learning about changes in environmental conditions and ecologic processes.

FARM–COMMUNITY ENTREPRENEURIAL PARTNERSHIP AND THE RISKY SHIFT FROM BULK COMMODITY PRODUCTION IN PIATT COUNTY

As early as 1974, a few farmers in Piatt County had acquired enough grain storage, transportation, and drying capacity to eliminate their need for services from the local elevators. Lynn Clarkson, a young farmer on his family's farm, saw no reason to pay the extra 14 cents per bushel that it cost for the local elevator to sell his grain to Tabor Grain, whose Decatur elevators then sold the grain to Archer Daniel Midland. Bernie Craft, who managed the Tabor elevator, told Lynn that he would like to buy directly from Lynn, but that he would be punished by the system if he did. However, he said, if Lynn had a grain dealer's license, then he could buy from Lynn directly. Soon thereafter, Tabor Grain was purchased by Archer Daniel Midland, and now handles identity-preserved grain for the company.

Lynn got the license, then used it to create Clarkson Grain. Working with some other farmers in the Cerro Gordo, he began investigating other ways to reduce the number of links between him and the end user. A German broker purchased a load of white corn from Clarkson Grain to ship to South Africa. The South African purchaser was so pleased with the grain that he sent Lynn a fax asking for more grain of the same quality. Lynn had seen the opportunities for marketing white corn internationally, and had found appropriate seed stock that produced a corn with good taste, texture, and process yields in the Central Illinois area. But he did not like the volatility of international markets. He knew that with a good crop year in South Africa, that market would no longer be available to him. Therefore, he began looking for domestic markets with a consistent demand for his differentiated product.

Armed with his experience as a foreign exchange student in Chile, he looked for current white corn users in the United States. Chicago's dynamic Mexican American community had a tradition of white corn use, particularly in making tortillas. Lynn met with many of the tortilla manufacturers in the city, learning exactly what they liked in corn and adjusting his seed accordingly. But he still did not make any sales. Most of his time was spent being rejected by potential customers. They were not particularly interested in getting the same thing from Clarkson Grain that they currently were getting from another supplier.

On his many visits, in which his interest and his Spanish language ability helped build up trust over time, Lynn noticed a number of system costs that he could reduce by the way he delivered white corn. The current suppliers sent the corn in 100 lb bags, which needed several workers to take them off the truck, rip them open, pour them into the bin, and then dispose of the costly sacking. He finally convinced one tortilla factory owner to let him bring in Clarkson Grain millwrights to build a grain-handling system for the factory in downtown Chicago. That new system allowed the grain to be transferred into the factory bins, increasing its quality even more by reducing the amount of handling and reducing system costs.

But taste, texture, and process yield were still the key. Farmers could no longer use their favorite seed varieties or their favorite pesticides. Lynn spent a lot of time with community farmers, explaining why the processor needed particular qualities and why certain practices resulted in the specific characteristics the particular manufacturers wanted. He also continued to spend time with the tortilla manufacturers to improve constantly the product he delivered to them in light of their demands.

With Lynn serving as a link between the producer and the end user, communities of interest among farmers began to grow. Farmers interested in producing differentiated products rather than growing "government grain" (grains supported by the farm programs and, after the Freedom to Farm Act of 1996, loan deficiency payments) sought out Clarkson Grain. A number of these producers were organic growers, deeply concerned about the environmental impact of conventional agriculture. Clarkson early identified markets for organic soybeans with specific characteristics. Specific kinds of organic corn followed. But to do true organic growing and make money, the entire rotation, which in Central Illinois includes corn, soybeans, and wheat, needed a market.

Working with organic farmers for particular characteristics of taste and texture, Lynn found a market for organic wheat. Once that alternative market path, with its premium, was in place, community growers found it much easier to move toward organic products. But even with the rotation cover, organic growth is not enough. The farm community entrepreneurial partnerships continue to work to meet the demands of end users and translate these demands into the genetics and crop management that most closely meet them.

The wife and husband team of Tracy Norcross and Allen Williams had a contract partnership with Clarkson grain to grow no-till white corn for the tortilla market (Figure 9.1). They produced the corn using no-till methods. In no-till farming, the soil is left undisturbed from harvest to planting. Planting or drilling is accomplished in a narrow seedbed or slot created by disk openers. Coulters, residue managers, seed firmers, and modified closing wheels are used on the drill or planter to ensure adequate seed to soil contact. In a properly designed no-till system, pest (weed, disease, and insect) control is accomplished primarily with the following cultural practices: rotation, sanitation, and competition. Judicious use may be made of herbicides to provide the crop with a competitive advantage over the weeds (USDA/Natural Resource Conservation Service [NRCS], 2000).

Tracy and Allen purchase their inputs from a local supplier (Piatt County FS Growmark) and a specialized pesticide consultant. Together with their neighbors Chalk Taylor and Rick Alan, Tracy and Allen provide labor for testing the soil, applying nutrients, spraying herbicide in the spring, planting the corn, harvesting the corn, and planting the cover crop. They market their corn to Clarkson Grain, providing spe-

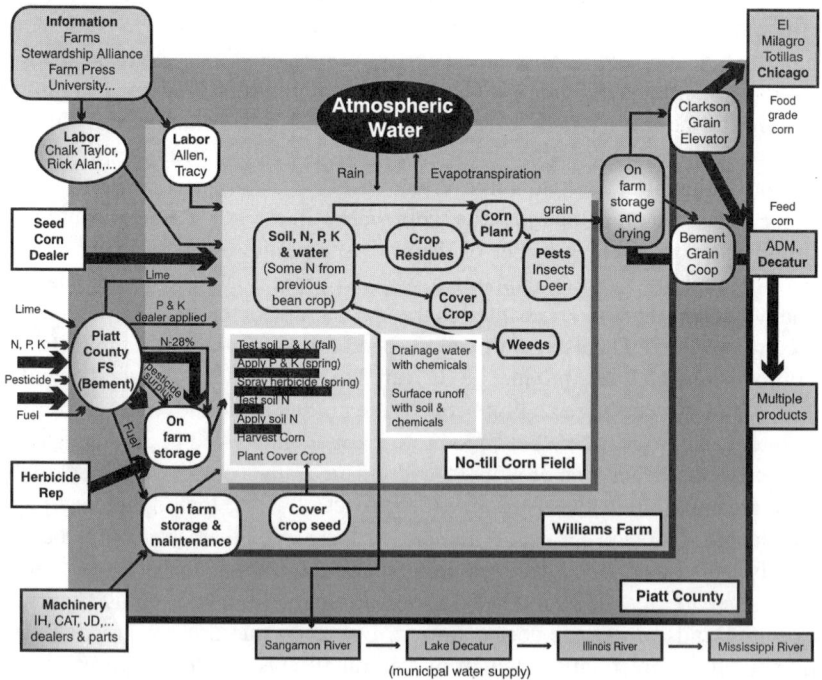

Thick lines represent relationships that will be broken by the new system.

FIGURE 9.1 Resource Flow Diagram for Tracy and Allen's No-till Corn Production Enterprise

cialized corn to a Chicago tortilla factory and to the Bement Farm Cooperative Elevator, which in turn sells to Archer Daniel Midland Co. (ADM) in Decatur. The chemicals and nutrients from the land flow into the Sagamon River, and hence into Lake Decatur, the municipal water supply whose quality is threatened by high levels of nitrate and sediment. That water then flows to the Mississippi River and the Gulf of Mexico, contributing to the hypoxia there. That linkage worried Tracy and Allen, who are concerned about ecologic health and social equity, as well as the economic vitality of their household.

Their concerns about the environment and food safety prompted them to experiment with organic farming on 93 acres of rented land starting in the early 1990s. The absentee landowners had a long-standing relationship with Allen and Tracy and shared many of their conservation concerns. They adopted an organic rotation that produced corn, soybeans, small grains, and a green manure crop. There are no livestock on the farm. Allen, a popcorn lover, discovered an heirloom variety with exceptional eating qualities that fit well into the organic rotation because it required less nitrogen than food-grade or feed corn. He located a supplier in Ames, Iowa.

Instead of spraying herbicides, they must till, rotary hoe, and cultivate to control weeds. This requires labor in addition to their no-till labor force and additional fuel. Furthermore, when the corn is higher, they must hand-weed, which they do not do with their no-till corn. This is a new labor need that makes organic popcorn a "risky shift." Now they use both migrant workers and high school students for hand-weeding at the peak of the growing season. One source of labor is a reserve labor force because it comes from outside the community and is occupied elsewhere during other times of the year. Another is a reserve labor force within the community: high school students. A community side effect is that this work also involves students with farming. However, hand-weeding has some potential community-level problems, particularly with migrant workers. They need housing and other infrastructure while they are there on the farm.

The corn is harvested by machine in both fields, and in both cases they store it on the farm. All the no-till corn is processed off the farm, and some of the organic popcorn is processed on the farm to be sold directly to a growing network of folks who have heard about the great popcorn. During the comparative field-based analyses, ways of breaking various bottlenecks were discussed. One alternative would be to keep the migrant workers in the community longer by engaging them as popcorn packers when they are not harvesting. Another alternative would involve working with the local school system and 4-H club to set up a youth entrepreneurship program. The youth could take over the packaging, managing the direct sales (particularly through the Internet and a school-based popcorn stand) and getting a cut of the profits in addition to payment for their current work of hand-weeding.

Another alternative would be to spin off a separate business within the community, which would be in charge of the packaging and marketing aspects of the corn sales. This business would be loosely networked with the farm, yet stay within the community. What then moved outside the community would not be high-volume, low-value corn, but low-volume, high-value corn that would not travel nearly as far or through as many hands. On-farm sales could still occur to bring people back to agriculture, back to visit the tall grass prairie that Tracy and Allen have planted on a part of their farm, which serves as a nature trail for local groups and visitors from urban areas, including Chicago. Another alternative might be to mechanize hand-weeding through development of appropriate technology, which could evolve into another local small enterprise.

New institutions built on new relationships have to be in place to handle the new resource flows identified in Figure 9.2. These new entrepreneurial partnerships could be made with high school students, migrant workers, or local business people. Some old institutions get left out: the pesticide distributor, the nitrogen and the nitrogen application from Piatt County Farm Service Company (FS) (Farm Services), Bement Grain Cooperative, and Archer Daniels Midland Co. Some of these institutions will be permanently eliminated, while others may change.

With the organic popcorn field, Tracy and Allen still service their farm machinery from local dealers. They still have on-farm storage and maintenance of machinery, and they still use fuel that they buy from their local Piatt County Farm Services dealer, which is their Farm Bureau Cooperative where they would have bought their pesticides and their fertilizers. The Piatt County Service Company lost input sales, but continues to sell fuel to Tracy and Allen.

Piatt County Service Company is adjusting to the new system. Because they no longer have a comparative advantage in grain storage, grain drying, or transportation, cooperatives such as the Piatt County Service Company need to seek new way to serve their members. For example, the coop might work with other businesses to pro-

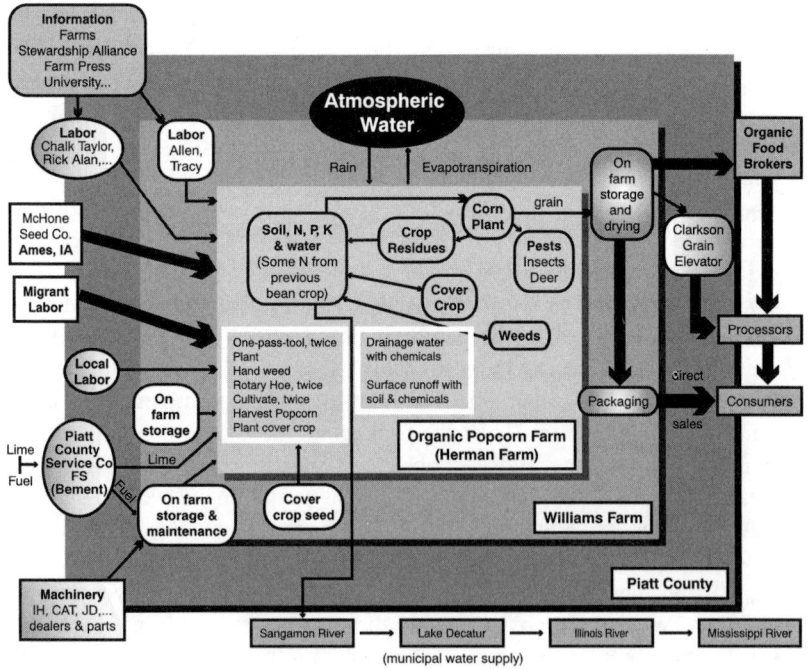

Thick arrows indicate new institutional linkages that need to be formed.

FIGURE 9.2   Resource Flow Diagram for Tracy and Allen's Organic Popcorn Enterprise

vide alternative employment to farm workers and their families so they can be available for seasonal labor. As a responsive, producer-owned input supplier, the coop would shift what it does in response to the changing conditions.

Additionally, a body of research is indicates that nitrate transport in drainage water from the organic production system tends to be less than from the conventional corn–soybean system (Drinkwater et al., 1998; Goldstein et al., 1998). Monitoring the tile drainage on Tracy and Allen's farm has produced similar results: nitrate concentrations in the drainage water from the organic farm have rarely exceeded the drinking water standard for nitrate (Mitchell, et al. 2000). This provides a water quality benefit to the city of Decatur, and may also provide a selling point for the popcorn. A study has been initiated recently at the University of Illinois to examine the economics of water quality benefits provided by organic farming to Lake Decatur and similar watersheds.

The comparative resource flow diagrams stimulate the farmer and other people in the community. The diagrams are not maps, but schematics of the process. Thus, everyone can begin to see what a different form of production requires both on the farm and in the community. A lot of work on sustainable agriculture is focused on a variety of on-farm techniques to improve the agroecosystem. Researchers have assumed that by aggregating costs, time used, and profits, the impact on the community can be determined. This approach is qualitatively different, because it focuses on the institutions that must change to make on-farm change possible. The only way this can be done is by comparing "standard" work to the innovation. The continued innovation of Clarkson Grain is based on understanding the standard work of conventional agriculture, then figuring how to make agricultural work better for ecosystem health, for farm households, and for the community. By constantly forming new relationships and strengthening communication along the value chain, new partnerships are continually made and remade as conditions change.

Changes are less likely to be made when too many changes must be made at once, particularly when the institutional structure is not in place to support these changes. Allen and Tracy have been very deliberate in moving into more organic and identity-preserved crops. Shifts become less risky if institutions change with them. Therefore, the personal contacts made by Lynn and other staff of Clarkson Grain have helped them to make the risky shifts they value for themselves and the community. This model can help providers of traditional inputs to rethink their role and help input providers move into biologic pest controls and farm management/consultant services. The Piatt County cooperative could change from a pass-through retail institution that adds wealth to the community only on their retail markup to being a service firm that provides more services locally, thus returning more of the dollars spent by the farmer to the community.

FOCUS ON RELATIONSHIPS

Skills needed for success in agriculture are changing rapidly. As in other parts of the economy, globalization and industrialization make agriculture a more complex enterprise. Furthermore, as in other industries, concern for environmental as well as economic outputs and outcomes increases the complexity even more.

CONCLUSIONS

Whereas commodity production has always been disadvantaged in an open system without supply controls, increasing global competition makes the situation today even more difficult. Threats to the economic sustainability of family farmers and the intergenerational transfer of farming as an occupation are thus high. Furthermore, high chemical input and commodity production systems designed to be managed by a single individual (often separate from the owners of the land) have led to serious deterioration of soil quality, water quality, and biodiversity, threatening environmental sustainability. Finally, the current system increases the distance between the haves and have-nots, threatening community and equity sustainability.

Farming alternatives require new mindsets by farmers and new institutions in rural communities and regions. New farm community entrepreneurial partnerships (FACEPs) can help to create both. As awareness of the multiple functions of farming are increased, FACEPs are a mechanism through which the risky shifts required to create new food systems more compatible with all three kinds of sustainability can occur, increasing the attractiveness of the occupation for coming generations.

REFERENCES

ACEnet, *Collaborative Cause Marketing Handbook for the Specialty Food Industry: Research and Recommendations,* Appalachian Center for Economic Networks (ACEnet) Publications, Athens, Ohio, 2000.

AIR-CAT, (Measurement of Consumer Attitudes and their Influences on Food Choices and Acceptability) *Consumer Attitudes Towards Meat,* AIR-CAT Project, Matsforsk, Norway, 1996.

Baskin, Y., *The Works of Nature: How Diversity of Life Sustains Us.* Island Press, Washington, D.C., 1997.

Berkes, F., Folke, C., and Gadgil, M., Traditional ecological knowledge, biodiversity, resilience, and sustainability, in *Biodiversity Conservation,* Perrings, C., Mäler, K. G., Folke, C., Hollings, C. S., Eds., Kluwer Academic Publishers, Dordrecht, The Netherlands, 1994, 269.

Chaundri, R. and Timmer, C. P., *The Impact of Changing Affluence on Diet and Demand Patterns for Agricultural Commodities,* World Bank Staff Working Papers No. 785, The World Bank, Washington, D.C., 1986.

Coaldrake K. F., Sonka, S. T., Sudharson, D., and Winter, F. W., *New Industries and Strategic Alliances in Agriculture: Concepts and Cases,* Stipes Publishing, Champaign, 1995.

Committee on Environment and Natural Resources. 2000. Integrated Assessment of Hypoxia in the Northern Gulf of Mexico. National Science & Technology Council, Committee on Environment and Natural Resources, Washington, D.C..

Council on Agricultural Science and Technology, *Gulf of Mexico Hypoxia: Land and Sea Interactions,* Task Force Report 134, Ames, IA, 1999.

Cronon, W., *Nature's Metropolis: Chicago and the Great West,* W.W. Norton, New York, 1991.

Drinkwater, L. E., Wagoner, P., and Sarrantonio, M., Legume-based cropping systems have reduced carbon and nitrogen losses, *Nature,* 396, 262, 1998.

Flora C. B. and Flora, J. L., Entrepreneurial social infrastructure: a necessary ingredient, *Ann. Am. Acad. Poli. Soc. Sci.,* 528, 48, 1993.

Flora, C. B., The sustainable agriculturalist and the new economy, *Greenbook,* Minnesota Department of Agriculture, Energy and Sustainable Agriculture Program, St. Paul, MN, 1995.

Flora, J. L., Social capital and communities of place, *Rural Sociol.*, 63, 481, 1998.

Flora, J. L., Sharp, J., Flora, C. B., Newlon, B., and Bailey, T., Entrepreneurial social infrastructure and locally initiated economic development in nonmetropolitan U.S., *Sociol. Q.*, 38 (4), 623, 1997.

Folke, C. and Berkes, F., *Understanding the Dynamics of Ecosystem-Institution Linkages for Building Resilience,* Biejer Discussion Paper Series No. 112, Beijer International Institute of Ecological Economics, Stockholm, Sweden, 1998.

Gefu, J. O. and Gilles, J. L.,1990, Pastoralists, ranchers, and the state in Nigeria and North America: a comparative analysis, *Nomadic Peoples,* 25, 34, 1990.

Goldstein, W. A., Skully, M. J., Kohl, D. H., and Shearer, G., Impact of agricultural management on nitrate concentrations in drainage waters, *Am. J. Alternative Agric.,* 23(3), 105, 1998.

Goolsby, D. A., Battaglin, W. A., Lawrence, G. B., Artz, R. S., Aulenbach, B. T., Hooper, R. P., Keeney, D. R., and Stensland, G. J., *Flux and Sources of Nutrients in the Mississippi-Atchafalaya River Basin: Topic 3 Report for the Integrated Assessment on Hypoxia in the Gulf of Mexico,* NOAA Coastal Ocean Program Decision Analysis Series No. 17, NOAA Coastal Ocean Office, Silver Spring, MD, 1999, p. 130.

Granovetter, M. S., Economic action, social structure, and embeddedness, *Am. J. Sociol.,* 91, 481, 1985.

Hamel, G. and Prahalad, C. K., Competing for the new economy: managing out of bounds, *Strategic Manage. J.,* 17, 237, 1996.

Hames, R., Game conservation or efficient hunting, in *The Question of the Commons: The Culture and Ecology of Communal Resources,* McKay, B. and Acheson, J., Eds., University of Arizona Press, Tucson, 1987, p. 92.

Harris, C. K., *Consumer Satisfaction with the U.S. Food System,* Michigan State University Department of Sociology, East Lansing, 1988.

Hassanein, N., *The Way America Farms: Knowledge and Community in the Sustainable Agriculture Movement,* University of Nebraska Press, Lincoln, 1999.

Havens, A. E., Capitalist Development in the United States: State, Accumulation, and Agricultural Production Systems, in *Studies in the Transformation of U.S. Agriculture,* Havens, A. E., Ed. Westview Press, Boulder, p. 26.

Hurt, R. D., *American Agriculture, A Brief History,* Iowa State Press, Ames, 1994.

Jellison, K., *Entitled to Power: Farm Women and Technology, 1913–1963,* University of North Carolina Press, Chapel Hill, 1993.

Jones, E., Consumer demand for carbohydrates: a look across products and income classes, *Agribusiness,* 13, 599, 1997.

Keefer, L. and Demissie, M., *Watershed Monitoring for the Lake Decatur Watershed, 1997–1998,* Illinois State Water Survey Contract Report 637, Illinois State Water Survey Champaign, 1999.

Kloppenburg, J. Jr., Social theory and the de/reconstruction of agriculture science: local knowledge for sustainable agriculture, *Rural Sociol.,* 56, 519, 1991.

Knutson, R. D., Penn, J. B., and Flinchbaugh, B. L., *Agricultural and Food Policy,* Prentice Hall, Upper Saddle River, NJ, 1998.

Kohl, D., Shearer, G., and Commoner, B., Fertilizer nitrogen: contribution to nitrate in surface water in a corn belt watershed, *Science,* 174, 1331, 1971.

Kreitzman, J. P. and Mcknight, J. L., Asset-based community development: mobilizing an entire community, in *Building Communities from the Inside Out: A Path Towards Finding and Mobilizing a Community's Assets,* Center for Urban Affairs and Policy Research, Northwestern University, Chicago, 345, 1993.

Lauenroth, W. K., Burke, I. C., and Gutmann, M. P., The structure and function of ecosystems in the central North American grassland region, *Great Plains Res.,* 9, 223, 1999.

Lejeune, A. and Cloutier, L. M., *Modelling the Knowledge-Creating Process: The Case of Strategic Alliances in Agribusiness,* Working Paper 07-96, Research Center in Management, University of Quebec, Montreal, March, 1996.

Malone, J. W., Consumer willingness to purchase and pay more for potential benefits of fresh food products, *Agribusiness,* 6, 163, 1990.

McDean, H. C., Dust Bowl historiography, *Great Plains Q.,* 6, 117, 1986.

Mintzberg, H., Musing on management, *Harvard Bus. Rev.,* 74, 61, 1996.

Mitchell, J. K., McIsaac, G. F., Walker, S. E., and Hirschi, M. C., Nitrate in river and subsurface drainage flows from an East Central Illinois watershed, *Trans. ASAE,* 2000.

Moscovici, S., Social influence and conformity, in *The Handbook of Social Psychology,* Lindzey, G. and Aronson, E., Eds., 3rd ed., New York, 1985, 347.

Nevis, E. C., DiBella, A. J., and Gould, J. M., et al., Understanding organizations as learning systems, *Sloan Manage. Revi.,* 37, 73, 1995.

Ohlendorf, G. W. and Jenkins, Q. A. L., Who cares about animal agriculture, *Rural Sociol. Soc.,* Washington, D.C., 1995.

Perkins, J. H., Reshaping technology in wartime: the effect of military goals on entomological research and insect-control practices, *Technol. Culture,* 19, 169, 1978.

Porter, M. E., What is strategy, *Harvard Bus. Revi.,* 75, 61, 1996.

Prichard, B., The transnational corporate networks of breakfast cereals in Asia, *Environ. Planning,* 32, 789, 2000.

Rhoads, B. L. and Herricks, E. E., Naturalization of headwater agricultural streams in Illinois: challenges and possibilities, in *River Channel Restoration,* Brookes, A. and Shields, D., Eds., Wiley, Chichester, 1996.

Ritson, C. and Hutchins, R., .Food choice and the demand for food, in *Food Choice and the Consumer,* Marshall, D., Ed., Blackie Academic and Professional, London, 1995, p. 43.

Sabatier, P.A. and Jenkins-Smith, H. C., *Policy Change and Learning: An Advocacy Coalition Approach,* West View Press, Boulder, 1993.

Sandburg, C., *Chicago Poem,* Henry Holt and Company, New York, 1916.

Sapp, S. G., Harrod, W. J., and Zhao, L., Socially constructed subjective norms and subjective norm behavior, *Soc. Behav. Personality,* 22, 31, 1994.

Sauer, C. O., Grassland, climates, fire and man, *J. Range Manage.,* 3, 16, 1950.

Schleister, K. H., Ed., *Plains Indians, A.D. 500–1500: The Archeological Past of Historic Groups,* University of Oklahoma Press, Norman, 1994.

Smith, S., Farming: it's declining in the U.S., *CHOICES,* 6(1), 8, 1992.

Sonka, S. T., Cloutier, L. M., and Banik, M. M., *Organizational Learning and Strategic Alliances,* presented at Using Firm-Level Models for Strategic Management Research in Agricultural Economics Symposium, American Agricultural Economics Association, Indianapolis, August 6–9, 1995.

U.S. Depatment of Agriculture (USDA), Table10: agricultural chemicals used, in *1997 Census of Agriculture, Piatt County, Illinois,* Available: *http://govinfo.library.orst.edu/cgi-bin/ag-list?10-147.ilc* Accessed: 9/19/99

USDA/Natural Resource Conservation Service (NRCS), Available: http://www.no-till.com// Accessed: 9/15/2000

Vileisis, A., *Discovering the Unknown Landscape: A History of America's Wetlands,* Island Press, Washington, D.C., 1997.

10 A Learning Approach to Community Agroecosystem Management

Clive Lightfoot, Maria Fernandez, Reg Noble,
Ricardo Ramírez, Annemarie Groot,
Edith Fernandez-Baca, Francis Shao,
Grace Muro, Simon Okelabo, Anthony Mugenyi,
Isaac Bekalo, Andrew Rianga, and Lynette Obare

CONTENTS

Introduction 131
A Learning Approach 133
    First Dimension: Organizational Structure for Learning 134
    Second Dimension: Process for Learning 135
    Third Dimension: Instruments for Learning 137
Learning Instruments for Future Visioning of Agroecosystem Management 138
Learning Instruments for Clarifying Requirements, Partnerships,
and Responsibilities 143
Learning Instruments for Clarifying Characteristics of Successful
New Partnerships 146
Learning Instruments for Reflecting on Agroecosystem Performance 148
Learning Instruments for Reflecting on Partnership Performance 150
Conclusion 151
References 154

INTRODUCTION

Farmers, local extension workers, and nongovernment organization (NGO) field staff play increasingly more important roles in "community-based" agriculture and natural resource management projects or programs. Ideas about priority problems and how they might be solved are expected to come from the community. A bottom-up, participatory approach to project design and implementation is the operational hallmark of these projects.

Communities rife with conflicts over the exploitation of farmland and natural resources are suddenly expected to work together to conserve nature. Suddenly, farming systems that degrade the soil and pollute water resources are expected to become ecologically sound and government and nongovernment institutions that have had little experience working together are asked to form partnerships.

Although no one doubts the desirability of these changes, little time and few resources are given to bring them about. Project participants are given no time to understand the perspectives of different communities about agroecosystems and their management. Indeed, little effort is given to finding out who the stakeholders are, let alone time for negotiating concerted action in the management of agroecosystems.

No one should really be surprised when project evaluators report that most farmers are not participating in the project, and that few will continue after the project ends. Equally, no one should be surprised when the expected farming or conservation improvements have not been realized. Farmers often are not impressed with the impact of so-called "improved" technologies. They complain that funds are attached to technological fixes that are inappropriate. Little effort is given to the development of knowledge systems for ecologically sound agriculture. Traditional knowledge about ecologically sound practices is rarely documented in a manner that is useful to other farmers. Project participants are rarely plugged into the growing international knowledge system of organic, ecologic, or alternative agriculture.

There is little room in projects for learning and change. External monitoring and evaluation, the main opportunities for adjustments, usually provoke defensive attitudes in which mistakes are hidden rather than used as opportunities for learning. Because local people do not have the capability or responsibility for evaluation, valuable lessons go unlearned. Because local people remain isolated from external knowledge networks, more appropriate technologies go unused.

To complicate matters further, many African nations are decentralizing and privatizing much of their public sector agriculture support services. District level staff are now asked to respond to farmer demands and to form partnerships with other service providers to meet those demands. The logic behind these policies is not only to save government money through sharing tasks and narrowing responsibilities, but also to provide better targeting of services and more efficient services. Building viable interinstitutional partnerships is hard, slow work. Moreover, responding to farmer demands requires considerable flexibility and dynamism of those organizations. How to create farmer demand for services and how to form viable partnerships rarely are subjects of study. There are few success stories and best practices on which to build. In these circumstances, the challenge confronting donors, central and local government officers, and project participants is to invent their own ways of working. This is partly because local conditions and complexities require a level of on-site innovation that cannot be satisfied by emulation of "best practices." This is not to say that the proverbial wheel must be reinvented everywhere, but best practices do need to be adapted to local conditions, or better still, re-invented by local people. A capacity to re-invent and innovate is essential to partnership building and community progress in agroecosystem management.

In this chapter, the authors describe how a learning approach provides communities, local government, and agriculture service providers with opportunities to learn their own way through to better partnerships and better management of agroecosystems. They describe an organizational structure for learning, a process for learning, and several key instruments to facilitate learning. The chapter is concluded with a look at what progress has been made in community development of agroecosystem management strategies, farmer demand for agriculture support services, formation of partnerships, and capacity for local innovation. The conclusion ends with a brief look at the constraints to further development of a learning approach to community-level agroecosystem management.

A LEARNING APPROACH

A learning approach builds joint capacity among community members, field-level development workers, and service providers for local on-site innovation. Learning facilitates innovation in the way local people work together and how they assess the performance of their partnerships and their agroecosystem management strategies. Enhanced innovative capacity sets the stage for improving the management of agroecosystems and the effective demand by farmers on agricultural support services. It also sets the stage for handing over more responsibilities to local actors at every stage of the project cycle from design through evaluation.

The learning approach developed in this discussion draws on four lines of research. Research in the area of farmer participatory development provides a great deal of experience in how to engage farmers in research and development projects (Chambers, 1997; Korten, 1980; Pretty et al., 1995). Insights gained from research on learning systems and soft systems provides a second foundation for this work (Bawden, 1991; Checkland and Scholes, 1990; Daniels and Walker, 1996). Operational details have been greatly informed by research in the areas of agricultural knowledge and information systems analysis (Engel and Salomon, 1994; Ramírez, 1997), multiple stakeholder management in forestry, and protected area management (Borrini-Feyerabend, 1996; Daniels and Walker, 1996; Ramírez, 1999). On the biologic side, operational details have been informed by research in the areas of agroecosystem analysis (Altieri, 1989; Conway, 1985; Lightfood and Noble, 1993; Lightfoot et al., 1993). The research methods of agricultural knowledge and information systems assessment have been woven with the research methods of agroecosystems analysis to capture both the learning about the way stakeholders are organized to respond to complex situations and the agroecologic aspects (Altieri, 1989; Conway, 1985; Lightfoot and Noble, 1993; Lightfoot et al., 1993).

This research provided the theoretical and methodologic inputs into the development of a learning approach that has engaged the International Support Group (ISG) and its local partners in Kenya, Uganda, and Tanzania for the past 3 years (Development Support Services, 1999; International Support Group, 1999; Shao et al., 2000). This development attempted to insert a learning approach into the ongoing development activities of a broad range of organizations at community, district, and national levels. It is from these experiences that examples are drawn to illustrate three dimensions of a learning approach. The first dimension described here is organizational. It permits a comparison of organizational structures between research and development projects. Process is the second dimension. This dimension elaborates a process for learning. Instruments for learning make up the third dimension. These instruments facilitate our learning about

- communities' future visions of agroecosystem management and opportunities for their realization
- partnerships and alliances needed if communities are to realize their visions
- negotiations to build partnerships and alliances for action
- reflections on agroecosystem management and partnership performance they have.

FIRST DIMENSION: ORGANIZATIONAL STRUCTURE FOR LEARNING

Although organizational linkages in research and development projects vary enormously, they have common patterns. Common patterns are, of course, oversimplifications, but they do make easier the kind of comparisons shown in Figure 10.1. Here the organizational linkages in research and development projects are top down. Many research projects build direct links between researchers working at the national level, whether in a university or government research organization, and farmers. On-farm experiments are found in many agriculture projects. Development projects, in contrast, involve extensionists from national and district levels and sometimes local NGOs. Development projects work directly with groups of farmers. In both cases few lateral links exist at the different levels. Where lateral links do exist, they tend to be between research and extension at the district level and increasingly between NGOs and extension at the point where the former are engaging the latter in their projects.

In the growing number of community-based natural resource management projects, things are different. Here one finds linkages built among all relevant stakeholders at the local, regional, and national levels. These projects tend to follow a

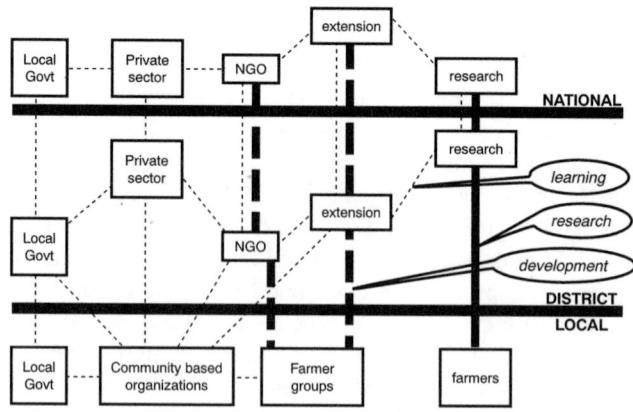

FIGURE 10.1   Linkages between participants in research, development, and learning settings.

participatory action learning approach to development (Röling and Wagemakers, 1998).

A learning approach forges lateral links among organizations and groups at each level and between levels. Over time, these linkages can result in the development of informal interinstitutional learning coalitions. Local- or village-level learning coalitions need to bring together farmer self-help groups, community-based organizations, and local government authorities. District-level interinstitutional learning coalitions need to bring together representatives from NGOs, government research and extension agencies, local government authorities, and private sector agricultural service providers. A similarly composed interinstitutional coalition also is needed at the national level.

The learning approach framework also requires that different levels link together. Local coalitions benefit greatly from linkages with district-level organizations. District- and national-level organizations wishing to respond to community demands benefit greatly from linkage with local learning coalitions. Opportunities occur from time to time within projects and within the operations of local governance for dialogue between organizations at all levels and, although less frequently, for dialogue between levels. However, without the commitment and resources of individuals within the organizations concerned to continue meeting and learning, little progress will result. Infrequent, random consultations or workshops are not enough to sustain a learning approach. Moreover, without a clear process for learning and instruments to facilitate learning, little progress can be made. Organizations seek new linkages not for the abstract notion of learning, but to pursue their own goals. Linking and negotiating are awkward, time-consuming efforts, but the organizations have come to realize that no better alternative, especially in this age of decentralization. These are the second and third dimensions, which are discussed next. Although the organization for learning can be started within projects, it must move from project to project in order to increase its skills and sustain itself.

SECOND DIMENSION: PROCESS FOR LEARNING

A number of elements were considered critical to effective joint learning:

- face-to-face accountability and group pressure to favor community influence
- farmer-led analysis, visioning, and planning
- reflection and adaptation of the instruments of learning

The process for learning has five distinct phases as shown in Figure 10.2. The learning process allows the partners to analyze the performance of the partnerships and find out where there is room for improvement. In this sense, "learning" embraces a process of reflection in terms of the partnerships and the agroecosystem's behavior and performance.

The first phase starts at the local level with farmer self-help groups and community-based organizations learning about the agroecosystems in their areas and their management 30 years ago and today. Viewing from the past, local people envi-

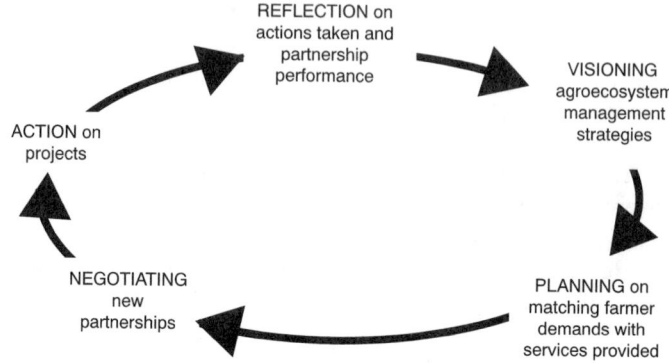

FIGURE 10.2   Phases in a process of learning.

sion how they would like to change their agroecosystems in the future. The desired changes form the basis for identifying the resources, services, and support needed to realize their future visions. Local learning about better ways to manage agroecosystems is enhanced by interactions with district- and national-level extensionists and researchers.

These visions and the requirements of community members are presented to district-level organizations in the second phase of the learning process. In this phase, communities' demands for agricultural support services are matched with the services offered. This also provides an opportunity for community members' visions to be informed by district-level extensionists and researchers. Thus local- and district-level organizations learn which services match demands and what new services should be created to meet demands. Where demands go unmet, policy issues of interest to national-level policy makers are raised.

When the resources, services, and support available match the community's requirements, there is a basis for negotiating partnerships between community members and public or private sector groups, entities, or enterprises. In this phase, the community and local organizations learn to develop partnerships that will increase the community's access to the resources, services, and support required to realize their future visions of how their agroecosystem should be managed. Providing opportunities for private and public services to present the objectives and mandate of their own organization and the constraints they face has proved to be an important aspect for the negotiation of a good partnership.

After rounds of negotiations between partners, concerted actions occur. Action, the implementation of improved agroecosystem management strategies, is the fourth phase in the learning process. The hallmark of these actions is that

- they are directed toward a vision of the future fashioned by local people
- local people are the key actors in the implementation of the actions
- indicators of performance are established as the partnerships are negotiated

These performance indicators are used to facilitate reflection, the fifth and last phase in the learning process. After action or the implementation of projects, partners at the local and district levels need to reflect on the performance of their partnership and that of the agroecosystem management strategies. After reflection, district- and local-level organizations revisit the community's future visions of how agroecosystems should be managed and what service partnerships are needed. Changes are made as a result of what has been learned, and another cycle is started. Learning is a continuous process that has no end, as indicated in Figure 10.2.

THIRD DIMENSION: INSTRUMENTS FOR LEARNING

Learning instruments facilitate each phase of the learning process. These instruments help the learners answer key questions. Each phase of a learning process has its own specific set of key questions.

Phase 1:

- Future agroecosystem management strategies are envisioned and the resources needed to implement them.
- What is the current status of our agroecosystem in comparison with the past?
- What would we like to see our agroecosystem look like in the future?
- With whom do we need to partner to realize our vision?

Phase 2:

- Comunity members' requirements are matched with accessible resources, services, and/or support.
- What opportunities do the communities have to gain access?
- What new opportunities need to be created?

Phase 3:

- Providers and communities negotiate partnerships.
- What conditions facilitate the negotiation of effective partnerships?

Phase 4:

- Partners design strategies of action around areas of mutual interest and implement their plans.

Phase 5:

- There is reflection on performance.
- What indicators will allow us to learn whether the improved agroecosystem management practices and the newly negotiated partnerships are performing well or not?

The following section contains a brief description of learning instruments useful for visioning, planning, negotiating, and reflecting on the phases of the learning process. Each instrument is described and illustrated by examples from work done in Ghana, Tanzania, Uganda, and Peru.

LEARNING INSTRUMENTS FOR FUTURE VISIONING OF AGROECOSYSTEM MANAGEMENT

Maps are instruments that help communities to learn how the quality, quantity, and use of the agroecosystems they manage have changed over the past 30 years, and to visualize how they would like to manage and use them in future. Matrices can be used to organize the information so that required resources, services, and support can be used to identify the partnerships that will be needed to make the envisioned changes. At the end of phase 1, a series of maps and matrices is available that provides information on the current state and the intended future state of natural resources as well as on the changes that need to be made and the kinds of partnerships that can help bring them about.

Maps facilitate learning by allowing different groups with different interest, to visualize how the agroecosystem has changed both for the better and the worse over the past 30 years. They also allow community members to discuss with each other and with outsiders, frequently government extensionists and researchers, what they want their agroecosystems to be like in 20 years. These maps use local categories to characterize farmland, forests, grazing areas, swampland and rivers, and other water resources. They also include major landmarks such as community boundaries, roads, houses, and other infrastructure. They indicate the major species of crops, livestock, fish, and trees. The map of the future is a vision of all the new roads, houses, markets, water supplies, and other infrastructure that communities would like to see. New agroecosystem management strategies for the forests, croplands, grass, and swamplands are represented on the completed map.

Matrices can help farmers organize the information on the current state of natural resources and contrast it with the changes needed to put improved management strategies in place. Working from the present and future vision maps for each change proposed, farmers identify the resources, services, and support they will need to implement each change. Then a clear link is established between a requirement and the kind of partnership needed to realize the improved agroecosystems management strategy.

The examples of agroecosystem maps in Figures 10.3 and 10.4, prepared by farmers from Soroti District in Uganda, clearly show an intention to intensify agroecosystem management in the future (Development Support Services, 1999; International Support Group, 1999). The current scattered farms with small plots of root crops, few animals, and almost no trees will be replaced in the future with more intensive farms that have a more diverse array of crops including coffee and upland rice, zero-grazing cattle operations, fish ponds, and citrus orchards. It should be noted that the expected rise in population is represented by a need for more houses,

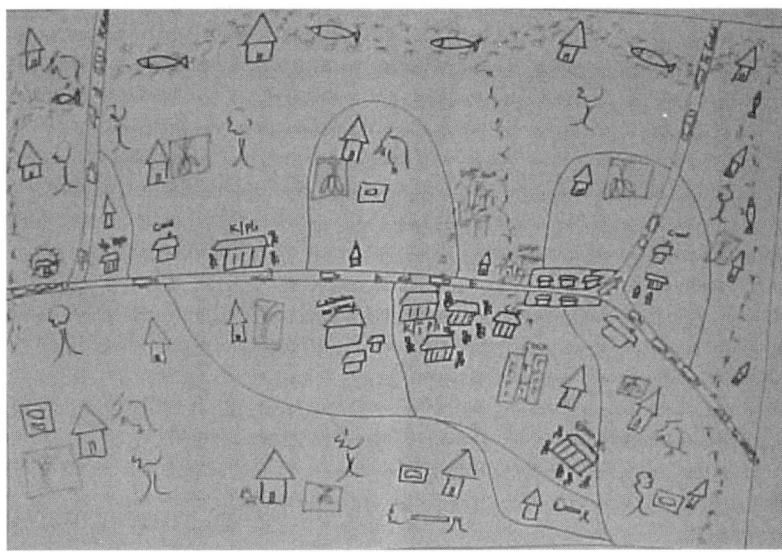

FIGURE 10.3   Agroecosystem map of the present situation, Soroti District, Uganda (*Source:* From Development Support Services, *Proceedings of on Orientation Exercise in Linked Local Learning,* July 2–3, 1999, DDS, Soroti, Uganda, p. 67.)

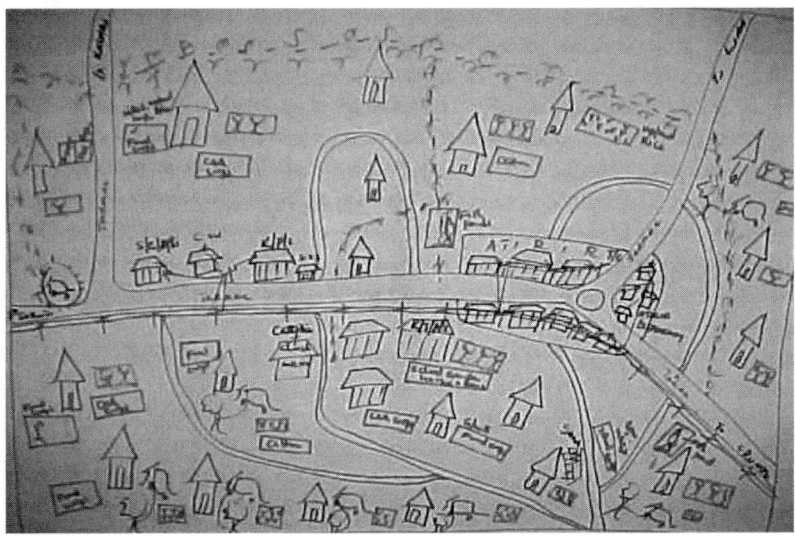

FIGURE 10.4   Agroecosystem map of the future vision, Soroti District, Uganda (*Source:* From Development Support Services, *Proceedings of on Orientation Exercise in Linked Local Learning,* July 2–3, 1999, Development Support Services, Soroti, Uganda, p. 67.)

TABLE 10.1

Future Changes in Agroecosystem Management and Key partners, Soroti District, Uganda

Natural Resource	Changes Management	Partners for Implementing Changes
Cropland	Implement soil erosion control measures, farm land consolidation	Community, local government, Department of Agriculture extension agents
	Introduce coffee, citrus, upland rice and fish ponds	Local government, Department of Agriculture extension, Department of Forestry, seed suppliers, credit institutions, marketing agents
Grazing land	Introduce improved pasture species, better livestock health, more zero grazing,	Community, local government, Department of Agriculture Extension, veterinary services, NARO, credit institutions
Wetlands	Implement conservation areas, water dam construction for irrigation.	Community, nongovernment organizations local council, Lands and Survey, Department of Agriculture Extension, credit institutions, marketing agents

Source: From *Proceedings of an Orientation Exercise in Linked Local Learning,* July 2–3, 1999, Development Support Services, Soroti, Uganda.

a school, a marketplace and bigger roads. The changes in the croplands noted on the map are captured in a matrix of future changes and the partners needed to implement them.

The matrix in Table 10.1 shows that grazing lands are to be rehabilitated by introducing improved pasture species and reducing the intensity of grazing with an increase in zero-grazing operations. The wetlands are to be designated as conservation areas with irrigation facilities to enhance productivity.

The matrix indicates that a wide array of partners will be needed to provide the necessary support and services. In addition to government services, there is a demand for seed suppliers, credit services, and marketing agents as well. The matrix also makes clear that to implement some changes, groups of community members will need to form partnerships with each other. For example, conservation of wetlands, management of grazing lands, and building of soil erosion measures all require intracommunity partnerships.

As in the case of Soroti District, communities in the Quilcas District, Peru, used participatory mapping to identify their desired future. Men's and women's groups constructed their separate visions, which then were joined to arrive at a common future vision for the community as a whole. Figure 10.5 shows the future vision of the community of Colpar in 1998. The vision is that of improved livelihoods achieved by way of sustainable agricultural production, good infrastructure, and tourism. The vision proposes that strong community organization and a clear development plan are critical to arriving at the vision. However, stronger organization and a clear plan depend on building capacity at family and community levels so that the over time, the elected authorities will be capable of facilitating collaboration among organizations in the community. The community also will need to be effective in negotiating part-

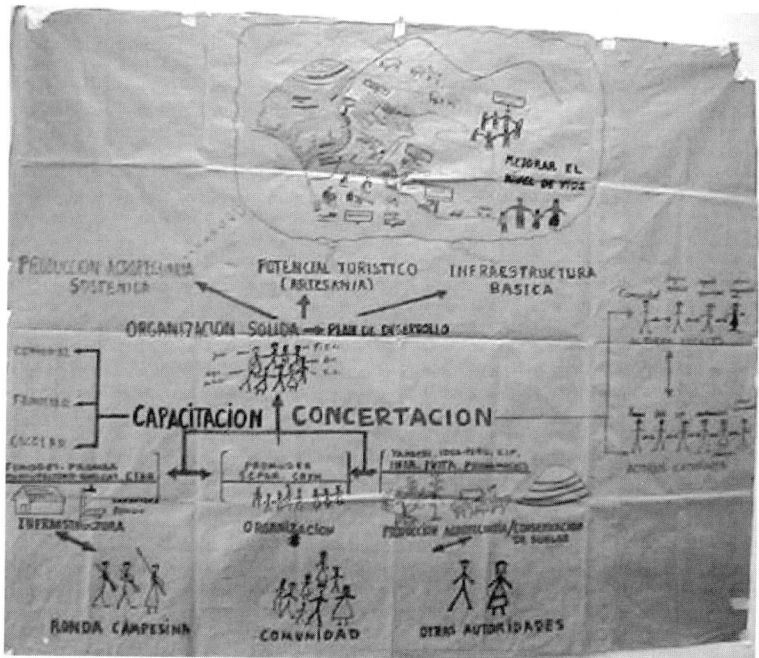

FIGURE 10.5   Future vision for the community of Colpar in 1998 (*Source:* From Fernandez-Baca, E. and Fernandez, M. E., *Evalvando de las Estrategias de Desarrollo de Comunidades Rurales,* Final Project Report to Red. de Investigación en el Manejo de Sistemas de Producción (RIMISP), Amersfoort, Netherlands, 2000; p. 67.)

nerships with government and nongovernment support and service organizations outside the community.

Future visioning can be used not only as a tool to identify requirements for resources, services, and support, but also as a tool to monitor shifts in priorities over time. Table 10.2 compares the vision of Colpar with that of Quilcas. The visions of these two neighboring communities are similar. For both Quilcas and Colpar, strong community organizations, sound and sustainable resource management, and infrastructure that facilitates physical and social well-being, communication, and networking among communities and with the market are the main components of the visions. However, priorities have shifted over the years. As goals are met through innovations and changes in strategy, new and more refined priorities have become part of the vision. Table 10.2 compares the future visions of two communities in the District at 2-year intervals between 1996 and 2000.

As with other participatory methods, these kinds of learning instruments can easily be misused. Maps of future visions can easily be hijacked by one interest group or another. For example, only so many people can physically be involved in drawing a map, so the task should be carried out by small groups of farmers with common interests. It is important to be sure that as many perspectives as possible are taken into account. For example, women often use natural resources differently from men, and they often are not included in community consultations.

TABLE 10.2

Desired Future Visions of Two Communities in the District of Quilcas Peru

Quilcas 1996

- Close coordination between the municipal and community authorities so that tasks are well defined and organization is improved
- Conserved soils and irrigated areas within the community to increment and diversify agricultural production

Quilcas 1998

- Member families in better socioeconomic condition
- Good roads and an irrigation system.
- Member families have the capacity to industrialize agricultural products
- Greater interest on part of financial institutions to support the district development plan
- Interinstitutional collaboration board strengthened by agreements on common goals and distribution of responsibilities among members

Quilcas 2000

- Improved soil conservation by way of infiltration ditches, reforestation, and pasture management
- Community animal production unit operating at full capacity and fish farm implemented
- Irrigation system–established irrigation so that every family can have a small vegetable garden
- Intercommunity boundary conflicts resolved
- Rules established for use of communal grazing areas
- Agreement on distribution of responsibilities with the district authorities

Colpar 1996

- Improved management of natural resources, reforestation, soil conservation, and production through the use of organic fertilizers
- Design and implementation of an integrated agroforestry and pastoralist system. Conserved fauna
- Reservoirs for small irrigation of vegetable gardens and forage crops, for drinking water, and for fish farming

Colpar 1998

- Sustainable agricultural production based on a solid community organization for land use
- Access to basic infrastructure by all member families
- Capacity-building opportunities for human development
- A plan to stimulate ecotourism as a means to improve the livelihood of member families in operation

Colpar 2000

- Erosion protection through reforestation, terraces, and infiltration ditches
- Strong and solid community organization
- Irrigation infrastructure forage and vegetable production
- Improved livestock production
- Organic agriculture production increased
- Better elementary education
- Negotiations for the establishment of an agricultural school
- Increased land areas under community management
- Pre-Hispanic remains restored and functioning fish farm to attract tourism
- Roads that connect Colpar with neighboring communities of Llacta and Casacancha
- A parabolic antenna to improve communication
- Sanitation infrastructure in the town center

Source: From Fernandez-Baca, E. and Fernandez, M. E., *Evalvando de las Estrategias de Desarrollo de Comunidades Rurales,* Final Project Report to Red. de Investigación en el Manejo de Sistemas de Producción (RIMISP), mimeo, Amersfoort, Netherlands, 2000; Red Internacional de Metodos de Investigación de Sistemas de Producción

International Support Group, *East African Seminar: Developing a Framework for Linked Local Learning,* Project Final Report to Danish International Development Agency, Mimeo, ISG, Amersfoort, Netherlands, 1999; and Fernandez-Baca, E. and Fernandez, M., 2000.

Caution also should be applied to farmers' visions of better ways to manage agroecosystems. Farmers' ideas are not always ecologically sound. Ecologic soundness is sacrificed not only because knowledge is lacking, but also because of a calculated trade-off between utilization and conservation. Balancing these trade-offs is a very difficult act for desperate communities. Community visions of how agroecosystems might be managed in the future benefit greatly from the exchange of information and ideas within the community as well as with extension officers and researchers.

LEARNING INSTRUMENTS FOR CLARIFYING REQUIREMENTS, PARTNERSHIPS, AND RESPONSIBILITIES

A series of tables can be used as learning instruments to help communities set out their requirements for services and support. The demands of the community are sorted into three categories: requirements that match the services and support offered, requirements that can be met by creating a new service, and requirements that cannot be met. In a second table, the responsibilities and actions needed to make services already offered available to the community are explored. Later, the new services needed, the partnerships that can leverage them, and the responsibilities this will entail are organized. The use of tables facilitates learning by helping to put the vision of the community and its requirements in a format that facilitates joint study and discussion with potential partners and service providers. These kinds of instruments guide partners and service providers into making decisions about how they will respond to the community in its quest for improved agroecosystem management.

Tables 10.3, 10.4 and 10.5 from Kilosa District in Tanzania illustrate how farmers, NGOs, and government extension officers at local and district levels responded to farmers' demands for services to meet their future visions for agroecosystem management (Shao et al., 2000). Services that matched farmers' requirements tended to be those offered by ongoing NGO and government extension programs that provided advice or training. Farmers also required advice and training in areas not covered by current offerings. New services were to be created for training in appropriate technologies and the formation of cooperatives and self-help groups. Government extension services and NGOs also agreed to create services for the introduction of improved livestock production and cash cropping. The requirements relating to deforestation, water use, and improved germ plasm went unfulfilled, as did the demand for credit.

Table 10.4 illustrates how farmers, NGOs, and government extension officers of Kilosa District in Tanzania proposed that the training of pastoralists on improved livestock production could be an effective means of reaching farmers' future visions. The Ministry of Agriculture and Cooperatives already offers training to pastoralists. However, for the pastoralists to be able to use the training received, land demarcation and group organization are necessary. Therefore, the actions to be taken include not only the delivery of the training program, but also the facilitation of groups among the pastoralists and the negotiation of a plan to demarcate available grazing land and water resources.

Table 10.5 illustrates how farmers and their partners brainstorm new services needed to meet farmers' requirements. The table guides the participants into making

TABLE 10.3
Response Capacity to Farmer Requirements, Kilosa District, Tanzania

Requirement Matches Existing Service	New Service Is Needed to Meet Requirement	Requirement Cannot Be Met
Advice on formulation and monitoring of bylaws on the management of natural resources	Introduce improved pastures, livestock breeds, cattle immunization, treatment dips, market improvements, cattle transfer permits and guaranteed grazing areas	Monitoring and control of deforestation and forest burning, assistance with tree planting
Training of pastoralists on improved livestock	Identify new cash crops to be introduced into different agroecosystems	Improved availability of seeds
Advice on proper use of land	Training on appropriate sustainable technologies that fit local environment	Monitor proper use of water sources
Training to ensure gender issues are properly addressed at village level and at work places	Advice on how to form farmer production groups, training on formation of cooperatives	Provision of financial assistance for inputs, working tools and farmer training

Source: From Shao, F., Mlay, E., and Muro, G., Eds., *Proceedings of a District Multi-Stakeholder Workshop on Linked Local Learning,* June 12–16, 2000, Mikumi, Kilosi, Farm and Natural Resources Management Consultants, Dar es Salaam, Tanzania, 2000.

TABLE 10.4
Improving Services That Match Farmer Requirements, Kilosa District, Tanzania

Service Matching Requirements	Division of Responsibilities	Actions to Be Taken
Training of pastoralists on improved management of grazing lands and production and maintenance of improved livestock breeds	Training Help in establishing livestock production groups (CBOs) Demarcation of grazing lands	Training program on raising good/productive livestock Program on establishment of livestock production groups Program to determine and demarcate available grazing land and water resources for livestock in grazing areas

Source: From Shao, F., Mlay, E., and Muro, G., Eds., *Proceedings of a District Multi-Stakeholder Workshop on Linked Local Learning,* June 12–16, 2000, Mikumi, Kilosi, FANRM, Dar es Salaam, Tanzania, 2000.

decisions about which partners are needed to create the new service and how responsibilities will be shared among the partners. In Kilosa District, farmers, NGOs, and government extension officers brainstormed a new service in the breeding of improved livestock to meet farmers' requirements. All involved realized that to establish this service, a complex of partnerships would need to be formed. These

TABLE 10.5
Characteristics of Breeding Service in Response to Farmer Requirements, Kilosa District, Tanzania

New Service	New Partners	New Responsibilities and Actions
Breeding and multiplication of improved livestock	Livestock keepers nongovernment organizations	Research on improved pastures and livestock breeds
	District councils	Repair and construction of cattle dips/spray races
	Researchers	Strengthening of markets for purchase and sale of improved livestock
	Central government	Issuance of livestock transfer permits

Source: From Shao, F., Mlay, E., and Muro, G., Eds., *Proceedings of a District Multi-Stakeholder Workshop on Linked Local Learning,* June 12–16, 2000, Mikumi, Kilosi, FANRM, Dar es Salaam, Tanzania, 2000.

partnerships would have to involve NGOs, national and local government authorities, researchers, and the livestock keepers. For the effort to be successful, a concerted effort among partners would be required to carry out the array of new responsibilities in research, infrastructure, marketing, and regulation the participants decided were necessary.

Inevitably, in any attempted response to farmer requirements for improving their strategies for agroecosystem management, some demands cannot be met. If these demands are critical to the realization of future visions, they become matters for policy. In the aforementioned example, credit to help farmers introduce new crops for the marketplace or to improve pastures on their grazing lands was not available. The farmers convinced the district and local government extension officers present that this lack of credit was a matter of concern to policy makers. The farmers would need to plant more cash crops if they were to increase income sufficiently to reverse practices such as overgrazing. In addition, they would need credit if they were to improve pastures as part of a strategy to reverse the degradation of their grazing lands. The government officers agreed to bring this policy matter to the attention of higher levels of government.

The use of learning instruments can be limited or enhanced by the context in which they are used. Community requirements often go beyond what government can provide, so the participation of other potential partners from the NGO and private sector is vital. For the morale of all participants, farmers and government officials alike, it is important to avoid a situation in which most of the farmers' demands go unmet. There often is a tendency in multistakeholder sessions for service providers, particularly those from the government, to become defensive if their services are criticized. Therefore, both farmers and service providers should adopt a learning attitude and see these sessions as an opportunity to learn which services are not meeting farmers' requirements and which new services might be more useful. The way communities manage their agroecosystems will change as their resource base improves, as will their requirements for support and services. It is therefore prudent for service providers to maintain a fair amount of institutional dynamism so they can respond to the changing requirements of farmers and communities.

LEARNING INSTRUMENTS FOR CLARIFYING CHARACTERISTICS OF SUCCESSFUL NEW PARTNERSHIPS

Learning instruments can aid communities and service providers in developing a set of attributes essential for successful partnerships. Ideas are generated in brainstorming sessions in which farmers and service providers recall the reasons why good partnerships worked and bad partnerships failed. The underlying reasons for success and failure become attributes around which to negotiate and build partnerships that work.

The use of a table facilitates learning because it helps participants come up not only with attributes of successful partnerships, but also with ideas of what negative things can happen if a partnership is not constructed on the basis of these attributes. Participants also identify the positive results from strengthening the attributes identified. The completed table guides communities and service providers to choose the criteria and conditions that are most relevant to a specific partnership. Knowing what

TABLE 10.6
Attributes of Successful Partnerships and Outcomes of Developing Them

Attributes	Negative Outcomes If Not Developed	Positive Outcomes If Developed
Trust between farmers and service providers	Services not readily accepted	Successful service delivery
Cooperation between farmers and service providers	Refusal to accept new technologies, poor participation	Proper joint implementation
Services delivered at the right time	Service delivered late, not worth using	Services delivered on time that can be used when and where intended
Awareness creation	Community not involved in development planning activities	Knowledge of development programs in their community
Knowledge of services needed according to capability	Lack of good environment for good service deliveries	Plans that are acceptable to all partners
Clear roles of each partner defined	Poor relationship between partners	Efficient and proper action by all partners involved
Communication among partners	Top-down directives	Sustainable commonly agreed-on plans
Transparency during planning and implementation	Unimplemented directives and orders	Trust between partners
Participation in decision making	Refusal to implement and, poor participation	Mutual benefits for partners
Creation of services	Unclear/unwillingness to implement plans	Information/knowledge spread to all

Source: From Shao, F., Mlay, E., and Muro, G., Eds., *Proceedings of a District Multi-Stakeholder Workshop on Linked Local Learning,* June 12–16, 2000, Mikumi, Kilosi, FANRM, Dar es Salaam, Tanzania, 2000.

might happen if either partner does not fulfill the conditions or help strengthen the identified attributes helps to reinforce commitment to the partnership. Finally, the attributes and outcomes will be used as criteria for assessing the success of the partnership.

Table 10.6, illustrates the kinds of ideas farmers, NGOs, and agriculture extension officers generated using this simple table (Shao et al., 2000). The number and range of ideas make it easy for each of partners to select some conditions and criteria that need to be improved in their partnership. In this instance, the participants

TABLE 10.7
Assessing the Value of Partnerships to Reach the Future Vision, Quilcas District, Peru

Key Institutions	Purpose	Present Status of Relationship
Community families	Give impetus to local development	Very active
Community of Quilcas	Organizes access to resources and is the legal representative for Colpar community to the government	Respects Colpar's autonomous development efforts
District of Quilcas	Is responsible for public services within the area designated	Little coordination and unclear division of responsibilities with the community
Interinstitutional coordination board	A space for dialogue and negotiation among institutions working in the area	Although efforts are under way, the space has not been consolidated.
Ministry of Agriculture PRONAMACH	Has provided ideas for soil conservation and reforestation that are being implemented by the community on its own	Has withdrawn from community
Ministry of Health	Responsible for district health services	Has interest but very few resources
Ministry of Education	Responsible for education and schools	Provides support to the parents' education group
Commercial agriculture establishments (local and regional)	Source of external inputs and advice for agricultural sector	Prioritizes advice and inputs for conventional production systems
Nongovernment organizations (Grupo Yanapai, SEPAR)	Follows local development process and provides support to local organizations and technical advice	Are accepted by community as far as they provide support and are transparent in their objectives
Research Institutes (INIA, CIP)	Source of alternative technologies and scientific knowledge	At present, provide support for agricultural experiments

Source: From Fernandez-Baca, E. and Fernandez, M. E., *Evalvando de las Estrategias de Desarrollo de Comunidades Rurales,* Final Project Report to Red. de Investigación en el Manejo de Sistemas de Producción (RIMISP), Amersfoort, Netherlands, 2000; p. 463

agreed that trust and communication between partners, transparency in decision making, and awareness building were the most critical attributes of partnerships. They also agreed that without trust between partners, the services might not even be requested or accepted. Communication between partners also was considered crucial because one-way top-down directives do not result in sustainable partnerships or effective learning.

In Quilcas District, Peru, farmers from Colpar used a table to assess the status of relationships they had with government and NGOs that already had a presence in the district. This exercise was carried out to help participants understand better what each support organization or service provider had to offer so they could begin to evaluate where partnerships could be strengthened. First, a list of institutions considered to be important to local development priorities was made, and the purpose of each entity was described. Then the current status of the relationship was described, with the following criteria taken into account:

- the degree of respect they show for the community and its interests
- the real contribution they make to local development efforts
- their potential for long-term support

The results of the assessment are given in Table 10.7. Community authorities used this assessment to design a strategy for strengthening partnerships with those organizations whose agendas would contribute most to reaching the community's future vision.

This kind of instrument is less prone to misuse than some others. Nevertheless, some caution is necessary in developing good attributes and outcomes. Bias can occur when the consequences of negative outcomes do not affect both partners equally. It is important to good partnerships that both positive and negative outcomes should equally affect both partners. Because a partner's response to negative outcomes or positive benefits from a partnership can change over time, criteria need to be revisited.

LEARNING INSTRUMENTS FOR REFLECTING ON AGROECOSYSTEM PERFORMANCE

Learning instruments that allow partners to reflect on how their partnerships are performing and on how their actions are affecting the "sustainability" of their agroecosystems use both indicators and scoring. In both cases, the initial task is to develop indicators for assessing performance. In addition, a scale on which to score each indicator is required. Indicator scoring facilitates learning because plotting the scores graphically demonstrates clearly to farmers and their partners where matters are improving and where they are not. Clear visual graphs and charts immediately direct the participants' attention to where progress has been made and where it has not. Clear graphs provoke discussions on the all-important issue of what might be done to improve poor performance. Clear graphs also facilitate transparent decision making, which is especially needed when decisions are made to change a course of action or terminate a partnership.

In assessing the performance of agroecosystem management strategies, it is important for farmers to choose indicators that not only make sense to them, but also are easy for them to measure. With regard to the performance of agroecosystem management strategies, farmers are interested in answers to questions such as these:

- How much more profit does the farm make for every dollar invested?
- How much more productive is the farm?
- How much more diverse is the farm?
- How much more recycling is going on?

It should be noted that farmer indicators assess both economic and ecologic dimensions. The profitability of the new agroecosystem management strategy assesses economic efficiency. Ecologic dimensions are assessed by determining increases in biologic production, biodiversity, and recycling of biologic wastes. Scoring for economic efficiency uses a simple profit–cost ratio. Scoring for biomass production capacity uses the weight of biomass produced by all species. Recycling of biologic wastes is scored using the number of recycling flows, and species diversity is scored using the number of species cultivated or collected. Performance scores for these four indicators are plotted on a simple four-way kite graph. It is important that each indicator be arranged on an increasing scale of improvement so that the larger the kite formed the better the performance. Farmers find it easier to use scores in which the larger number indicates better performance.

Figure 10.6 shows how a group of farmers from Mampong District in Ghana assessed the performance of their "future" agroecosystem management strategy after 1 year of implementation (Prein et al., 1993). The farmers' future vision of an improved agroecosystem management strategy included the rehabilitation of a previously neglected wetland area by improving water flow and impounding water in a

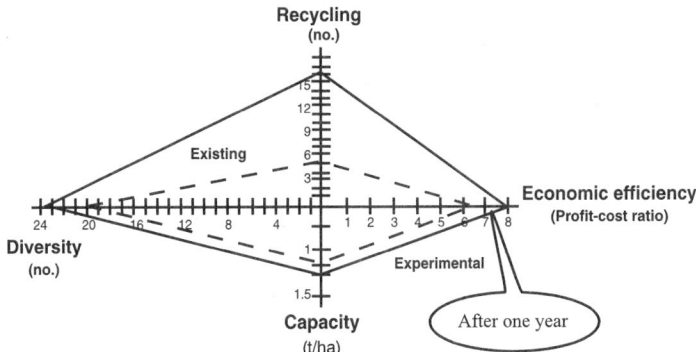

FIGURE 10.6   Four-way kite graph of agroecosystem management performance indicators from Mampong District, Ghana (*Source:* From Prein, M., Ofori, J., and Lightfoot, C., *Research for the Future Development of Aquaculture in Ghana,* Conference Proceedings 42, ICLARM, (International Center for Living Aquatic Resources Management), Manila, Philippines, 1993, p. 94.)

small pond. The pond provided sufficient water to stock fish and irrigate newly established vegetable plots. Wastes from all the animals were recycled to the pond and vegetable plot to substitute for chemical fertilizer. After 1 year, with savings on external
inputs, increased products going to market, increased meat and vegetables consumed
at home, and increased internal recycling of wastes, farm gross and net incomes went
up by 50%. Cash profits rose to an $8 return on every dollar invested. On the ecologic
side, species diversity increased to 24 species, biomass capacity to 1.4 tons per
hectare, and waste recycling flows from 5 to 16.

LEARNING INSTRUMENTS FOR REFLECTING
ON PARTNERSHIP PERFORMANCE

In assessing performance of partnerships, it is important for partners to choose the
indicators from the attributes developed for building and negotiating the partnership.
Transparency in decision making, trust between partners, access to information,
awareness building, and mutual benefits all are criteria that make good performance
indicators for partnership performance. Once indicators are chosen, a scale on which
to score performance is developed. Given the subjective nature of most partnership
indicators, performance often is seen only as good, okay, or bad. At a very basic level,
performance can be scored with just these three levels. However, better feedback is
provided when more levels of scoring are used. Each partner scores the other.
Farmers score the performance of the service provider on the delivery of an agreed-
on service, whereas service providers score the performance of farmers on the responsibilities they agreed to take on in the partnership. Thus the instrument captures the
perspectives of each partner. Once the scoring has been done, all the partners should
reflect on both the positive scores and, more importantly, the negative scores.

Moreover, visualization and also documentation ensure a transparency that helps
in decision making. This is particularly true when projects are not working well
because of poor relationships between partners. Understanding the perceptions of others often helps partners to discover that others are not satisfied with the way they are
working. Such understanding can be the start of negotiating desired changes to meet
expectations better in future, or the start of a mutual decision to end the partnership.

In Kilosa District, Tanzania farmers assessed the performance of the partnerships
in which they were involved. Of the two projects assessed, the dairy goat project was
a success story, whereas the honeybee project was a failure (Shao et al., 2000). To
assess these two projects (Figure 10.7), farmers chose four indicators: trust, information access, technical success, and mutual benefit. They also developed a 5-point
scoring scale as follows: very bad, bad, average, good, very good. The dairy goat project succeeded because it was technically viable. The farmers were given good information on the project and on goat rearing technology both before and during the
project, and the farmers felt they had benefited greatly from the project. They felt that
the service providers had benefited as well, but the farmers did not develop more than
an average amount of trust for the service provider during the project. The farmers
felt that although they had received adequate information during the implementation
of the honeybee project, the partnership had failed. It failed because raising bees was

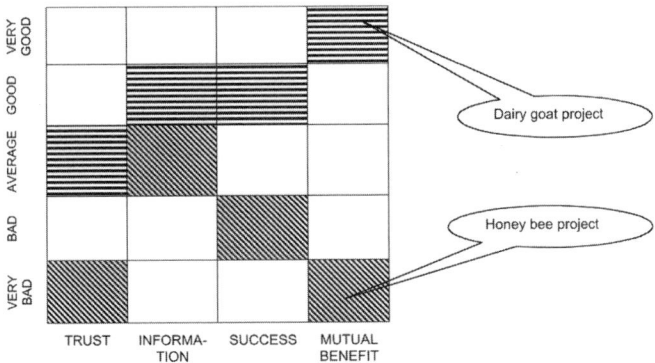

FIGURE 10.7 Assessment of two projects using partnership quality indicators (*Source:* Shao, Mlay, and Muvo, 2000)

difficult technically. The farmers were able to produce very little, so they did not benefit much. Finally, throughout the project, the partners were not able to build trust for each other.

Perhaps more than any of the other learning instruments, those scoring partnership performance raise emotions and heated debates. Sometimes debate is a good thing, and sometimes it is not. A common and useful debate often occurs over the choice of the most important indicators. Men and women may propose different indicators because they may have different concerns regarding agroecosystems management. Other groups in a community will have specific interests, and thus will suggest particular indicators. Time spent negotiating agreement on indictors is useful because it helps to forge understanding between disparate groups.

Unhelpful debates often arise over the subjectivity of scoring. When highly subjective indicators are scored using "feel good" scales, the learning instrument can break down in needless heated argument. When this happens, it is best to revisit the indicator and scoring scale rather than pursue an agreed assessment. However, one has to be careful about suppressing perceptions. This is particularly true in the assessment of partnerships. Partnerships are based on relationships that often work on perceptions. Each partner scores partnership performance because all partners want to discover how the others appreciate their contribution to the common effort.

CONCLUSION

The learning approach in agroecosystem management improvement is based on the view that an equal partnership between rural communities and private and public sector organizations and entities can increase opportunities for ensuring environmental sustainability and dignified livelihoods for those who manage marginal and degradable land areas. The learning approach focuses on enhancing community capacity to require resources, services, and support from government and other service providers. With a clear vision of how their future should look, and supported by information of what is needed to achieve that vision, community members are in a better position to present their case to district authorities and other potential partners. However, this is not just about the community asking, but also about service providers responding. Instruments that can help community members and service providers work together on a response to farmer-identified requirements have been exemplified here in this discussion.

The use of the instruments presupposes that potential partners, community members, and service providers have been convened. District authorities often are the most appropriate convenors if the task is to determine which of the farmers' requirements match with existing services, what new services need to be created, and what requirements will have to go unmet. Moving from knowing what to do to doing something about it requires that both government organizations and NGOs have the dynamism needed to pick up needed new services by dropping old and unwanted ones. Many local governments are not prepared to put the necessary effort, time, and resources into getting consensus in their organizations to make the required changes. Local government managers are often unclear about just how flexible their organizations are and too insecure of their own employment to risk changing or dropping services or entering into the kinds of new service partnerships that farmers are demanding.

Progress has been made in changing attitudes, realizing the necessity of partnerships, and developing partnership formation skills. The creativity and knowledge of communities regarding agroecosystem management often surprise agricultural service providers at the district level who hitherto thought of small-scale farmers as backward and resistant to modern agriculture. For their part, farmers have changed their attitude toward government as well. Farmers now realize that funds are no longer available for handouts and free inputs as in the past. They are more aware that officials face their own difficulties of understaffing and insufficient financial resources. All concerned recognize the need to build partnerships.

It has become clear that there is no going it alone in agroecosystem management. Government, farm communities, and service providers must work together in partnerships. Some progress has been made in the development of instruments for building partnerships. There are instruments for identifying potential partners, and there are instruments that can help communities and service providers to work together in developing and clarifying roles and responsibilities for each partner. Also, there are instruments for identifying the attributes of desired partnerships and conditions that can be negotiated. Finally, instruments are available for reflecting on partnership performance. The development of these instruments represents local innovation that has occurred as a result of using the learning approach.

Progress in enhancing capacity for local innovation is another important effect of working with a learning approach. The ability of local people to invent ways of doing things that work in their contexts is vital to community progress in agroecosystem management. Local capacity to develop learning instruments facilitating consultations at a community or district level that take people beyond the "talk shop" has emerged. Operational procedures for deciding when, where, and how consultations should take place and who should be there are being prepared. Capacity to invent ways of visualizing, recording, and documenting what has been learned from the use of the learning instruments is increasing. Documentation not only captures outputs for all to see, but also ensures transparency. Transparent impact assessments and decision-making practices are vital if investors, whether local governments, charities, or donors, are to became more committed.

Progress in developing a learning approach for improved agroecosystem management, in which partnerships among communities, local government, and service providers are necessary, faces many constraints. In the offices of the local government and service providers, constraints are everywhere. Officers are not prepared to pay the high transaction costs of establishing learning frameworks. It takes time, effort, and resources for the concerned organizations to meet and develop learning instruments and to agree on the appropriate operational process for learning. It takes time and effort to build enough consensus within the organization to support a learning approach. This is particularly true where departments within an organization are warring with each other. Finding the necessary funds to support the learning process is another constraint because sharing the budget across organizations is difficult. Government agencies are seldom willing to allocate their scarce funds to NGOs, or even to buy the services of NGOs for what are seen as low-priority activities, such as staff learning.

A more difficult set of constraints is the complexity of relationships between local and central government and donors. There are questions of confidence. Central managers often doubt local capacities. Donors often attach heavy-handed control over funds for fear of local misuse. There are questions of empowerment. District governments, although empowered to decide, often wait for "go signals" from the central government and donors. Then there are questions of interpretation. Central governments and donors may say fund allocation is a local decision, but their veto on budgets undermines this interpretation from the local staff's point of view. Technical assistance often is viewed in a similar fashion. Centers may say technical matters are local decisions, but their insistence on providing technical advisors contradicts this affirmation. It is very rare for a local initiative or innovation to override that of the technical advisory team. Learning approaches that facilitate and rely on local innovation cannot thrive if local decisions are not respected.

Learning approaches thrive in situations where the learners want to learn and where the learners bestow legitimacy on the convenors of multiorganization coalitions. Learning approaches cannot be imposed in a top-down fashion by project implementers. Rather, those wishing to support learning approaches should seek invitations from the learners to facilitate learning. They should get mandates from the learners. Ideally, the learners should own the funds for learning as well as the responsibility for making progress. Meeting these "ideals," however, is difficult both within special projects and within regular work programs of centralized organizations.

Special projects have the ability to implement actions and get participants working. They are, however, less able to respond to participants' changing needs, and they rarely have mandates to give up economic control of the project to the participants. They are not equipped to support local innovation, especially when technical advisors are preoccupied with doing things a specific way. Although building a learning space into ongoing development agency work avoids many of the weaknesses of projects, it does have its own problems. Undoubtedly, costs in time and money are small when the learning activity is substituted for existing consultations or evaluations, and when it is spread out over a long period. Certainly, the learners can empower the facilitators and bestow legitimacy on the process. However, a learning activity is almost impossible to package so that human and financial resources can be allocated to it. Although monitoring and the measurement of impact need to be tailored to local processes, the authors are convinced that the learning approach opens new opportunities for taking control of agroecosystem management improvement processes.

REFERENCES

Altieri, M. A., *Agroecology: The Science of Sustainable Agriculture,* Westview Press, Boulder, 1989.

Bawden, R. J., Systems thinking and practice in agriculture, *J. Dairy Sci.,* 74, 2362, 1991.

Borrini-Feyerabend, G., *Collaborative Management of Protected Areas: Tailoring the Approach to the Context,* International Union for the Conservation of Nature, Gland, Switzerland, 1996.

Chambers, R., *Whose Reality Counts? Putting the First Last,* IT Publications, London, 1997.

Checkland, P. and Scholes, J., *Soft Systems Methodology in Action,* John Wiley and Sons, Chichester, UK, 1990.

Conway, G. R., Agroecosystems analysis, *Agric. Adm.,* 20, 31, 1985.

Daniels, D. and Walker, G., Collaborative learning: improving public deliberation in ecosystem-based management, *Environ. Impact Assess. Re.,* 16, 71, 1996.

Development Support Services, *Proceedings of an Orientation Exercise in Linked Local Learning,* July 2–3, 1999, Development Support Services, Soroti, Uganda.

Engel, P. H. and Salomon, M., Rapid Appraisal of Agricultural Knowledge Systems, a participatory action research approach to facilitating social learning for sustainable development, in *Systems-Oriented Research in Agriculture and Rural Development,* Papers of AFSRE International Symposium, Montpellier, France, 1994.

Fernandez-Baca, E. and Fernandez, M. E., *Evaluando la Sostenibilidad de las Estrategias de Desarrollo de Comunidades Rurales,* Final Project Report to Red de Investigación en el Manejo de Sistemas de Producción (RIMISP), mimeo, Amersfoort, Netherlands: ISG, 2000.

International Support Group, *East African Seminar: Developing a Framework for Linked Local Learning,* Project Final Report to Danish International Development Agency, mimeo, International Support Group, Amersfoort, Netherlands, 1999.

Korten, D., Community organization and rural development: a learning process approach, *Public Adm. Rev.,* 40(5), 480, 1980.

Lightfoot, C., Dalsgaard, J. P., Bimbao, M., and Fermin, F., Farmer participatory procedures for managing and monitoring sustainable farming systems, *J. Asian Farming Syst. Assoc.,* 2(2), 67, 1993.

Lightfoot, C. and Noble, R., A participatory experiment in sustainable agriculture, *J. Farming Syst. Res. Ext.,* 4(1), 11, 1993.

Prein, M., Ofori, J., and Lightfoot, C., *Research for the Future Development of Aquaculture in Ghana,* International Center for Living Aquatic Resources Management Conference Proceedings 42, International Institute for Environment & Development, Manila, Philippines, 1993.

Pretty, J., Guijt, I., Scoones, I., and Thompson, J., *A Trainer's Guide for Participatory Learning and Action,* International Center for Living Aquatic Resources Management, London, 1995.

Ramírez, R., *Participatory Learning and Communication Approaches for Managing Pluralism: Implications for Sustainable Forestry, Agriculture, and Rural Development,* Pluralism and Sustainable Forestry and Rural Development Conference, December 9–12, 1997, Rome, 1999, p. 117.

Ramírez, R., *Understanding Farmers' Communication Networks: Combining PRA with Agricultural Knowledge Systems Analysis,* International Institute for Environment & Development Gatekeeper Series No.66, International Institute for Environment & Development, London, 1997.

Röling, N. G. and Wagemakers, M. A., Eds., *Facilitating Sustainable Agriculture: Participatory Learning and Adaptive Management in Times of Environmental Uncertainty,* Cambridge University Press, Cambridge, UK, 1998.

Shao, F., Mlay, E., and Muro, G., Eds., *Proceedings of a District Multi-Stakeholder Workshop on Linked Local Learning,* June 12–16, 2000, Mikumi, Kilosa, mimeo, Farm & Natural Resources Management Consultants, Dar es Salaam, Tanzania, 2000.

11 Bridges to Sustainability: Links between Agriculture, Community and Ecosystems

Lorna Michael Butler and Richard Carkner

CONTENTS

Introduction 157
Agroecology within a Social Dimension 159
The Concept of Foodshed 159
The Problem 160
Purpose 161
Case Studies 162
    Food Bank Farm, Hadley, Massachusetts 162
    Tolt Farm, Carnation, Washington 164
    Thompson Farms, Boring, Oregon 166
Sustainability Connectors 169
Conclusion 171
References 172

INTRODUCTION

There is often a mental disconnect between agriculture, community, and the environment. Rarely do political agendas, academic researchers, government programs, or nonprofit organizations' goals address the crossover between agriculture, the community, and the surrounding natural resource system. Most often, one component is treated separately from the others. These divisions do not reflect the way that nature is organized, nor the way in which human communities adapt to changing natural environments, technologies, and available resources.

The trend in North American agriculture toward fewer and larger farms exacerbates the disconnect. More vertically integrated agriculture means that processing and marketing firms acquire greater control of commodity production and marketing. They guarantee supply by contracting with farmers. However, contracts do not guarantee any commitment to farmers, communities, environment, or consumers. The trend is for corporate firms to exercise centralized control of technology and production processes, leaving the farmer with little flexibility in deciding how to manage his or her relationships with nonfarm neighbors or more distant customers. The farmer also is the person who takes care of the natural resources.

The industrial approach to North American food production distances farms and farmers from consumers. If you eat, you participate in a global food system. All of us can walk into a supermarket and find foods from throughout the world—apples from New Zealand, grapes from Chile, melons from Mexico, kiwi from New Zealand, cucumbers from Canada—yet we know little about how these foods are grown. Do the growers use chemical pesticides? Are the farm workers fairly rewarded for their efforts? What is the true cost of transportation? How environmentally friendly is the production process? How much profit goes to the farmer? The other side of the equation is equally disturbing. Even in our own community, it is often difficult to find fruits, vegetables, and other products that originate in our own region.

Early anthropologists debated the impact of environment on social and political behavior. Kroeber (1939) emphasized the collection of ecologic information to explain social phenomena, and Steward (1955) used a cultural ecology paradigm to explain relationships between certain subsistence systems and their environment. Early studies of indigenous agriculture systems drew on ecology. For example, Rappaport's (1967) seminal New Guinea study, *Pigs for Ancestors,* argued that a human population was a species within an ecosystem, and that the total environment could be understood as a system. Major cultural processes such as ritual, fallow cycles, pig population, war, and peace were shown to be interrelated with the natural ecosystem. Rappaport's work on subsistence agriculture contributed in a major way to our knowledge of the relationships between agriculture, environment, and social systems.

Although early ecologic approaches had their limitations, they had a great influence on the work that followed. Geertz (1963) was one of the first anthropologists to note the usefulness of the ecosystem as a unit of analysis to explain the internal dynamics of total systems. Goldschmidt's (1978) well-known book, *As You Sow* (1947/1978), probably had the greatest influence on contemporary studies about the links between farm structure and community life.

In spite of our knowledge that cultural factors have an impact on ecologic dimensions such as population size and resource use, the cognitive dimensions of human behavior have been relatively overlooked. We are now beginning to acknowledge the impacts, often negative, that individual management practices have on the landscape. For example, Andreatta's (1998) political ecology study of small holders in the Windward Islands, West Indies, documents the negative environmental impacts of the increase in cultivated acreage and the associated loss of vegetative cover. The fragile nature of the physical environment coupled with a relatively unsupportive political structure and the impact of the global agrofood sector has forced farmers into less sustainable tillage systems that require chemical inputs and contribute to soil erosion. Centuries of human and environmental interaction are now threatening the future of local farming systems, yet it is the farmers who are charged with natural resource stewardship, and natural resources that attract much needed regional tourism dollars.

According to increasing evidence, traditional agriculture systems that adopt a multifunctional agricultural strategy are effective at achieving food security, income generation, and environmental conservation. These are the systems that generate a variety of products, manage a continuum of agriculture and natural resources, combine intercropping and agroforestry, combine stable and diverse production, and and produce for both subsistence and market needs. Multiple-use strategies that enhance the multifunctional nature of agriculture may be an important concept for the future health of farms and rural regions throughout the world (Altieri, 2000).

Agroecology, which looks at the multifunctional nature of agriculture, emphasizes holism and relationships between people and their environment as people manage natural, financial, and social resources. Resource management is site specific. Social, political, legal, and historical dimensions are every bit as important as environmental dimensions. Knowledge of internal household dynamics (as opposed to total populations) and the social relationships of production, consumption, and, distribution are essential pieces of the total picture (Moran, 1990).

AGROECOLOGY WITHIN A SOCIAL DIMENSION

Agroecosystem analysis, or the interplay between external and internal social, economic, biologic, and environmental processes of the agricultural system, helps to explain a particular production system either at the spatial or temporal level (Altieri, 1987).

An agroecosystem is a hierarchical system containing both ecologic and socioeconomic components, in which one level feeds into the next. An agroecosystem has recognized goals and strategies for attaining goals that satisfy human or social needs. These can be assessed by examination of productivity, stability, sustainability, and equitability (Conway, 1994). Viewed this way, agroecosystems can be assessed for their contributions to the sustainability of the community as a whole, including the manner in which they protect the natural resource system. Agroecosystems generally look at watersheds, as determined by natural geography and hydrology.

THE CONCEPT OF FOODSHED

Foodshed has been used as the social and economic analog of watershed. The concept of the foodshed, first introduced in 1929 (Hedden, 1929), and more recently expanded by Kloppenburg et al. (1996), offers a way to decrease the distance between disciplines, and between consumer, farmer, community, and nature. The term refers to the cultural connection provided by "food" to the natural ("shed"). Foodshed serves as "a unifying and organizing metaphor for conceptual development that starts from a premise of the unity of place and people, of nature and society" (Kloppenburg et al., 1996, p. 34). While providing a broad interdisciplinary mechanism for reflecting about the place where we live and eat, foodshed serves as a tool for action that can move us toward a more sustainable society. Food becomes a place to start collective action.

Foodshed analysis involves asking directed questions about the food system and collecting useful information for the purpose of education and action to reestablish a sense of community. Kloppenburg et al. (1996) proposed that the concept of foodshed could be used to include the elements and properties of a preferred food system, thereby contrasting it with the existing global food system. The concept of foodshed encompasses the following five principles:

1. The foodshed is embedded in a moral economy that influences market forces. Food production is reembedded with people's needs rather than with "effective demand." Food becomes more central to quality of life and human relationships. Food production and consumption can strengthen family, community, and civic culture.
2. The foodshed is a commensal community. That is, it is shaped and expressed through human communities. "Commensalism," an ecology term, refers to a relationship between two different organisms in which one gets food from the other without causing damage to the other. Foodsheds are sustainable relationships without harm between people who eat together, and between people and the land. For example, new consumer–producer linkages are evident in cooperative relationships between restaurants and farmer suppliers, consumers and farmers in community-supported agriculture (CSA) arrangements, and the growth of farmers' markets.
3. Foodsheds are created by self-protection, secession, and succession. People in them work toward an alternative food system by gradually disengaging from the dominant market-based system to create an alternative food system that has potential for social change.
4. Foodsheds are characterized by their proximity to needed social and environmental resources such as native species, land, water, labor, processing, energy, markets, transportation, and the like. Self-reliance is tied to the foodshed's reliance on local or regional connections—a response to the commensal community. Social welfare, resource conservation, and energy efficiency are priorities. The community is responsible for the stewardship of the land and the people.
5. The foodshed is a human activity embedded in the natural resources of a region. Natural conditions represent a measure of limits to be respected rather than overcome. Nature can suggest a regional palate of locally and seasonally available food (Kloppenburg et al., 1996, p. 34)

THE PROBLEM

It is the authors' contention that many of the current agricultural and environmental (agroecosystem) conditions, such as salinization, water pollution, soil erosion, soil compaction, genetic homogeneity, genetically modified seed, and so on, are strongly rooted in prevailing social, cultural, and economic systems. Entrepreneurial farming and marketing systems that are demonstrating models of sustainability need to be examined in the context of an agroecologic environment that includes people and their relationships with each other. The authors contend that it is important to examine the human, ecologic, and economic dimensions of agricultural businesses to identify particular management strategies used to foster linkages with the total community. This is particularly revealing where farm businesses are located in urbanizing regions such as the Pacific Northwest or New England. Many communities are made up of nonfarm residents who are expressing their appreciation more and more for fresh, safe, wholesome produce, pleasing surroundings, and a vibrant natural environment. Increasingly, urban and suburban residents have little to no contact with their food sources or with the land.

Do farmers, their families, or their employees intentionally build "bridges to sustainability" through their selection of particular agronomic practices, management strategies, or marketing models? Do farm operators add to community and ecosystem sustainability through their community and organizational activities? How important is the operator's personal and management philosophy to long-term sustainability of the agroecosystem and community?

PURPOSE

This chapter argues that local community and agricultural sustainability is greatly dependent on the establishment of common denominators or "connectors" between farm system units and their nonfarm neighbors, which are of mutual benefit socially, economically, and environmentally. The illustrations presented highlight the importance of sustainability connectors that have been established between people involved in urban edge agriculture and their local communities, organizations, and the natural environment. The authors contend that the underlying goals and philosophies of the farm operators make this exchange possible because they have been able to identify certain common denominators appreciated by all participants. In fact, they have been able to design a system compatible with their own personalities. This has led to a win-win situation.

Primary attention is given to the impact of rapid growth and urbanization on small-to-midsize farming systems and communities. Highlights are presented from three case studies: one from Massachusetts, one from Oregon, and one from Washington State. The first case, *Food Bank Farm,* is based on a partnership involving paid interns, food bank volunteers, and consumers. The second case, *Tolt Farm,* is a sole proprietor organic farm adapted to fit the management and labor capabilities of the owner-operator and his decision to market through regional farmers' markets. The third case, *Thompson Farms,* is an intergenerational family farm that puts a high priority on family participation in the business and the maintenance of strong community relationships. Each of the cases studied represents a distinct "personality" that evolved in response to unique regional opportunities identified by the operators. The personality is the product of operators' own special personalities, goals, and philosophies and of the assets found in the particular environment.

The three cases are presented to illustrate ways that small-to-midsize farm operators have been able to capture available ecologic and human resources and direct them to the achievement of business, community, and environmental sustainability. These cases illustrate a number of practical strategies that might be used by farm and market operators to build bridges to sustainability between their own agroecosystem and the surrounding community.

Connectors include such things as the following:

- capitalizing on individuals' talents and interests
- taking advantage of location
- learning what customers (and potential customers) want and appreciate
- identifying "hidden" community and ecosystem assets
- making efficient use of human, physical, and technical resources
- paying attention to people's passions
- being clear about one's own goals and values

These connectors form "bridges to sustainability."

CASE STUDIES

FOOD BANK FARM, HADLEY, MASSACHUSETTS

Food Bank Farm consists of 55 continuous acres along the Fort River, one half mile from the confluence of the Fort and Connecticut Rivers. Evidence gathered from archeologic research "indicates that the land has been used for some form of agriculture for the last 1000 years" (Food Bank Farm, n.d.). The river bottom soil is prime vegetable growing land.

The farm is a project of the Western Massachusetts Food Bank. The food bank, which owns the land, is jointly managed and operated by Linda Hildebrand and Michael Doctor. Linda has a farm background, and Michael is from an urban background. The farm managers are paid a salary plus a dividend based on the farm's net income. They commit half of the farm production to the food bank and the other half to nearly 500 families through a community supported agriculture (CSA). The farm managers believe in what they are doing and realize that their incomes could be higher if they were not connected to a food bank. However, they are committed to their community.

A CSA is a marketing system directly connecting consumers with farms. This is a relatively new marketing system spreading rapidly across the United States. The movement began in Europe and Japan more than 30 years ago. The first CSAs appeared in the United States less than 15 years ago and have since sprung up throughout North America.

A CSA is an arrangement whereby consumers buy shares of a local farm's harvest. The shareholders' fees support the farm operations, and, in return, the farm supplies shareholders with weekly shares of fresh produce. The fee structure is based on the value of the farm share as measured against local supermarket food prices. The purpose is to provide fresher, not necessarily cheaper, quality organic produce and to provide a connection to the food source.

In addition to farm-produced food, the farm also retails locally produced food not grown on the farm, such as maple syrup, apples, cider, eggs, and poultry, to CSA shareholders. These products are purchased from local farmers and sold to CSA shareholders.

Food Bank Farm uses organic production practices, thereby reducing purchased inputs such as synthetic pesticides and fertilizers. Cultivation, hand-weeding, and crop rotations are used to control weeds and reduce plant disease. The operators direct seed most of their crops, but start the season with purchased transplants. Because a local greenhouse produces transplants, this represents another connection between local food production and related businesses within the community.

Farm labor consists of the farm operators and three paid interns. The interns receive food and lodging as well as a stipend. The farmers have developed highly efficient production and harvest priorities that allow them to operate on a large scale with a small labor force. Whereas they work long hours during the summer, they take many Saturdays and every Sunday off. Initially, interns were recruited through advertisements directed toward college students. Now word of mouth provides the contacts and the interns for the farm. Many interns have liberal arts backgrounds, and with their farm experience, some are now farming on their own.

Crops are available on a pick-your-own basis. This reduces labor costs and involves shareholders. Customers pick up their shares on one of three identified pickup days, and have unlimited access to u-pick items such as flowers, snap peas, and beans. Items are available to customers any day of the week, thus building customer goodwill.

The connections that Food Bank Farm provides between agriculture and the community are multifaceted. On the farm-input side, the farm accepts organic yard waste from the Hadley, Massachusetts community, which is composted and applied to the farmland. Because the farm is within the city limits of Hadley, it provides a convenient location for recycling yard waste. The recycling program is an informal system in which the farm provides residents with guidelines on what materials are acceptable. This recycling of organic waste reduces deliveries to local landfills and provides valuable nutrients and organic matter to replenish the soil, thus making a contribution to the sustainability of the local ecosystem.

The CSA process also contributes to the ecosystem by reducing the need for packaging and refrigeration. Food is harvested and provided the same day to customers, so there is no need for refrigeration. Customers bring their own containers, reducing the need for packaging. Community volunteers provide labor to support the farm, which in turn supports the Western Massachusetts Food Bank Program. Volunteers, who come to the farm through the food bank program, make a commitment to their community by supporting the local food banks. In turn, these volunteers develop a greater understanding of local food production and further their own sense of community. Approximately 10% of the farm's labor requirement is met through volunteers.

On the farm-input level, the farm is making a commitment to train interns who want to become farmers on their own. The farm managers seek interested people who are serious about learning sustainable food production practices.

Connections are established between the land and its CSA customers in a number of ways. By picking up their food on the farm each week, the customers experience the changing mix of food production throughout the season. Many customers bring their children to experience the farm, and they may also assist in harvesting pick-your-own vegetables and flowers. By doing so, they establish a direct connection between the land, the food they eat, and the flowers they enjoy.

Another connection the farm provides for the community is making available to CSA shareholders other local food grown not on the farm but within the community. The farm provides seasonal items such as apples, cider, and maple syrup, in addition to a regular supply of eggs, fresh poultry, and bread from a local bakery. This process provides expanded markets for locally produced food. It also appeals to CSA customers because they can get more of their needs met in one trip.

The farm hosts local gatherings for its CSA members on the farm several times a year. These events involve sharing farm-produced food, live music, and family entertainment. The socialization that occurs helps to develop a bond between share members and the farm. The farm also is made to be an inviting place for children by providing play equipment for their enjoyment.

Evidence that Food Bank Farm is making connections is suggested by a number of other observations. Through word-of-mouth advertising, there is a waiting list to become a CSA shareholder. Interns serious about learning sustainable agriculture seek Food Bank Farm. The farm is in the enviable position of being able to choose among applicants. Other farmers frequently contact the farm for information about production, marketing, and community involvement. To these requests, the farm operators willingly respond by sharing their experiences.

TOLT FARM, CARNATION, WASHINGTON

The second case study farm, operated by Steve Halstrom, is located near the Tolt River close to Carnation, Washington. This farm, like Food Bank Farm, is on alluvial soil.

Steve Halstrom, a 55-year-old certified organic farmer, developed a working familiarity with food production early in life. He grew up on 5 acres of land, gardened, and participated in 4-H activities as a youth. After graduating from college, he worked for 27 years in the computer industry. He bought his current farmland 20 years ago, and began farming 5 years ago after taking early retirement. On a personal basis, he began to promote local agriculture, and currently is working with King County Growth Management on local agricultural policy issues. Halstrom indicates that he does not need additional income. Rather, he farms because he wants to show that agriculture is feasible using organic methods and to help people discover fresh food. He also enjoys gardening and marketing, meeting customers, and the reward of customers' appreciation.

The farm crop mix includes corn, lettuce, and a wide variety of vegetables. Tolt Farm consists of 2 farmed acres and 8 acres of forest on a 10-acre base. The farm is in a narrow river valley that has not yet felt the impact of major urban expansion. The adjacent forest is upland and provides a spring for irrigation and a buffer isolating the farmland.

Production practices are organic. Composted manure is applied for fertilizer, and cultivation and hand-weeding are used for weed control. Manure is purchased from livestock producers in the community and composted on the farm. Direct seeding is used, and attempts are not made to extend the season by using plant starts, plastic row covers, or cold frames.

Steve Halstrom prefers to avoid the use of petrochemical-based plastics. For this reason, he uses glass rather than plastic greenhouses (fewer petrochemical products are involved). He also direct seeds, as opposed to using greenhouse-produced plant starts. The former method involves less plastic.

In Halstrom's first year of farming (1995), he expanded his home garden, raising mixed vegetables and corn. He sold produce in partnership with another small farmer at the Seattle University District Farmers' Market. At the same time, he sold corn to local grocery stores and from the back of his pickup truck. Halstrom thought about trying a CSA but opted not to, primarily for philosophical reasons. For example, in terms of fuel consumption, he feels that it is "better for me to drive into Seattle than have customers drive here." His decisions reduce energy consumption by consumers even though this is not a farm expense for him. This suggests that he may have a more holistic view of local agriculture through his consideration of the impact his decisions have beyond the farm gate.

During the second year (1996), Halstrom expanded the garden and made adjustments according to his newly acquired sense of "what would sell." Based on his selling and growing experience and what people were willing to pay, he adjusted his choices to increase his profitability. For example, he reduced the amount of corn, added lettuce, and sold produce at his own stall in the University District Farmers' Market. He sold the excess at Everett Farmers' Market, and donated produce to the food bank. Halstrom found a fair amount of spoilage with this marketing system.

Halstrom's third year of farming (1997) was similar to the second, incorporating more refined growing techniques and expanded lettuce production. He sold more frequently at the Everett Farmers' Market on Sundays. The farm was able to sell more of what was grown and experienced somewhat less crop loss.

During the fourth year (1998), Halstrom concentrated his selling efforts on the University District Farmers' Market. By participating in Seattle's Columbia City Market on Wednesdays, Halstrom was able to vastly reduce losses. Tolt Farm nearly doubled its income and, with low expenses for few purchased inputs, proved profitable. Profits increased with Halstrom's knowledge of what to grow and how to display and sell at farmers markets, as well as his expanding relationships with customers. Steve Halstrom prefers taking his produce to farmers' markets and dealing with customers directly. He enjoys meeting his customers and knowing who is consuming the food he produces.

The owner provides all the labor and does most of the marketing. He receives some assistance with selling at Saturday markets. Students have offered to help, but he has kept their participation at a minimum. This is part of his commitment to develop a small farm system based on the owner-operator's labor and management.

The farm operator, Steve Halstrom, encourages small farmers to consider specialty produce: "Look for flavorful varieties that are easy to package and will survive." He observes that if you have a product "failure," customers always can find that product elsewhere. He advises small farmers not to go into debt and to keep the farm "something you can handle yourself, with your own labor." He warns farmers not to "push the season" or to go against to the natural growing season of the plants. In Halstrom's words, "The small farmer must enjoy growing and selling."

THOMPSON FARMS, BORING, OREGON

Thompson Farms, 20 miles southeast of Portland, are located near Boring, Oregon. The farming system has changed a lot since Victor and Betty Thompson started farming here in 1947. Today, Thompson Farms are operated and managed by Larry and Kathy Thompson; Larry's mother, Betty Thompson, an active partner; Larry and Kathy's son, Matt, and daughter, Michelle; and several other family members. These farms are a family affair, and according to Larry, "that may be the secret of the farms' success. " Over two, and approaching three, generations, the family has adapted the farming system to the goals of family members, and to the interests of the nearby urban population.

Larry is clear about the need for the farms to fit with his own goals as "caretaker of the land." He states, "When I leave this ground, I want it to be in better shape than when I arrived." He is adamant about not using harsh chemicals that could leach into rivers and streams, and about creating a safe food system for the local community that does not harm the environment. "In return," says Larry, " the community realizes and believes in what we are doing, and they have embraced us with their support."

The farms got their start when Larry's father, Victor Thompson, originally from South Dakota, and Betty came to Oregon to operate a strawberry farm. "At that time," according to Betty, "a 5-acre strawberry farm would provide enough income to support a family of four."

Larry grew up helping his father raise strawberries, raspberries, and broccoli. Every summer the family worked in the fields, harvesting and hauling the produce to the cannery. This would go on into the fall as Larry and his brother continued to work after school until the harvest was complete.

Today's farms consist of 100 acres on which more than 30 different crops are grown, including the flowers that Betty manages. The system has changed dramatically over time, in response to changes in the community and the family and the need to become more efficient. Housing developments began to surround the farm in the 1970s. About 10 years ago, processor prices dropped. This prompted the family to try selling broccoli directly to local grocery chains. This worked for a while, until they were outpriced by produce from Mexico and California.

In the mid-1980s the family began to rethink their farming and marketing strategies by taking advantage of the things they knew best. They also put a greater priority on goals that matched their own personal values. More emphasis was put on the customer, the community, and the environment. Marketing shifted from a processor orientation to a fresh market orientation. "My mother and I liked people," said Larry, "so we decided to create our own image and take advantage of having customers close by. . . . Everyone loved the fresh-picked taste of our strawberries, raspberries, marionberries, and broccoli."

Today, the farms have 10 harvesters, all members of one extended family. They are employed year-round, have regular time off with benefits, and are well paid. Larry says, "We need to ensure they make a good honest living . . . and that they get a good return for their work." The farms fly the staff from their home base to Oregon in the spring and home again in the fall. A lot of attention is given to family celebrations such as birthdays and Mother's Day, and staff opinions are sought to improve practices and organization. As a result, the family of harvesters has been with the farms for 11 years. Additional harvesters are hired at peak periods, and a farm manager who speaks three languages is hired during the harvest season. She serves as liaison between Larry and the harvest crew, and is key to a u-pick following from Portland's Russian community.

The secret is an involved and enthusiastic crew and family, each of whom has an important and valued responsibility. For example, during the summer, Larry's son and daughter are paid to run the farmers' markets. "This," according to Larry, "instills a sense of pride about agriculture among the younger generation." He feels that customers like to know that there are good wholesome kids involved in the farm and in serving the public. Besides six regional farmers' markets, the family operates two roadside stands.

Larry and his family have adapted their farming system to their own personal qualities and resources. "If you are going to look at a direct market approach," advised Larry, "you have to be comfortable with people coming on to your farm. They are going to run over your irrigation plots once in awhile. But in spite of the occasional problem, we have a great time. We have learned that it is important to our customers to . . . talk to a farmer."

The farms cultivate a high number of beneficial insects through the use of cover crops. For example, Larry points to the strawberries they are preparing to plant and says:

> That's the cereal rye that we use in our crop rotation pattern. We will be disking that in, subsoiling fairly deep, then rotatilling and planting as quick as we can. I do not like to see bare soil on my farms. You won't find bare soil on my farms at any time of the year. . . . We keep something planted so that the soil microorganisms have something to eat during the wintertime. You need to have natural insectaries available from the beneficial insects. . . . That's sustainability.

Larry stresses the value of the humus provided by the cereal straw. "It helps suppress the weed populations in the strawberries," he says, "so we can avoid the use of insecticides in all of our berries." Dandelions, another source of beneficial insects when left in the aisles between the raspberries, are mowed just as the raspberries begin to bloom, forcing the beneficial insects into the canopy of the raspberries where they search out nectars. Larry notes, "The theme here is sustainability, not only of cover crops, but of the crops as they are growing, and what we can do to enhance nature, and what she will do naturally for us."

The family pays attention to what the consumers like to eat. Larry says, "We raise Hood and Rainier strawberry varieties because this is what the public likes to eat. . . . Our market is the community. People come out and pick their own." Larry contends that other varieties would produce a higher tonnage, and they would be more attractive. However, "maximum yield is not the key to economic profitability. . . . rather a berry that the customer loves."

Another unique characteristic of Thompson Farms is the role they play in the community. The Thompson family likes people and kids. They have identified ways to encourage young people to visit the farm. At least 10 acres of pumpkins are planted for fall tours. Fliers are distributed through the schools, and young people come for tours to learn about the farm, pick pumpkins, and have fun. Preschool children who come out by bus are charged $2 each. This guarantees them a ride on the trailer, a talk about the crops that are grown and the beneficial insects, a taste of fresh broccoli, and the biggest pumpkin they can carry home. After they are treated to a little floret of fresh broccoli, they unanimously agree that broccoli is good. This is a switch from what they thought before. Most have never tasted a vegetable fresh from the field.

Larry shares a chickweed story to illustrate the value of the children's farm visits. "Chickweed," according to Larry, "works well as a cover crop because the bees use it as a source of nectar, and when the weather gets hot, it dies." During the tour Larry tells the kids about chickweed, which normally is considered a weed. However, Larry tells them about the useful role the chickweed plays in attracting the bees and beneficial insects in early spring, and that it dies when the weather gets hot. Larry explains why this "weed" is a useful cover crop. When one of the kids got home, his Dad said, "What did you learn today?" The kid said, "About chick weed, Dad." Dad says, "What the heck is chickweed?" You can imagine what came next.

Larry says, "Not only is this valuable science education, but the kids go home and tell their parents, and who do you suppose comes out to the farm the next weekend?" Larry tells about one mother who came out to the farm and said, "My kid says we have to get broccoli here. . . . where is it?" At least 27 schools took part in farm tours last year.

Every year there is at least one story about the farms in the local paper. Larry makes a point of calling the papers, and may even take a crate of strawberries in when he places the annual strawberry ad. Recently, by using the newspaper, Larry promoted a 3-day celebration at the farm stand in honor of Betty's 75th birthday. This included free berry shortcake for a 3-hour period. The story provided valuable publicity. He also likes to encourage university faculty to visit the farms. As he says, "They learn and we learn. Sometimes they do research based on what they observe at our farms."

Another link to the community is through Larry's participation on the board of the Gresham and Hollywood Farmers' Markets. The family takes an active role in the communities where they sell their produce, and in the region. This includes an annual float in the Gresham Teddy Bear Parade with bales of hay, pumpkins, vegetables, and kids. Larry's leadership roles also extend to serving as regional chair of the U.S. Department of Agriculture Western Sustainable Agriculture Research and Education Administrative Council.

In conclusion, Larry says:

> We farmers have an opportunity to take a direct role in the way agriculture is going. The pendulum has gone as far as it can go towards large-scale farming. . . . The communities that surround our farms. . . are ready to say "we have good agriculture in our area." Agriculture can be environmentally conscious. It is economically rewarding to the community. It creates good community spirit. . . . The future of our local economies and our food security systems begins with our own communities.

SUSTAINABILITY CONNECTORS

The case study farms and the communities they serve are woven together in a mutual interdependence that sustains them both. Among benefits to consumers are the freshest possible produce in its most nutritious form. The benefits to the environment include food production without potentially toxic chemicals, reduction in fossil fuels used to produce chemicals and fertilizers, and a decrease in long-distance transportation costs. Benefits to farm operators include a sustainable livelihood and the opportunity for farmers to build appreciative relationships with customers and their families.

On the basis of the three preceding cases studies and the authors' observations of similar small-to-midsize farms situated in urbanizing regions, seven *sustainability connectors* have been identified that seem to bridge urban-edge agriculture with local communities, organizations, and the environment. The authors' research suggests that these seven connectors are important bridges to sustainable agroecosystems in which people are actively engaged.

1. *Use urban opportunities:* All of the cases illustrate the opportunities available to farms situated close to urbanizing areas. Farm operators who can take advantage of the proximity of urban consumers have a distinct advantage when it comes to marketing produce and products that are in demand. Not only does this generate a profit, but there also are other spinoffs such as valuable relationships between farm and nonfarm people, and for nonfarm people, an understanding of the science of farming, and the food production process. One farm provides food in return for labor; another entertains and teaches children.

2. *Recognize human interdependency:* All of the cases take advantage of the innate need that seems to exist between urban and rural people, farm and nonfarm people, and adults and youth: to reach out to each other to experience the food culture of the other. For farmers, this relationship is partly tied to profit, yet with this comes an appreciation of the opportunity to engage with and learn from other people. This suggests that human relationships built around the food system and the land may rekindle the human spirit and offer the possibility of new food-related structures. This

might be likened to Stevenson's (1998) notion of competencies, which he suggested may be useful for negotiating alternative agrifood systems. Relational competencies focus on new forms of food citizenship, especially new organizational patterns. These may involve reciprocal relationships between farmers, marketers, and ordinary people. These are truly interactive relationships, as those between the Thompsons and their customers, or between Food Bank Farm and their volunteers or interns. Farmers' and consumers' awareness of the interdependence of rural and urban areas opens up opportunities for joint problem solving. According to Castle (1998), solutions to rural problems may be rooted in the knowledge of rural–urban interdependence. The viability of rural communities, businesses, and the environment must be examined in the context of the interrelationships between rural and urban areas, not in isolation of one place from the other.

3. *Embrace natural resources:* The farmers in the case studies have clearly made a commitment to practices that are environmentally friendly. Steve Halstrom purchases manure from nearby livestock operators and composts it for his fields. He intentionally avoids the use of petrochemical-based plastics, and to reduce energy consumption, he prefers to drive to farmers' markets rather than have customers drive to his farm. The Thompsons place value on natural cover crops to enrich the soil and attract insects, rather than apply toxic pesticides. Food Bank Farm provides a convenient location for their urban neighbors to recycle yard waste, and in doing so, obtains a free supply of organic matter. All of the farms avoid the use of packaging and refrigeration by marketing directly to the consumer.

4. *Know yourself:* In all three of the cases, the operators and their family members seem to have a clear sense of their own personal qualities. In the case of Tolt Farm, Steve Halstrom has designed a system in keeping with his desire to operate farming and marketing activities himself. Larry and Betty Thompson love to talk with their customers and with young people. They have devised ways to ensure quality human relationships that they enjoy, and which bring in business. The same can be said for the satisfaction gained by Linda and Michael of Food Bank Farm through relationships with shareholders, volunteers, and interns.

5. *Know your customer:* On each of the case farms, management has made an effort to learn about customers' preferences. The Thompsons plant strawberry varieties that their customers like, despite the fact that yields may be lower. Food Bank Farm makes other local products available on their farm to facilitate "one-stop shopping" for CSA shareholders. Food Bank Farm entertains its customers with music, children's play equipment, and other family activities, whereas the Thompsons offer fun educational activities for school children. They also take advantage of their customers' family values by having the whole family participate in marketing activities.

6. *Incorporate equity considerations:* An equitable food and agricultural system fosters a decision-making system that is democratic, identifying the costs and benefits and the way they are distributed among farmers, farm workers, customers, nonfarm rural and urban residents, future generations, and residents throughout the globe (Allen et al., 1991). Although this sounds idealistic, the goal is worth examining. Two of the cases presented, Tolt Farm and Thompson Farms, are clearly committed to ensuring consumers' access to fresh, safe, high-quality food. Fair and equitable treatment of farm labor is high on the list of Thompsons' values. Similarly, the participation of Food Bank volunteers, customers, and interns in Food Bank Farm reflects a democratic ideal.

7. *Engage in community participation:* All of the farms play an active role in the communities where they expect to do business. The Thompsons participate in the annual community parade, encourage school visits to the farm, provide human interest stories in the local paper, and serve on local and regional boards and committees. Steve Halstrom participates in local growth management issues, and Linda and Michael build community goodwill through an intern and food bank volunteer program.

CONCLUSION

The agrarian model in North America is rapidly being transformed by a food system in which technical control of biologic process and management control of food manufacturing and distribution is wielded by a few multinational firms. The degree of this system's momentum suggests limited opportunities to redirect it to strengthen links between agriculture, communities, and ecosystems. At the same time, an alternative paradigm is developing that offers a stark contrast to the industrial food system. The case studies in this chapter show that changes can be made that connect consumers with farms and the ecosystem. The idea that farming is a core component of the total community ecosystem has come full circle, with trendsetters recognizing the importance of sustainable agriculture to a community. The case studies illustrate that if farmers take a consumer-oriented approach to production and management practices, everyone wins. The producer–consumer disconnect is decreased. In addition, it is apparent that education is one of the keys to change and survival. Each of the farms has mechanisms in place to ensure that members of neighboring communities have opportunities to learn about and appreciate food production practices, whether through publicity, direct-marketing activities such as CSAs, school tours, or internships to train the next generation of farmers. They also exemplify the importance of a multifunctional approach to food, agriculture, and the environment that may be of great importance to communities in the future.

It is apparent that the systems discussed in this chapter are not going to survive by competing on the basis of price, the concept that drives the industrial food system. In all likelihood, they will be underpriced, either by those systems that are using conventional methods, incorporating economies of size and chemicals or imports that have access to lower input prices, labor in particular. What differentiates the farms discussed in this chapter is that they are moving beyond the competitive treadmill trap model and attempting to differentiate their produce and services on factors other than price. They are forging mechanisms to become price makers. For them, the keys are the consumer and the community. All the case study farms astutely recognize the fact that consumers' needs and communities' interests are changing. There are opportunities for creative multidimensional strategies that enhance sustainability.

This nation has never before been wealthier. An affluent society is one that can afford to be more discerning in its choice of food, and one that can afford to be more environmentally conscious. Maslow's (1968) hierarchy of needs illustrates how one's priorities change as different needs are fulfilled. Perhaps one of these evolving dimensions is locally grown, high-quality, fresh produce for attaining a healthier population as well as a more sustainable community and environment. The farmers discussed in this chapter are attempting to meet these new needs. They are using practices friendly to the environment, healthy to the consumer, and good for communities.

Small- and medium-size farms, represented by the case studies, are changing the rules of the game for at least part of the food system. The findings suggest that farmers such as those described in this chapter are questioning the industrial paradigm that relies on distant control of the food system. They are taking control of the marketing process by making direct connections with consumers in their communities. At the same time, they are making a positive contribution to the ecosystem. Changes can take place by questioning the status quo, thereby stimulating a new personality of farmer and new relationship-based structures.

REFERENCES

Allen, P., Lundy, J., and Gliessman, S., *Sustainability in the Balance: Expanding the Definition of Sustainable Agriculture,* Issue Paper 3, University of California Santa Cruz Agroecology Program, Santa Cruz, 1991.

Altieri M., *Agroecology: The Scientific Basis of Alternative Agriculture,* Westview Press, Boulder, 1987.

Altieri, M., Multifunctional dimensions of ecologically based agriculture in Latin America, *Int. J. Sustain. Dev. World Ecol.,* 7, 62, 2000.

Andreatta, S. Transformation of the agrofood sector: lessons from the Caribbean, *Hum. Organ.,* 57, 414, 1998.

Castle, E. N., *Agricultural Industrialization in the American Countryside,* Henry A. Wallace Institute for Alternative Agriculture, Greenbelt, 1998.

Conway, G. R., Sustainability in agricultural development: trade-offs between productivity, stability, and equability. J. Farming *Syst. Res., Ext.,* 4, 1, 1994.

Food Bank Farm , Baseline data, gathered by R. Carkner September, 1999.

Geertz, C., *Agricultural Involution,* University of California Press, Berkeley, 1963.

Goldschmidt, W., *As You Sow,* Harcourt, Brace, New York, 1947, reprinted by Allanheld, Osman, Montclair, 1978.

Hedden, W. P., *How Great Cities Are Fed.* D.C. Heath & Co, Boston, 1929.

Kloppenburg, Jr., J, Hendrickson, J., and Stevenson, G. W., Coming into the foodshed, *Agric. Hum. Values,* 13, 33, 1996.

Kroeber, A., *Cultural and Natural Areas of Native North America.* University of California Press, Berkeley, 1939.

Maslow, A. H., *Toward a Psychology of Being,* D. Van Nostrand, New York, 1968.

Moran, E. F., Ecosystem ecology in biology and anthropology: a critical assessment, in *The Ecosystem Approach in Anthropology: From Concept to Practice,* Moran, E. F., Ed., University of Michigan Press, Ann Arbor, 1990, chap.1.

Rappaport, R., *Pigs for Ancestors,* Yale University Press, New Haven, 1967.

Stevenson, G. W., Agrifood systems for competent, ordinary people. *Agric. Hum. Values,* 15, 199, 1998.

Steward, J., *The Concept and Method of Cultural Ecology: Theory of Cultural Change.* University of Illinois Press, Urbana, 1955.

12 Rural Community Leadership in the Lake Benton Watershed

Wayne Monsen

CONTENTS

Introduction 175
The Agroecosystem 176
The Rural Community 177
Building Social Capital for a Healthier Agroecosystem .177
Community Mobilization for Improving Ecosystem Health 178
The Community Vision 180
Inclusiveness Helped the Coalition Work Together 184
References 185

INTRODUCTION

Community sustainability is a concept that refers to long-term economic, environmental, and community health. In the document *Sustainability Checklist,* Bauen et al. (1996) stated that sustainability helps communities in two ways. One way is by "considering the long-term consequences of today's decisions: Do they enhance or detract from the community's ability to prosper into the future? What will their effects be on later generations?" (Bauen, et al., 1996; Monsen and Toren, 2000).

A second way of considering sustainability is by "thinking broadly, across issues, disciplines, and boundaries. Sustainability suggests that creating economic vitality, maintaining a healthy environment, and meeting human needs are closely related rather than separate tasks" (Bauen et al., 1996).

The Whole Farm Planning Program at the Minnesota Department of Agriculture (MDA) developed a leadership model, rural community leadership (RCL), that encourages the participation of citizens in planning for the sustainability of their community. Through a series of workshops, residents of Lake Benton, a southwest Minnesota rural community with different backgrounds and interests, came together to discuss and take action to improve their community. They worked together, whereas historically they seldom had talked or worked together.

0-8493-0917-4/01/$0.00+$.50
© 2001 by CRC Press LLC

The MDA recognized that leadership must come from the people themselves, and not from agencies outside the community, to engender community sustainability. Consequently, the MDA began a pilot project designed to foster proactive citizen leadership. The process tested in this project allows the local residents and agency staff to inquire and plan for the type of community they want for their future. The results are better relationships, people understanding each other's interests, and implementation of government programs that the people want. This approach is a win-win solution to program development and implementation.

The project in Lake Benton, Minnesota, exemplifies a community of residents with diverse backgrounds and worldviews actually working together. Such working together helps government programs to be more effective both in dollars spent and community support for the type of work implemented.

THE AGROECOSYSTEM

The Lake Benton watershed is located in Lincoln County in southwest Minnesota. The watershed is situated on Buffalo Ridge, a unique ridge that goes from northwest to southeast from Saskatchewan through the Dakotas and into southwest Minnesota. The area is the second highest location in Minnesota. The ridge is approximately 1,700 ft above sea level, whereas the elevation of the prairie not on the ridge is about 1,000 ft. This change in elevation makes the climate harsher than in the nearby lower elevations, that have stronger winds and lower temperatures. The area is quite hilly, which allows for more soil erosion during rain events and snow melt.

The landscape is predominantly farmland. Corn and soybeans make up most of the crops grown, and livestock are present on many of the farms. A growing number of large livestock facilities occupy the area. Because the landscape is rolling and hilly, the soils have a tendency to erode, carrying nutrients into the surface waters and ultimately into Lake Benton Lake.

Lake Benton is 6 miles long and 1 mile wide. It is the largest lake in southwest Minnesota, noted for great fishing and recreation. The lake is ranked in the top 10 of best lakes for walleye in Minnesota. In the past 10 years, the lake has been infested by an exotic weed, the curly-leaf pondweed. This invasive weed covers about three fourths of the lake. It is thought that the nutrients added to the lake from farmland and feedlot runoff are contributing nutrients for the explosive growth of the curly-leaf pondweed, which does not have any natural enemies to help keep its proliferation in check. In June and July the weeds are so thick and so near the surface of the water that boating and fishing are nearly impossible. In August, the weed dies back, but the old plant growth washes up on shore, rots, and smells. The shoreline needs cleaning to remove the rotting plants. Consequently, fishing and other water-related activities are on the decline. Residents of the area, especially people living on the lakeshore, are very concerned about the lake quality and the negative impact on the community.

In addition to the damage to recreation aspects of the lake, the amount of economic revenue brought into the area is greatly diminished. The Lincoln County Economic Development Office reported that the number of fishing licenses purchased in Lincoln County in 1997 was down 60% from 15 years earlier. Based on the

1991 National Survey of Fishing and Wildlife-Associated Recreation by the U.S. Department of Interior and adjusted for inflation, it is estimated that the lake generated $669,480 of income in 1996 for the area from the anglers' purchases of food, tackle, equipment, fuel, and the like. If the lake were not infested with curly-leaf pondweed, the income generated would greatly increase.

THE RURAL COMMUNITY

The population of the Lake Benton watershed consists of about 1,500 residents: 700 farmers and rural residents, 700 people living in the city of Lake Benton, and 35 families living on the shore of Lake Benton Lake. The predominant ethnic background of the population in the city of Lake Benton is German and Scandinavian. In contrast, the farm and rural residents of the watershed are considered a "melting pot" because many ethnic backgrounds are represented without a dominant heritage.

The average age of the residents is approximately 55 years. This includes both the farm and nonfarm population. As in most rural communities in Minnesota, there is concern that the population is getting older and not enough young people remain in the community or want to move into the community.

With one half of the population involved in farming, the economy in the area is heavily dependent on agriculture. There is discord between the farm community and the nonfarm community that leads to little communication between the two sectors. When there is a history of mistrust between different sectors of communities, it is difficult to institute projects that move communities to where they want to go.

Historically, when natural resource problems arose, it was easier for concerned citizens to work directly with natural resource agencies than to bring all parties together to work through the problem. This was the situation in the Lake Benton watershed with the concern about the curly-leaf pondweed problem in Lake Benton Lake. The members of the lakeshore association, Lake Improvement for Everyone (LIFE), developed a working relationship with local natural resource agencies, but relationships with other sectors of the community were not fostered, especially those with the farm community. The agencies and LIFE wrote a watershed management plan designed to clean up the lake if implemented. The plan called for changes in farming practices, but did not include the farmers in designing the plan.

BUILDING SOCIAL CAPITAL FOR A HEALTHIER AGROECOSYSTEM

Members of LIFE wanted to work to clean up the lake and make the area a better place to live. Because LIFE members were not part of the farming community, they often were viewed by the farmers as antifarming. The curly-leaf pondweed problem became so bad that the LIFE members wanted to do something to help the problem. They approached the problem the only way they knew how, with the help of the local Soil and Water Conservation District in Lincoln County and the Redwood Cottonwood Rivers Control Area. Together they designed a watershed plan. They knew that they had to include farmers in the process, especially to implement changes in farming practices that would reduce nutrients and sediments entering the lake.

To encourage farmers to take advantage of farm programs, Carolyn Brinkman, a LIFE member, called the Minnesota Department of Agriculture (MDA) in September of 1996 to see if the MDA could get grant money to area farmers to help them keep livestock out of the lake and streams. As the discussion progressed, it was noted that plans were written about farming, but without involving farmers in the planning. The lake association was searching for ways to involve farmers in the planning but did not know how.

At this same time, the MDA was developing a whole-farm planning program to assist farmers in taking a holistic approach to farm planning. The MDA offered to work with the farmers in the watershed to help them take a holistic approach, an endeavor that looks at the farm as a system. With this approach, farmers are encouraged to include the broader community in their planning. A vision for the community will help them plan for the actions they will incorporate into their farm operations.

COMMUNITY MOBILIZATION FOR IMPROVING ECOSYSTEM HEALTH

The RCL model is designed to enhance the leadership abilitities of the residents in rural communities so they can be proactive in planning for their community's sustainability. Planning and government decisionmaking often are criticized as being reactive more often than proactive. Critics state that there is a lack of vision or insight as to how decisions should be made and how these decisions will affect the future. In the RCL model, citizens are invited to find how they are able to influence community processes and to help establish sustainability policy. Citizens also learn how to work comfortably within community systems. Through this project citizens learn how to collaborate with government and nongovernment organizations.

Implementation of RCL is through workshops in which people of diverse interests and experiences come together to focus on issues of mutual concern. The main objectives are to involve the public in making decisions and to have citizens take ownership of their decisions. The RCL method assumes that the combined knowledge and wisdom of citizens can be used to make effective decisions and implement effective programs. This process was selected by MDA because (1) the decision is workable, meaning it can be implemented; (2) the information is used before the decision, not after; and (3) the decision is perceived as fair and equitable, with all the interested parties participating in the development of the decision (Adams et al., 1995). Once residents feel they really have a say and can do things, trust can be built between community members and agency personnel.

A series of RCL workshops were presented. The workshops were designed to help participants work together and learn from each other about the variety and commonalities of vision for the community. They built on the latent values, the aspirations, and the skills that local residents already had, using theories and methods about systems thinking, learning organizations, driving forces and trends, future search, and visioning (Schwartz, 1991; Senge, 1990; Van Der Heijden, 1996; Weisbord and Janoff, 1995; Wilson et al., 1990). The broadest public involvement was generated when emerging issues became the basis for management and decision making. These workshops were designed as consecutive meetings, but the meetings did not need to be held on consecutive days.

The first workshop was designed to help the participants recognize the aspects of the community they wanted to keep and enhance: the social, human, and natural capital. To get at these aspects, the participants were asked, "What do you want your landscape to be like in 50 years?" and "What do you want your community to be like in 50 years?" These questions created the commitment for long-term action and thinking. They also helped the participants to brainstorm possible actions with no time frames or commitments attached. The list of possible actions they generated was used in later workshops in which they actually designed action plans.

The goal for the second workshop was to develop a shared vision statement and commitments to action plans. This session helped the participants to see why it is important to have farmers and nonfarmers planning together. In both small and large groups, the participants worked together, and through consensus developed a shared vision statement. A vision statement is important in community planning because it is developed and shared by many. The statement can change the relationship between the participants and the community, giving them courage to pursue the vision (Senge, 1990).

After the vision statement was written, each participant listed two important issues in the community that needed to be addressed. These issues were categorized, and the participants worked in small groups to design actions that could be taken to address the issues. This activity helped the participants to see that they were having a say in the type of activities they wanted, and that carrying out these activities would be their responsibility.

The third workshop helped the participants to see the trends and driving forces in the community and throughout the whole world with which they would have to deal. There are forces in society, technology, economics, politics, and the environment that will change how things will be done in the future. Awareness of the trends and forces helps communities to plan for action and to be aware if things are taking a different path than originally planned. Once forces and trends have been identified, this knowledge can be applied to the action plans, which will help the actions to be more affective in achieving their planned outcome.

This can be a somber exercise, because numerous trends in the world may have effects on the community that the participants do not want. Ignoring them, however, will not make the trends go away. By acknowledging them and planning for them, perhaps they can be averted or changed to help benefit the community. Often, people feel helpless in changing trends that seem beyond their control. The trend toward a global economy is an example of a trend that often is seen as beyond the local community's ability to change. A global economy has some negative effects on rural communities, but it is seen too far removed from the community for the community to do anything about it. But, by not addressing it, the community will let the trend happen to the community rather than take a proactive approach to have the trend be a benefit to the community.

The fourth and final workshop was designed to put visions into actions. Using the work of the previous sessions, the participants put time frames and projects into action. They revisited the categories of actions developed in the second session, considering the driving forces and trends discussed in the third session to see whether the categories needed to be changed, added to, or modified. Then the participants broke into small groups. Each group took a category and designed an action plan for that category. The small groups reported back to the large group and got feedback and buy-in from all of the participants. This put the seal of approval on the projects by the participants. This process also helps natural resource agencies to provide services that the citizens helped to design and therefore accept.

THE COMMUNITY VISION

Originally, the RCL workshops were to be conducted for farmers in the Lake Benton watershed (Figure 12.1). At their workshop, the farmers wanted to attend the series of workshops with a broader community because they felt it was important to include everyone in the community in this type of planning. They also felt they were blamed for all the environmental problems of the lake.

During the first workshop with the larger community, the participants decided that they should organize into a coalition. They called this new coalition the Lake Benton Watershed Holistic Management Coalition. They felt that a coalition would give them the impetus to do things in the community and would show other residents that the many organizations represented were working together.

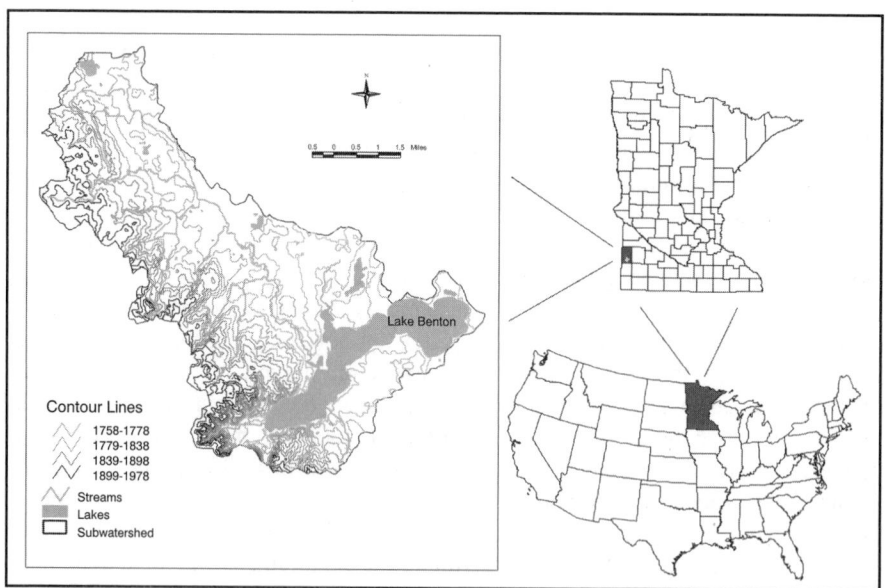

FIGURE 12.1   The Lake Benton watershed, Lincoln County, Minnesota.

The RCL process enabled the residents of the Lake Benton watershed to want to take leadership in their community. This process showed them that they had the power to lead and make decisions to better their community. The MDA staff provided facilitation for the process, but the actual leadership and commitment came from the participants themselves.

When the participants decided to form a coalition, they wanted three representatives from different sectors of the community to be coordinators. A farmer, a lakeshore resident, and a business person from Lake Benton were chosen as co-coordinators. These three sectors were selected because they represented the major portion of the community. The farmer represented the largest geographic area and nearly one half of the population, with about 700 people. The business person represented businesses and the residents in the city of Lake Benton, with a population of about 700 people. The lakeshore association member represented about 100 people living on the lake, made up of retirees and people who worked in town. These three leaders were responsible for keeping the coalition together, bringing new people to the group, and keeping the goals and visions of the coalition as the driving force for keeping the group together.

The coalition set up a monthly meeting schedule and wanted the MDA to be there to help run the meetings. They decided that they wanted to broaden the facilitated workshops and hold education activities throughout the workshop series. This was different from the original plans of the MDA because the original series of workshops were to be limited to four meetings. The MDA supported this development because it gave the coalition an opportunity to show its leadership.

The RCL model helped the residents of the Lake Benton watershed to develop a vision for what they want their community to be like in 50 years. The vision statement the participants agreed on was as follows:

A strong and diverse agricultural base providing opportunities for future generations. Maintain a viable community that values business, employment opportunities, services, schools, and churches. Decisions are made locally through positive communications and participation.

The visioning process led the residents to identify the aspects of the community that they wanted to maintain and enhance. Working at two levels, the natural resource and environment side and the social and community side, the people identified nine aspects of the community that were important to them. These included aspects that were more involved with the natural landscape, and others that were more civic in emphasis. The nine aspects were as follows:

1. *Local schools:* The people had a strong desire to have enough students to keep the school open and not consolidate with other schools.
2. *Health care:* The participants wanted affordable and adequate health care within the community.
3. *Maintained parks:* There was a desire to have plenty of park space for promoting outdoor recreation.

4. *Updated highways:* The participants wanted to ensure good roads for transportation of goods and services.
5. *Clean lake:* The participants wanted Lake Benton to be fishable and swimmable so people would want to visit this natural resource. They recognized that proper management of all the natural resources would benefit the lake.
6. *Identity with community:* There was a strong identity and pride among the people for the region of Lake Benton, and the coalition wanted to keep this identity strong.
7. *Healthy farms:* Lake Benton relies on farming for its economic base. The participants realized that a strong, vital farm economy was important for the well-being of the whole community.
8. *Prosperous businesses:* The success of nonfarm businesses was important to the participants.
9. *Tourism:* The promotion of tourism would provide economic benefit to the community.

This activity inspired a desire in the participants to identify some principles that they wanted to live by and implement as they took action either as individuals or in cooperation with the larger community. They wanted to base their decisions on the following principles:

- *Local control:* The residents of the community wanted to have control and a say in the programs and policies designed for their community.
- *Property rights:* Any decision that was made had take into account the property rights of the individuals affected. Policies could not be allowed to impair property rights.
- *Economics:* Any policy or program needed to be economical for the individual and the community. It was not right for programs to cause economic hardship.
- *Quality of life:* The residents were proud of the quality of life in their community. Programs and policies had to take into account the quality of life of the community.
- *Open communication:* There had to be open communication between individuals, organizations, and sectors in the community to prevent bad feelings and factions.
- *Flexibility:* All programs needed to have flexibility to fit the desires of the residents and the culture of the community.

These principles were important because they gave the residents some guidelines for any policies and actions taken. These principles sent a message to policy makers as they designed new policies. Individuals and businesses could use these principles to help them determine whether their plans for their businesses, farms, or residences followed these principles to guide their decisions.

The participants identified five priority issues that needed to be addressed by the coalition:

1. *Ground and surface water quality of Lake Benton and the watershed:* The coalition planned to develop action plans that addressed lake weed control, soil erosion, and nutrient runoff from the fields, city, and home site.
2. *Community and communications:* The coalition wanted to have a wide range of community involvement in their planning and actions. They would do this by networking and providing educational programs and tours.
3. *Livestock and manure management:* The coalition wanted to provide input on a local zoning ordinance that would protect the environment while providing adequate income potential for the farm operation, with design incentives for the installation of effective waste management facilities for livestock operations.
4. *Stewardship and biodiversity:* The coalition wanted to promote a wise approach to use of the natural resources in the area by analyzing social and economic impacts of their actions. They wanted to explore alternative uses for the resources.
5. *Recreation:* The coalition members were proud of the numerous recreation opportunities in the area. They wanted to work with tourism groups and encourage continued growth of outdoor recreation and the arts.

The participants thought some activities along the way would help them create and implement efficient projects. They held a tour of Lake Benton Lake with representatives from the U.S. Natural Resource Conservation Service, Soil and Water Conservation District, the Minnesota Department of Natural Resources, the sportsmen's club, and LIFE. A tour of farm conservation structures showed nonfarmers what farmers were doing to protect the lake. A nutrient management person from MDA met with the coalition and the community to talk about the nitrate problems in the rural water systems in Lincoln and Pipestone counties. That meeting was well attended because the quality of drinking water was of primary concern. One meeting included a discussion with the Lincoln County Commissioners for feedback to the commissioners about different issues. In another meeting there was a presentation by an aquatic weed specialist from the Minnesota Department of Agriculture discussing the weed problem and experiments being done to create different methods for controlling weed growth and spread.

The RCL model is an inclusive planning model designed to work for residents from all aspects of the rural community. It is important to be inclusive in community visioning and planning so that all sectors buy into the planning effort. Omitting sectors will leave the process open for complaints or the development of other strategies that counteract efforts.

In the RCL model, deciding whom to invite and include in the workshops is the first step. Local resident leaders are selected by MDA, and they are the ones who generate the invitation list for the first workshop. At the first workshop, the participants also identify others, either as individuals or organizations, to invite. The workshops also are advertised in the local press so anyone interested can participate. It is important for the local residents to invite participants personally, because this helps residents feel they have a vested interest in the success of the workshops.

INCLUSIVENESS HELPED THE COALITION WORK TOGETHER

The attitudes of the participants in the coalition changed over the duration of the project. Initially, farmers and the lakeshore residents did not talk to each other. There was no trust between them. The lakeshore residents thought the farmers were damaging the lake quality because of their farming practices, especially by allowing cattle direct access to the lake and tributaries. The farmers thought the lakeshore residents were concerned only for their personal recreation and aesthetic values and not with the economic viability of the farmers.

Through the workshop process, lot of progress occurred, with farmers and lakeshore residents working together. One farmer commented that he used to call the lake residents "the lake people." Now he called them "lakeshore residents." Trust was built because each group learned that the other was concerned and doing projects to help the lake and community. Throughout the duration of the project, the coalition wanted to provide education activities for the residents of the watershed. They thought that education would help them to work together and network as a community. Residents would learn the issues and desires of people from other sectors of the community with whom they normally did not interact.

An important strength of the RCL model is that it empowers individual participants to realize that they have a say in the future of their community. Sensing this empowerment, the participants involve themselves in designing programs and projects that will help their community achieve desired goals.

In addition to the action of the coalition, there was evidence of changes taking place outside the coalition. Farmers took advantage of the Environmental Quality Incentive Program (EQIP) payments to cost-share 75% of the installation costs of conservation structures. (EQIP is an incentive program of the U.S. Department of Agriculture.) Farmers in the watershed installed sediment basins to retain runoff water in their fields longer, thus retaining the sediment in the fields, and keeping it out of the lake. Otherwise, sediments would carry the fertilizer and pesticides farmers applied to their crops to the lake water, causing problems.

Another program the farmers used with EQIP was the installation of buffer strips along drainage ditches and streams. Farmers got permanent easements for the land on which they planted grass. This grass filtered and buffered the draining systems so that fewer nutrients ran off and entered the surface water systems.

Bridging social capital was created through the coalition to improve first natural, then financial, capital. The farmers, primarily interested in whole-farm planning to become eligible for federal programs, were pleased to see how they could contribute to achieving the community vision while not suffering economic hardship. The businesses and lakefront residents created a mechanism to improve natural capital, which in turn contributed to human capital (human resource education), social capital, and financial capital.

REFERENCES

Adams, E., Gray, K., and Baril, K., *Ground Rules Equalize Power as Governmental Agencies Manage Citizen Involvement,* Western Region Extension Publication, Pullman, March, 1995.

Bauen, R., Baker, B., and Johnson, K., *Sustainable Community Checklist,* Northwest Policy Center, Graduate School of Public Affairs, University of Washington, Seattle, 1996.

Monsen, W. and Toren, B., *Community Sustainability through Citizen Leadership,* M.A. Thesis, Augsburg College, Minneapolis, MN, 2000.

Schwartz, P., *The Art of the Long View: Planning for the Future in an Uncertain World,* Currency, Newfolk, 1991.

Senge, P. M., *The Fifth Discipline,* Doubleday. New York, 1990.

Van Der Heijden, K., *Scenarios: The Art of Strategic Conversation,* John Wiley & Sons, Chichester, England, 1996.

Weisbord, M. R. and Janoff, S., *Future Search: An Action Guide to Finding Common Ground in Organizations and Communities,* Berrett-Koehler Publishers, San Francisco, 1995.

Wilson, K., Moren, G., and Moren Jr., E. B., *Systems Approaches for Improvement in Agriculture and Resource Management,* Macmillan Publishing Company, New York, 1990.

13 The Winnebago Tribe's Agroforestry Project: Linking Indigenous Knowledge, Resource Management Planning, and Community Development

Lita C. Rule, Marcella B. Szymanski, and Joe P. Colletti

CONTENTS

Prologue 188
Introduction 188
The Winnebago Tribe of Nebraska 189
Tribal History of Land Use 191
Agroforestry as a Development Tool for the Winnebagos
of Nebraska 191
The Initiative 192
Determining Best Land Use: First Step to Development 193
    Planning for Best Land Use: A Feasibility Study 193
    Methods 194
        Forming Objectives 194
        Assessing Resources 195
        Evaluating Alternatives 195
    Feasibility Study Results 195
    Seven Alternatives and Fifteen Management Regimes 196
    Effects and Decision Criteria 197
    Ranking of Alternatives Based on Four Weighting Schemes 200
    Feasibility Study Conclusions 201

0-8493-0917-4/01/$0.00+$.50
© 2001 by CRC Press LLC

Continuing Community Efforts toward Development 201
    Development of the Agroforestry Demonstration Project 201
    Community Participation in the Effort 202
    Use of the PRA to Link Cultural and Other Resources to the Project 204
Experiences Gained and Lessons Learned 204
    The Tribe 205
    The Researchers 206
    The Larger Community 207
Conclusion 208
References 208

PROLOGUE

The development of the Winnebago Tribe's agroforestry project in Nebraska show-cases how a rural community can enhance an agroecosystem. First, it illustrates the unification of the varied community interests of a native nation, which contributed to its search for means to uphold its traditional land stewardship ideal. Second, it presents a holistic look, by the tribe itself, at its composition, needs, and resources in drawing up plans for land resource use for tribal development. Third, it provides ways of successfully getting community support in developing plans to meet community needs, ensuring representation of the whole community in the process. These are summed up by the immediate past leader of the tribe:

> I am pleased with the process that was developed to further changes near and around the Missouri section of the land. It has been nearly 10 years since the tribe made the commitment to change the way the land has been utilized for the past 50 years (constant farming) to something that is more conservation based. In addition, I am satisfied with the participatory methods that were used in determining the future use of our land. This community should be commended for their insights in the deliberation about their lands. It is from these findings that we confirm the need to change many of the ways we view our lands as leaders of our nation. (John Blackhawk, Past Chairman of the Winnebago Tribe.) Aug, 1996, Winnebago, NE.

This chapter describes how cultural values played a part in the Winnebago Tribe's decision making, community dynamics, and land use planning and management for a sustainable future. It also explains how the decision-making process became a unique learning experience for both the tribe and the nontribal researchers involved.

INTRODUCTION

Many Native American societies have looked on the land as a resource that needs to be used and managed with utmost care. This land stewardship philosophy is based on the realization that present-day users are only temporary occupants and caretakers of the land. Therefore, they should manage the land and other natural resources for future generations. The forest is a source of food, medicine, and shelter. It also is an important setting for many of the tribe's cultural and religious activities. Some tribes are looking at a more holistic, integrated management of the whole set of resources, not just forests, within the reservation as a fulfillment of various tribal objectives.

Managing land resources to achieve group objectives requires evaluation of the products of such management activities. Priority setting is a function of the value system of a social group. Many indigenous people regard land as a means to sustain human society, with the environment as an extension of themselves (Adamowicz et al., 1994). Sociocultural differences in value systems, especially when indigenous groups arrive at their values using Euro-American methods, yield contradictory evaluations of goods and of land use systems. What is viewed by Euro-American culture as "indifference to land ownership" is in fact a difference in values. For many indigenous peoples, the predominant value of sharing among themselves results in an indifference to the accumulation of individual wealth and property (Adamowicz et al., 1994).

Individual preference structures also are defined by the cultural aspects of a society. For instance, Euro-American society emphasizes individuality and financial success, whereas many Native American societies place the emphasis on family and spiritual harmony (Smith, 1994). Additionally, problems occur in assigning nonmarket values for objects, practices, or places that have sacred or revered values but no monetary or substitution goods (Adamowicz et al., 1994). These defining elements make it difficult to assign price valuation for natural resources and land-use decisions based on Euro-American constructs.

Unfortunately, however, the concepts of incorporating the so-called nonmarketable goods and services (e.g., aesthetics, wildlife, settings for religious and social rites, and culture) are often overlooked. The imposition of Euro-American concepts of measurement is still the norm rather than the exception in dealing with indigenous groups.

THE WINNEBAGO TRIBE OF NEBRASKA

The Winnebago Reservation of Nebraska is located in the northeastern corner of the state bordering the Missouri River (Figure 13.1). The Winnebago Tribe has strong extended family connections and a strong connection to the earth. Most individual land ownership is in small heirship holdings. The tribe's present population on the reservation is approximately 1200 people. Additionally, there are some members who live off the reservation because of outside employment.

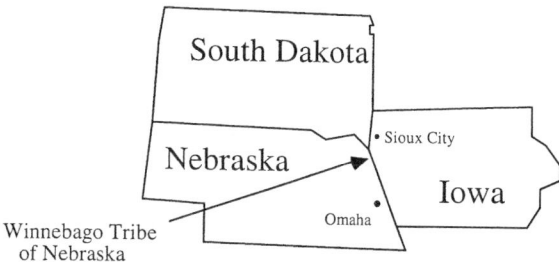

FIGURE 13.1 Location of the Winnebago Tribe of Nebraska lands. Big Bear Hollow is located east of Winnebago, Nebraska, on the floodplain of the Missouri River.

The Winnebago people called themselves "Ho-Chungra," or "People of the Parent Speech," or "Big Fish People." According to the tribe's historian, the Ho-Chunks (Winnebagos), along with other Siouan tribes, may have migrated from the Olmec civilization in Middle America around 1000 BC. Similar characteristics have been observed between the old Olmec religion and the tribe's traditional religion, the Medicine Lodge. Settling at a place called Indian Knoll in northwestern Kentucky around 500 BC, the Ho-Chunks and three other sister tribes left the knoll by AD 500 and entered Wisconsin. Their villages stretched from Green Bay in Wisconsin to northeastern Iowa. These people were responsible for thousands of effigy mounds through northern Illinois and southern Wisconsin built during the Effigy Mound Building Era of the Woodland Cultural Period (Smith, 1996).

Wars with other Indian tribes and their involvement in the French, British, and American wars caused the Winnebago to lose most of their homeland and to move across the midwestern states of Illinois, Iowa, Minnesota, and South Dakota before finally settling in Nebraska. In 1865, after that series of relocations, the U.S. government purchased 40,000 acres from the Omaha Tribe of Nebraska to provide the Winnebagos with a reservation of their own (Sultzman, 1998). There are now two separate Winnebago nations: one in Wisconsin and one in Nebraska. After land was given to the Winnebagos in Nebraska, additional purchases were made. The current Nebraska reservation base is 120,000 acres, with 30,450 acres currently owned communally by the Winnebago Tribe (Whitewing, 1997).

The Tribal Allotment Act of 1887 placed ownership of the land for the Winnebago Tribe with individual tribal members. The act was a continuation of a bad, albeit well-meaning, policy that attempted to turn the Winnebagos into settlement-type farmers. The effects of that policy still reverberate to this day. Currently, two thirds of the original land allotment have passed from Winnebago tribal ownership because of tribal members selling their individual allotments to nontribal members (Office of Native American Programs, 1998). Because the Winnebagos did not usually leave wills, lands owned individually by tribal members were handled by the Bureau of Indian Affairs (BIA), with land subdivided among individual heirs. Land ownership on the reservation today is a complex issue of multiple heirs (in some cases, hundreds of heirs) from single allotments and ownership by nontribal members, with only one fourth of the reservation actually belonging collectively to the members of the tribe.

Economically, the early self-sufficient farming and hunting economy was replaced by a temporary affluence caused by the 1887 Allotment Act that allowed Indian ownership, lease, and sale of reservation lands. An extremely complicated issue, the removals, loss of lands, and cultural challenges helped to generate many social problems and challenges. Despite these social problems, some aspects of reservation life based on tribal values of sharing and caring continued to grow and have been sustained, as manifested in the current tribal activities of powwows, feasts, gift giving, and other family and community events.

Currently, the tribal economy is basically dependent on the gaming industry. The tribe's Winne Vegas Casino, located in Sloan, Iowa, provides direct employment to more than 400 and indirectly to 100 members. The Tribal Council also has set aside additional revenues in other investment accounts to help prepare for the future of the Winnebago people (Anonymous, 1996). The revenue from its gambling casino has allowed the tribe to expand its land base by reacquiring some of the lands within the reservation previously lost to nonmembers.

TRIBAL HISTORY OF LAND USE

A history of land use and land use issues is very important in explaining and under-standing why decisions were made by a particular society, and why that society pos-sesses what it does now. The Winnebagos were a woodland tribe, and horticultural activities in a forest setting have always been a part of their tradition. Before the com-ing of the Europeans, the Winnebagos were hunters and gatherers of natural products for their food, shelter, clothing, tools, and weapons, and some bands of the tribe also became horticulturists (Smith, 1996). They were one of the northernmost agricultural tribes, and even with Wisconsin's limited growing season, they were able to grow three types of corn together with other products such as beans, squash, and tobacco. Fishing and hunting supplemented their agricultural produce. In the fall, they used dugout canoes to gather wild rice from the lakes in the area (Smith, 1994).

The Winnebago Reservation in Nebraska is a rural agricultural area. The 1887 Allotment Act attempted to convert the Winnebagos from the traditional horticulture-type of land use to the ways of the white farmers, (e.g., more large-scale farming). Eventually, the tribe lost almost all of the good farmland to the white settlers. The remaining agricultural land for the most part has been leased out. Most of the remain-ing tribally owned land is forested, not good for agriculture, and located on the east-ern side of the reservation near the Missouri River. The tribe once lived near the river and started migration toward the town of Winnebago only in the 1940s. In a 1992 study (Rule et al., 1994a), as part of a scoping process to obtain information about tribal feelings regarding the use of a piece of tribal land near the Missouri River, members of an Iowa State University interdisciplinary research team talked with elders and other members of the tribe. Some tribal elders reminisced about the time when certain species of trees or grass grew aplenty within the reservation. Others dis-cussed how medicinal herbs and other plants had been cultivated before and were in abundance for the people. Although it was suggested that land cessions and removals after short settlement in several areas may have caused the tribe to move away his-torically from horticulture to hunting as their primary endeavor (Smith 1996), results of this later study strongly indicated the tribe's strong affinity with and attachment to the land.

AGROFORESTRY AS A DEVELOPMENT TOOL
FOR THE WINNEBAGOS OF NEBRASKA

The traditional philosophy of land stewardship and a society's special relationship with forestry favored a certain type of possible developmental activities that are desir-able to the Winnebagos. Developments associated with the use of their land resources are acceptable only if they contribute to the improvement of tribal life socially, economically, politically, and environmentally. One of these developments is agroforestry, which belongs to a more comprehensive realm of activities called "social forestry," described by Gregersen (1988) as a broad range of forest-related activities that provide products and services and income for the local community. On tribally owned lands, agroforestry systems may be used to address broader social needs of the tribe, such as the creation of avenues for the tribe to use in passing down intergenerational knowledge, projects for the youth, and opportunities for small family and school garden projects.

Agroforestry is a land use option that has been increasingly identified as an environmentally sound and potentially sustainable system. Several factors determine the sustainability of agricultural land use practices, including agroforestry. York (1988) classified these factors into three groups: biologic, physical, and socioeconomic and legal. The interaction of these three groups of factors often constitutes the basis of all production systems. The desirability of a land use system may be evaluated using a simple rule often associated with accepting an innovation: the system must be biologically and technically feasible, economically viable, socially acceptable, and politically responsible (Rule et al 1994b).

Agroforestry systems provide many opportunities for meeting social, cultural, and economic needs. A main objective in agroforestry projects, as Mercer (1993) described, is to increase the efficiency in rural resource use through reduction or elimination of ecologically destructive land use practices and by introducing new or improved agroforestry enterprises to produce increases in income and living standards that are sustainable. Agroforestry may have its roots in the early 1900s, when Smith (1914) started to advocate greater production through mixing agricultural and forestry practices. However, only in recent years has agroforestry been recognized as a valid land management practice in North America. Although all three traditional types of agroforestry systems (agrisilviculture, silvipasture, and agrisilvipasture) exist (Rule et al., 1994a), the most common, especially in the Midwest and the Great Plains, are riparian buffers and windbreaks designed to meet soil conservation needs (Schultz et al., 1995). On tribal lands, such as those of the Winnebago Tribe, agroforestry systems could provide more than income and resource conservation. They also could offer opportunities for the social needs of these societies, such as settings for passing on intergenerational knowledge, including avenues for preserving and perpetuating indigenous knowledge and information.

THE INITIATIVE

One of the most important factors for the exploration of possible different types of land use for many native nations in North America has been the change in the status of sovereignty that has occurred in the past 20 years. In the early 1990s, revenues from the gaming industry helped to revive the Winnebago Tribe's ailing economy. In addition to other newly created investment accounts, the revenue from its gambling casino has allowed the tribe to reacquire some of the lands within the reservation previously lost to nonmembers, thus expanding its land base. This precipitated community-based decision making for the tribe concerning how and for what purpose its lands shall be managed.

In 1989, the tribe contracted the Johnson-Trussel Company to direct a series of workshops to aid the tribe in developing an interim land use plan. Several land and resource goals were articulated specifically addressing the protection and development of the tribe's renewable resources including forests, water, and the Missouri River corridor; the development and enforcement of tribal environmental standards; the restoration of tribal involvement in agriculture and the maintenance of agricultural values in crops, and the use of reservation lands to develop an economic base for the tribe.

DETERMINING BEST LAND USE: FIRST STEP TO DEVELOPMENT

PLANNING FOR THE BEST LAND USE: A FEASIBILITY STUDY

Starting in 1990, an interdisciplinary team (IDT) of forest biologists, rural sociologists, forest soil scientists, and forest economists studied the feasibility of converting a large tact of intensively cropped tribal land to an alternative use. (For more complete information and results, see Rosacker et al., 1992 and Rule et al., 1995.) The property, known as Big Bear Hollow, is a 1255-acre agricultural area that has been leased to a farmer, who uses it for irrigated and dry-land corn and soybean production. The area is located on the Missouri River floodplain about 7 miles east of Winnebago, Nebraska (Figure 13.1). Big Bear Hollow is part of a 2372-acre parcel of land owned by the Winnebago Tribe of Nebraska. This parcel is the largest contiguous unit of land owned by the tribe. The Glover's Bend area to the east and the bluff land to the west of the agricultural land make up the remainder of the parcel. The area in agricultural production is approximately 1 mile wide from east to west and 2 miles long from north to south.

The study focused on comprehensive planning, assessment, and evaluation of alternatives for conversion of this 1225-acre Missouri river bottomland site currently in agronomic production. The study complemented the Winnebago Tribe's strategic planning efforts as stated in their 1989 land use plan. The study was designed to assess and develop feasible alternatives for converting the site to the best land use that allowed for the highest attainment of the Winnebago Tribe's social, economic, and environmental goals.

The feasibility study had eight specific objectives:

1. Identify the issues and concerns of the Winnebago Tribe of Nebraska and the Winnebago Agency
2. Establish the decision criteria to be used for evaluating alternatives that will be developed for the conversion of the 1200-acre site to forest crops or a combination of forest and agronomic crops, over a 10-year period
3. Assess the capabilities of existing natural and human resources, as well as the socioeconomic and environmental aspects of current and alternative

agronomic and forest crop production practices, with special consideration
of the loss in revenue that occurs from any site conversion activity

4. Determine the ceremonial and cultural use of the site
5. Assess the nature of "rights to land" in the site, including allotted lands
6. Develop alternatives for the conversion of the 1200-acre site to forest crops
   or a combination of forest and agronomic (agroforestry) crops, including
   short-rotation woody crops (SRWC) for energy and fiber, over a 10-year
   period
7. Evaluate the economic trade-offs, benefits, and costs, and describe the
   social and environmental consequences associated with each alternative
   considered, including a study of the market potential for SRWC as energy
   and fiber resources
8. Recommend the best alternatives from the set of feasible alternatives

METHODS

A comprehensive land-use planning process was used. The following steps were
completed with input from the Tribal Council and the Winnebago Agency-Bureau of
Indian Affairs (BIA):

1. Identify issues and concerns to determine primary goals
2. Determine decision criteria
3. Collect data
4. Assess resource capabilities
5. Formulate alternatives
6. Evaluate alternatives (determine effects including benefits and costs,
   impacts, and trade-offs)
7. Recommend best alternative(s) to the Winnebago Tribe of Nebraska
8. Select the best alternative
9. Implement the best alternative
10. Re-evaluate the alternative

The last three steps of the decision-making model were to be completed by the
Winnebago Tribe of Nebraska and its governing body, the Tribal Council.

The study was divided into three phases, as discussed in the following sections.

Forming Objectives

Understanding the needs and concerns of the Winnebago people was the starting
point in determining the appropriate objectives for the study. The IDT met with the
Winnebago Tribal Council and personnel from the BIA to discuss and refine the state-
ments of concerns. This phase actually "framed the study" by identifying the objectives. Some of the objectives were competing and others complementary. The
objectives were those things the Winnebago people hoped to achieve to the maximum
extent possible given the constraints on the natural and cultural resources available.

The criteria to be used for measuring the achievement of the objectives were formulated during this phase as well.

Assessing Resources

Various resources of the Winnebago Tribe, the BIA, the Winnebago tribal elders, the Nebraska Department of Conservation, the U.S. Department of Agriculture-Soil Conservation Service, and the U.S. Army Corps of Engineers were used to determine past and present agronomic use at Big Bear Hollow. Further more, various research reports related to forestry, agroforestry, and farming were reviewed. A scoping assessment allowed for the development of the initial set of alternatives to be considered. The scoping assessment entailed assembling the objectives, decision criteria, resource capabilities, and other planning considerations into an initial set of alternatives.

Evaluating Alternatives

The evaluation of alternatives explicitly considered (1) a 10-year, phased-in implementation schedule, (2) establishment and production of a black walnut plantation, (3) the foregone annual crop output and cash rent from the current agronomic activities, (4) a 55-acre tree border and filter strip planting installed during the spring of 1991, (5) short-rotation woody crop plantation establishment and production, (6) agroforestry plantation establishment and production, (7) the impact of the 1990 Food and Agriculture, Conservation, and Trade Act (FACT Act), the Forestry Incentive Program (FIP), Agricultural Conservation Program (ACP), and state-level policies on agronomic, agroforestry, and forestry crops, and (8) the impacts of the Army Corps of Engineers' Missouri River management plan.

Each alternative was evaluated and compared in terms of qualitative and quantitative effects. Risk and uncertainty were explicitly considered. These effects were transformed to allow for summation of all effects from alternatives and comparison of all alternatives.

FEASIBILITY STUDY RESULTS

The IDT made three formal presentations to the Tribal Council. The first formal meeting was held to solicit their issues and concerns for the purpose of establishing the objectives. The second meeting was convened to gather input from the council on the scope of the initial set of alternatives, and to report on some of the initial resource assessment results. During the summer 1991, the IDT visited the Big Bear Hollow site several times to gather site-specific data on soils, cropping practices, and surrounding natural vegetation, and to consult with BIA personnel and tribal members. A third meeting in December 1991 involved elders and Tribal Council members in assigning weights to previously identified objectives.

A draft resource assessment report was presented to the Tribal Council in June 1991. After editorial and review comments, the IDT completed the resource assessment report 2 months later. Thus, the first two phases (forming objectives and assessing resources) were completed.

Working with the Tribal Council, the IDT developed a set of 20 objectives grouped into four categories: economic, environmental, social, and institutional/political. The Tribal Council expressed the desire to achieve a combination of social (employment), environmental (reduced impacts from agronomic activities), and economic (annual net cash flows) objectives. The economic output from the existing agronomic cropping of corn and soybeans was considerable. The Winnebago Tribe had (and has) been renting the agricultural land at Big Bear Hollow to a non-Indian farmer under a long-term contract. The rental agreement was worth more than $100,000 of annual income for the tribe. The council and elders expressed an interest in production of black walnut sawlogs and veneer, production of Indian corn and berries, and production of hybrid cottonwood trees for woody biomass yielding energy and fiber. The potential also existed for enhancing the developments planned by the Army Corps of Engineers and the Papio-Missouri River Natural Resources District for the river-edge portion of Big Bear Hollow called Glover's Bend. Both recreational and wildlife opportunities existed on site, especially given the other forested lands along the bluffs and the flood plain of the Missouri River.

During the third meeting in December 1991, the IDT presented an overview of the assessment report and then led the council and elders through a nominal group process (NGP) to establish weights for the previously articulated 20 tribal objectives. The council and elders elected to add another objective entitled "building continuity increasing the total to 21." The NGP of developing weights for the 21 objectives entailed several rounds of assigning weights by each member present at the meeting, discussing the values, and obtaining an overall group ranking for all 21 objectives. Social and environmental objectives topped the list of ranked objectives. Table 13.1 presents the 21 objectives and their nominal group rankings.

SEVEN ALTERNATIVES AND FIFTEEN MANAGEMENT REGIMES

Seven alternatives were developed representing a set of feasible alternative land uses for the Big Bear Hollow study area. Starting with the alternative 1, the current cropping situation or status quo, the remaining six alternatives represented a progress toward natural forest plantings and away from intensive agronomic cropping. Alternative 2, a slight variation on alternative 1, would retain the current cropping situation except for about 400 of the 1200 acres converted to black walnut, green ash, cottonwood, and silver maple trees. Alternative 3 would retain two center-pivot irrigation units south of the main blacktop access road for corn and soybean production. It would remove 400 acres in the northern part of the area for agroforestry and short-rotation woody crop production. Alternative 4 would retain only one center-pivot irrigation unit. The agronomic crop would be alfalfa, which would be shipped to the pelletizing plant on the Santee Reservation north and west of Winnebago and be converted into fish food. Approximately 400 acres would be planted to bottomland forest species including black walnut, and the remaining 400 acres would be devoted to agroforestry, Indian corn, a containerized tree nursery, and short-rotation woody crops. Alternative 5 would remove the alfalfa production and retain the agroforestry, Indian corn, containerized tree nursery, and short-rotation woody crops production. More land would be converted to bottomland forest, with outputs of timber, wildlife,

TABLE 13.1

Nominal Group Rankings for the Objectives Associated with the Winnebago Forest and Agriculture Alternatives Feasibility Study

Objectives	First Group Ranking	Final Ranking
To minimize health risks associated with farm chemical use	75	87
To build continuity	87	87
To enhance the water rights of the Winnebago Tribe of Nebraska	69	79
To enhance attainment of the goals of the Winnebago land use plan	66	78
To enhance the quality and quantity of water coming from the area	64	76
To enhance wildlife habitat	63	75
To enhance the complementary nature of the project with Glover's Bend	64	75
To enhance the diversity of plants/animals of the bottomland ecosystem	62	75
To foster the tribal philosophy of land use	61	72
To reduce soil loss	57	71
To enhance soil fertility	57	70
To foster long-term support of tribal goals and objectives by the BIA	59	69
To maintain about the same cash flow as from leasing the property	75	68
To minimize loss of income from converting the site to other land uses	58	68
To enhance the transfer of intergenerational tribal knowledge	57	66
To foster hunting and fishing opportunities for the tribe	58	66
To promote long-term opportunities for adult employment	54	61
To foster educational opportunities for the entire tribal community	50	60
To promote opportunities for seasonal youth employment	46	58
To foster long-term support of tribal goals and objectives by the Army Corps of Engineers	55	47
To develop recreational opportunities for tribal members	43	40

BIA-Bureau of Indian Affairs.

and recreation. Alternatives 6 and 7 would provide predominantly nonmarket goods (recreational and wildlife opportunities and outputs), with some marketable crops. Alternative 6 would establish two demonstration plots in a cooperative agreement with the Center for Semiarid Agroforestry located in Lincoln, Nebraska. Alternative 7 would call for the complete conversion of the 1200 acres back to native bottomland forests.

Combined in various ways, 15 management regimes were used to create the seven alternatives, which were then analyzed using a common set of 25 decision criteria. Each management regimen had an estimated number of acres and a specific set of activities and timing. Shelterbelts planted in 1991 on 55 acres of the 1255-acre site were included in each alternative.

EFFECTS AND DECISION CRITERIA

The 25 decision criteria were subdivided into four broad categories (economic, environmental, social, and political/institutional) for purposes of discussion. Table 13.2

TABLE 13.2
Matrix of Effects for All Criteria Across the Seven Proposed Alternatives

Criteria and Unit or Type of Measure Used	Alternatives						
	1	2	3	4	5	6	7
Yearly cashflow ($) (undiscounted)	90,000	67,500	45,000	104,440	104,440	0	0
Present net worth ($) (discounted @6%)	−601,120	−876,030	−900,698	−238,858	−385,864	−1,540,983	−1,604,630
Full-time jobs (no. of job opportunities)	0	0	1	2	2	2	1
Part-time jobs (no. of job opportunities)	0	1	2	10	6	4	2
Jobs for the youth (no. of job opportunities)	0	10	20	67	60	28	28
Well yield (acre-feet per year)	450	450	300	287	0	0	0
Crop water use (acre-feet per year)	2,219	2,334	2,430	2,901	2,601	2,598	2,609
H_2O from Missouri River (acre-feet per year)	80	0	0	0	0	0	0
Soil loss (index)	−10,800	−8,400	−6,421	−2,308	−1,012	−1,051	−600
Organic matter (index)	3,000	4,914	5,725	8,497	10,692	10,772	11,875
Bulk density (index)	2,260	4,377	5,477	7,866	10,528	10,625	10,635
Nitrogen fertilizer (pounds)	−72,000	−54,000	−36,000	−4,920	−4,920	−4,493	0
Pesticide danger (index)	−8,871	−6,657	−4,480	−1,094	−15	−496	−17
Species richness (no. of representative plant/animals)	6,000	12,000	12,225	17,947	29,017	29,280	30,000
Game/wildlife habitat (habitat value)	6,992	5,584	5,690	5,830	6,556	6,613	6,771
Educational opportunities (index)	0	4	5	10	10	7	8
Complementary developments (index)	0	2	3	8	8	8	8
Intergenerational enhancement (index)	0	1	4	10	10	8	8
Recreational use (no. of people participating)	0	352	665	957	1,270	1,265	1,304
Tribal control of land use (index)	2	4	5	10	10	10	10
Hunting activities (index)	290	270	274	157	92	53	14
Match with Tribe's land use plan (index)	2.5	4	5	6	6.5	7	6
Funding, BIA ($)	0	48,210	184,701	214,581	185,287	186,412	192,840
Funding, others (index)	0	5	6	10	10	10	10
Building continuity (index)	10	9	5	0	2	6	9

BIA-Bureau of Indian Affairs.

presents the summary of the "raw" effects (resource outputs, criteria values, responses, and magnitudes), as determined by the IDT, based on the possible outputs, responses, and contributions for each alternative in the attainment of specific objectives.

The environmental criteria measured the outputs and impacts for the seven alternatives. Farming activities in alternative 1 would have high well yield, high soil loss, more nitrogen fertilizer, and more pesticide use. The last three criteria certainly would imply more possible pollution and erosion. On the other hand, conversion of the whole area to trees (alternative 7) would have very low environmental impacts and high effects or values for species richness, game/wildlife habitat, organic matter, and bulk density criteria. From the standpoint of a natural environment, high values for these criteria and low values for the pollution/erosion-related criteria are better.

The economic criteria measured net economic benefits and annual cash flow. More emphasis was placed on elements or activities that would provide returns to the people such as farming income and jobs. On the basis of these criteria, alternatives 4 and 5 would provide the largest magnitudes in terms of annual income and jobs than any other alternative. The expected high economic returns were predicated primarily on projected income from the Indian corn, the nursery, and the berry patch. Many jobs would be generated by these activities and related forestry activities. Alternatives 1 and 2 would have lower annual cash flows than alternatives 4 and 5, not generating any jobs for the people of Winnebago. The benefit/cost ratio would be high for alternative 1 mainly because of the lower cost projected with respect to this alternative. Generally, the alternatives offering more diversity (more management regimes) tended to be better than other alternatives in terms of the economic criteria.

The social criteria measured responses dealing with educational and recreational concerns as well as internal cultural interactions. Except for the two criteria on hunting and building continuity, the alternatives with forestry activities scored better (had higher magnitudes) for this particular group of criteria. For example, alternatives 4 and 5 would have both natural environment and limited agricultural activities, higher magnitudes for educational and intergenerational enhancement, and better complementarity with the other projects (i.e., Glover's Bend) in the vicinity. These alternatives, however, scored low in building continuity, primarily because of the higher level of complexity and technology requirements implied by the many management regimes included in these two alternatives.

The political and institutional criteria measured the opportunities for outside funding and technical support for the alternatives. Hence, these options, especially alternatives 3 to 7, scored high on the funding criteria. Funding sources included the BIA and several other programs under the FACT Act of 1990. The two specific programs under the FACT Act that seemed to offer both cost share and annual rent income for forestry management regimes were the Conservation Reserve Program and the Environmental Easement Program. Alternatives emphasizing the development of a natural forest environment achieved better values; the Winnebago land use plan was considered. Alternatives 4, 5, 6, and 7 scored high in terms of matching the land use plan and the funding criteria. As with the social criteria, the complexity and new technology requirements of alternatives 4 and 5 caused lower scores as far as the criterion of building continuity.

To allow for a comparison across all seven alternatives, and considering all 25 decision criteria simultaneously, all raw effects were converted to z scores (Canham, 1990). This allowed all criteria effects for each alternative to be summed, yielding a total composite score for that alternative. The unitless z scores for a specific criterion, say present net worth, describe how far that effect is from the mean of all effects for that criterion relative to its standard deviation. (For procedures and results of converting the raw effects into z scores for each criterion, see Rule et al., 1995.)

RANKING OF ALTERNATIVES BASED ON FOUR WEIGHTING SCHEMES

Table 13.3 shows the ranking of the seven alternatives using the four weighting schemes. These different schemes were chosen to highlight the different perspectives that could be taken to evaluate the set of proposed alternatives relative to the 21 objectives. These schemes were nominal group process weights, equal weights, environmental priority, and economic priority. Several criteria were established to measure the attainment of the objectives identified by the tribe as important to them. In most cases, only one criterion was used to measure achievement of an objective. However, in some cases, two criteria were used to measure one objective. In these cases, each criterion was assigned one half of the weight for that objective. For these four schemes, the "best" land use alternative would be that with the greatest total weighted score.

Using the NGP scores, alternative 5 was ranked first and alternative 4 second. Using equal weights for all decision criteria provided rankings similar to those with the NGP. This result was somewhat surprising. Using a weighting scheme with emphasis on the environmental criteria, alternative 7 was ranked first and the status quo (alternative 1) last, whereas with emphasis placed on economic criteria, alternative 4 ranked first and alternative 5 was a close second. This was expected because of the expected cash flows and employment opportunities for both alternatives.

TABLE 13.3
Ranking of Seven Alternatives Using Four Weighting Schemes

Weighting Schemes	Ranking						
	1st	**2nd**	**3rd**	**4th**	**5th**	**6th**	**7th**
Nominal group[a]	Alt. 5	Alt. 4	Alt. 7	Alt. 6	Alt. 3	Alt. 2	Alt. 1
Equal weights[b]	Alt. 5	Alt. 4	Alt. 7	Alt. 6	Alt. 3	Alt. 2	Alt. 1
Environmental priority[c]	Alt. 7	Alt. 5	Alt. 6	Alt. 4	Alt. 3	Alt. 2	Alt. 1
Economic priority[d]	Alt. 4	Alt. 5	Alt. 1	Alt. 3	Alt. 2	Alt. 6	Alt. 7

[a] Tribal weights as determined from the nominal group process (NGP) are applied.

[b] All objectives are given equal weight.

[c] Environmental objectives are given a weight of 10, all others 1.

[d] Economic objectives are given a weight of 10, all others 1.

Alt. = alternative.

On the basis of the four weighting schemes, alternatives 5 and 4 seemed to be the best land use alternatives. Both alternatives were expected to provide a diverse set of agronomic and forestry products while providing important social and environmental benefits. In short, these agroforestry alternatives provided greater attainment of the 21 objectives than the other five alternatives.

Formal risk consideration was limited to the current net worth criterion with a 6% real discount rate. Informal risk and uncertainty elements were incorporated into each alternative by downward adjustment of expected yields and prices for risky management regimes. For example, in alternatives 4 and 5, the nursery regime was considered risky because it required high-level technical and managerial skills, large front-end costs, and uncertainty associated with the market for nursery products. Thus, the cash flows and current net worth were reduced by the assumed risk. No sensitivity analysis was applied to the effects.

FEASIBILITY STUDY CONCLUSIONS

As an initial attempt to determine the best land uses for Big Bear Hollow, a piece of tribal land owned by the Winnebagos of Nebraska, a multistep decision process was applied. Seven feasible land use alternatives were developed and evaluated. Development of tribal objectives and assigning of weights were both very important events in the decision process. It was clear that the tribe desired a complex set of social, environmental, and economic objectives to be achieved from Big Bear Hollow. On the basis of transformed effects and the four weighting schemes that varied the importance of the decision criteria, the best land use alternatives for Big Bear Hollow involved agroforestry production (alternatives 4 and 5) that yielded tree (wood, fiber, and nut), grain (Indian corn), forage, berry, wildlife, educational, and social/political benefits that addressed the tribe's expressed goals. Although no alternative from this set was adopted, the study led to the identification of a smaller-scale agroforestry system, which was begun in 1993.

CONTINUING COMMUNITY EFFORTS
TOWARD DEVELOPMENT

DEVELOPMENT OF THE AGROFORESTRY DEMONSTRATION PROJECT

In 1993, the development of an agroforestry project became the joint project of the Winnebago Nation, the U.S. Department of Agriculture (USDA) Forest Service National Agroforestry Center, the BIA, and Iowa State University. The project was to determine the site(s) for an agroforestry demonstration and then to design and develop the system. Specifically, the project was to establish an agroforestry demonstration to identify feasible agroforestry systems based on social, economic, and environmental criteria with a special emphasis on rural development. Site identification was facilitated through a workshop and a series of meetings between Winnebago administrative tribal personnel (land management, natural resources, water resources, and the Tribal Operating Procedure for Land Acquisition [TOPLA]), the Tribal Council chairperson, and outside project cooperators (USDA Forest Service and county extension, BIA, and Iowa State University). The final site selected was a 55-acre triangular section of land located in the southern portion of Big Bear Hollow.

A series of informal and formal meetings with key individuals (Winnebago Tribal members and personnel from land management and Little Priest Community College) and focus groups was used to define the goals for the demonstration project and to direct the overall type of demonstration desired by the tribe. Goals for the Big Bear Hollow demonstration followed guidelines set up by the tribe's interim land use plan (Winnebago land use plan, 1991) while incorporating alternative land uses. Specifically, the agroforestry system was designed to (1) protect and aid in the further development of the tribe's natural resources, (2) provide opportunities for the restoration of tribal member involvement in a horticultural/forestry-related system, (3) allow for sustainable land management systems to provide economic opportunities for the tribe and its members, and (4) facilitate intergeneration opportunities and employment for youth. Various demonstration designs were presented first to a focus group of tribal members and then to a larger group, including Winnebago land management staff, to help define the type of system desired. Results from the focus group sessions brought out two very important elements that were considered in the development of the agroforestry demonstration. One element was the recurrent interest expressed in growing blue/Indian corn as part of the agroforestry demonstration, and the other was the involvement of the school or youth.

The resulting tentative design for the agroforestry demonstration was a black walnut intercropping system, which would be put into place slowly over a 5-year period to allow for change. Black walnut, an indigenous bottomland area species, would be planted at a spacing of 66 ft, with sweet clover, blue/Indian corn, and organically grown soybeans rotated in the alleyways. The resulting system would provide a nut crop, high-value wood, Indian/blue corn for educational and cultural needs, organic soybeans for soil improvement, sweet clover for soil improvement, and fodder for the Winnebago bison herd. All of these were also income-generating opportunities. The demonstration also provided a mechanism for a desired conversion of the area back to forested land.

COMMUNITY PARTICIPATION IN THE EFFORT

To link culture and the community to the agroforestry demonstration, and to determine how the agroforestry system would fit into other land use needs, a participatory rural appraisal (PRA) was conducted in 1995. Studies using rural appraisal for a wide range of natural resource purposes can be found in the literature for resource economics (Pretty and Scoones, 1989), resource planning (Scoones and McCracken, 1989), and community forestry (Messerschmidt, 1991; Molnar, 1989), with participatory approaches such as PRA now becoming basic approaches in rural development.

The PRA took place over a 2-month period, with planning occurring on site. The PRA team consisted of five members: four Winnebago tribe members (two members from the Winnebago Land Management Department and two interns from a local Indian community college) and one non-Indian member from outside of the community. Five main geographic areas were the focus of the PRA: (1) newly acquired lands in the western portion of the Winnebago reservation (currently leased or in the USDA Conservation Reserve Program), (2) the village area of Winnebago, (3) an area located along the Missouri River (4) the wildlife refuge, and (5) the bison refuge.

During the initial stages of the PRA, brainstorming was used to determine the information needed and the means for obtaining this information for each area, and to generate a list of participants from existing Winnebago community groups. Seven groups representing different gender and age classes were contacted to determine participation interest. Representatives from each group and the Winnebago community were contacted personally and invited to participate in a community survey planned over a 5-day period. Each day during this period, certain activities for the community survey were to be carried out at each site. Participants were to work in groups and use diagramming, flowcharts, and pie charts to examine issues of concern for each land area. An orientation day before the 5-day study period was held to inform potential participants about activities planned and to answer questions.

Only one participant came to the orientation. After an impromptu brainstorming session, a more informal and less intimidating format for the community survey was chosen to increase community participation. The community survey was "retooled" as "Tour the Rez," with 4 to 12 participants visiting all five of the study sites for each of the 5 days. At each site, a PRA team member would give a brief description of the site and moderate any discussions that followed. Tape-recorded comments and participant feedback from the tour were used in planning questions for use in the second phase of the PRA.

For the second phase, because there was a reluctance to share opinions in a public format, an informal questionnaire titled "Continuing the Circle" was chosen. The focus of the questionnaire was on determining ways to incorporate indigenous knowledge into a decision-making model for land use preference. The informal survey, conducted at four different locations and times on the reservation, had roughly equal numbers of men (41%) and women (59%) participants. A total of 246 participants took part in the informal survey, representing about one fourth to one fifth of the Winnebago community: 69% Winnebago tribal members, 24% other tribes, and 7% non-Indian. Participants at each location were asked voluntarily to fill out a short survey comprising 33 questions that represented current and future land use. For the Winnebago community, informal personal interviews overlaid with an informal questionnaire format provided a means for participants to share in the process of land use planning.

The last phase of the PRA used a matrix to rate preferences for crop plants and trees and link these to social and nonmarket values. Various plants, trees, and horticultural products were scored on a scale ranging from 0 (not important) to 10 (most important) according to their importance for spiritual and cultural values, as a food source, and for teaching youth. Matrix rankings were collected from a total of 30 participants during informal interviews that included Tribal Council members and community members. The PRA revealed a strong preference for keeping the Missouri River corridor for wildlife purposes. Wildlife purposes ranked highest as first preference of land use for the river area, with spiritual uses ranking only slightly lower. There was a strong connection between wildlife and spiritual values that related to use of the river area. Whereas wildlife had a direct connection with forestry and recreation uses for the Big Bear Hollow area where the demonstration was located, it also had strong cultural and spiritual connections. Embracing this perspective, 70% or 160 respondents chose to remove this area from agriculture and return it to a woodland type of land use.

Overall interest in horticultural activities was moderately strong (59%), with more interest in family type gardens (84%) and cottage industries (70%). Trees were rated as strongly important for both cultural (61%) and wildlife reasons (79%), and only moderately important for economic reasons (55%). For horticultural crop preference, flint/Indian corn ranked very high (68%). Indian corn's strong cultural values were reflected in the strong to very strong importance attached to its cultivation (67%), and in preference rankings that were highest for cultural and spiritual values as well as opportunities to teach the youth.

The highest priorities of land use for the tribe/community were housing (78%), education (51%), and health facilities (51%). The PRA identified each individual tribal member's need for information on current land issues and individual land ownership (30% were unaware of the type of land use lease for their lands). In 1996, the data and results were presented to the Winnebago Tribal Council and community through the Tribal Department of Land Management.

USE OF THE PRA TO LINK CULTURAL AND OTHER RESOURCES TO THE PROJECT

The interpretation of information varies from one community to another. Use of the participatory process helped to incorporate indigenous or local knowledge into planning, and it allowed for culture and belief systems to direct the ways in which information could be collected and used. The PRA provides a mechanism for connecting culture to the biologic and economic realities that communities such as Winnebago face when making land use decisions. Linking resources, such as infrastructure with a community-based agroforestry system, means looking at the culture fabric of how these types of systems fit together. It helps in understanding the linkages between indigenous or local groups (e.g., drum groups, Language and Culture Department, Native American Church, Winnebago Bison Project) with respect to the agroforestry demonstration, and also provides input into the ongoing changes to the design of the demonstration.

EXPERIENCES GAINED AND LESSONS LEARNED

Planning for the Winnebago Tribe's agroforestry project provided some learning opportunities as well as some insights into the context that land use systems such as agroforestry have within the larger context of land planning issues. Issues of housing, education, and medical facilities are challenges that face all communities and the agroforestry system, challenges that must be addressed within the context of these other and sometimes competing land use issues.

The NGP initially involved the tribal members in identifying and weighing the importance of their collective goals with regard to the uses of a particular piece of tribal land. The PRA helped a lot in balancing two big items, considering the tribe's culture: (1) the collective interests in acceptable (sustainable and usually long-term) land use alternatives for their tribal lands and (2) the day-to-day shorter-term problems in the community that required attention.

For the Winnebagos, the whole planning process provided a forum that gave tribal and community members a learning opportunity and a new awareness of the tribe's land management and acquisition activities. During the same process, the Winnebago Tribe and community also taught a valuable lesson to outsiders, giving researchers, who were nontribal members, a new awareness of land use viewed in the context of the community's unique sociocultural and historical fabric. History, ecologic use, spiritual values, and cultural adaptations all connect to aid in providing the Winnebago Tribe with a direction and a vision of desired land use. This connection can perhaps be best and most eloquently addressed in the words of Rueben Snake Jr., past Tribal Chairman at the Winnebago 130th Homecoming Pow-Wow Celebration Program, 1996:

> The Indian today is faced with a unique situation. On the one hand, a dominant overwhelming culture permeating the life of the individual with its rules and ideals, and on the other, a meaningful philosophy, and culture vitally necessary to his existence as an Indian. When most people talk of re-establishing an Indian culture, the immediate response is, "Shall the Indian go back to living off the land?" That is hardly possible. Society, even if it wanted to, could not afford or allow that to happen. That land as it stands today could not support the Indian. Re-establishing Indian culture does not mean wearing braids and feathers. It does not mean demanding concessions from a society that will not grant them anyway. Being Indian is not merely a physical appearance or material gains but a way of life, a philosophy, a state of mind, a spiritual fulfillment which makes an Indian an Indian.

THE TRIBE

The PRA and the development of this agroforestry project can be viewed as part of an ongoing process of changing land use. By addressing issues surrounding land acquisition and by bringing an awareness of land management activities to tribal members, the Tribal Department of Land Management was able to address a broad spectrum of interrelated concerns that, at the same time, took account of the community's diverse interests in terms of land uses.

Perhaps the best positive impact on the tribe from the appraisal was an awareness of the land acquisition process and the identification of land use issues. For instance, the NGP brought together, for group discussion and evaluation, a collection of tribal objectives regarding the use of Big Bear Hollow. Through the NGP, some kind of goal prioritization for Big Bear Hollow had been established. Then the PRA identified individual tribal members' needs for information on current land issues and individual land ownership. For many community members, it was their first chance to see some of the different types of land use being considered on the reservation. Because of an inward migration into the Winnebago town area over the past 50 years, and because three fourths of the reservation land had been lost to non-Indian people, some community and tribal members were surprised at how far the actual boundary of the reservation extended. Among the PRA activities conducted was the "Tour the Rez," which offered many members an opportunity to visit the reservation as a whole and learn directly about current and proposed types of land use, including the agroforestry demonstration area.

In essence, the PRA provided the tribe's land management unit a way to engage the tribe and community actively in the process of land use planning. Although much of what the PRA revealed was not new information to the tribe's land management department, it did support a lot of issues that they dealt with on a daily ongoing basis. One issue, for example, was the need for increased awareness of location and leasing purposes of individual tribal member ownership on the reservation. The PRA did not provide a specific action mechanism for the tribe to use in adopting an agroforestry project. Rather, it provided the tribe a venue for its inclusion and its consideration in land use planning and decision-making processes.

THE RESEARCHERS

From the researchers' perspective, the feasibility (initial) study, the development of the agroforestry demonstration, and the PRA provided unique learning experiences on how a proposed agroforestry system could fit into the Winnebago Tribe and community, considering the tribe's sociocultural-historical framework. More directly, it helped to bring awareness to the researchers of the possible fusion of land use needs, community values, and evaluation methods with criteria to be used. One of the biggest lessons learned was that it is not only a question of how market and nonmarket benefits of an agroforestry system should be measured, but also a question of choosing the appropriate yardstick with which those benefits must be measured for results to reflect truly the tribe's value system and to fit their needs.

Earlier in this chapter, it was suggested that the cultural aspects of a society define individual preference structures. It also was indicated that sociocultural differences could give rise to valuation problems, especially when another system (e.g., Euro-American) is used to assess the worth of the benefits of a land use system (e.g., agroforestry) to an indigenous people such as the Winnebagos. If indigenous peoples consider land as a means to sustain human society and the environment as an extension of their being, it is not surprising then that to the Winnebagos the importance of the extended family is well placed in the context within which decisions involving the larger family are made. For instance, the value for certain agricultural commodities is no longer dollars but a way to provide generosity because the Winnebagos are indifferent to the accumulation of individual wealth and property based on their tribal values of sharing and caring. In contrast, mainstream society has placed more value on individuality and financial success. The scoping process that involved nontribal researchers talking with elders and other members of the tribe, and the more formal NGP used to elicit the members' feelings as to what was important to them provided significant and very valuable information to the nontribal researchers, particularly in regard to identifying and gauging acceptable alternative land uses for Big Bear Hollow in the feasibility study initially.

Learning to listen or learn is another valuable lesson for outsiders participating in PRA (Chambers, 1997). One of the most striking examples was the understanding of how indigenous knowledge is linked to the tribe's descriptive and learning processes. For instance, during the first few phases of the PRA, the researcher conducting the process wrote lists and recorded information in a linear fashion regarding how and what to include in the PRA during the initial planning stages. It was not until after community participation in this discussion failed that the researcher, sitting and surrounded by linear posted newsprint lists, realized that not one person from the tribe explained information in this vertical manner. What happened was that both elders and children used circles to illustrate examples of these relationships graphically. For the researcher, this new awareness required a shift in the thinking process, resulting in the emergence of a very different world view.

For the Winnebago community, the relationships between components (e.g., wildlife and Indian corn, opportunities for tribal members to be involved in land use, culture, and basic human needs such as housing) in a land use system are seen as an interconnected circle. These relationships are not represented in a linear fashion. Termed as "continuing the circle," the circular symbol represents a way of relating indigenous knowledge to preferred land use and of relating to the world that is Winnebago. For outsiders (e.g., the researchers) who are accustomed to linear processes of problem solving and decision making, failure to recognize the importance of these indigenously linked decision-making processes means finding solutions for which there are no problems.

THE LARGER COMMUNITY

The use of the scoping mechanisms (NGP and PRA) involved tribal participation and inputs into the planning process. These research processes guided the researchers in their quest to provide the best assistance to the tribe in terms of what the tribe needed. Use of these processes was based on the view that only the tribal members themselves, with their accumulation of past and present knowledge, could know best what they want, what they have, what is good for them, and what their preferences are in going about the business of getting what they want.

The PRA, in particular, brought forth the adaptability of PRA tools and their use in the research process (Szymanski et al., 1998). For the Winnebago community, informal personal interviews overlaid with an informal questionnaire format provided a means for participants to share in the process of land use planning. For the PRA process, although some techniques requiring direct participation in a group environment are good and applicable in some communities, the total approach may not be appropriate or comfortable for every community. Therefore, some modifications may be necessary. For the Winnebagos, an indirect approach to participation through an informal survey worked best for the community in general, but could be overlaid with group participation techniques for younger members of the community.

Because the PRA was used for both community development and research purposes, matching direct participatory techniques with other research methods could offer the best world to both the researchers and community. With the Winnebago PRA, an informal survey provided a wider range of community involvement when direct participation was not an option. The PRA was a vital step in the research process, offering a chance to learn how to learn from the community. One of the greatest values of using a PRA is its adaptability as a tool to fit the cultural dynamics of a particular community.

Techniques used to evaluate agroforestry projects must account for differences in economics and environmental issues and show how these may be combined to fit a particular culture. Linking indigenous knowledge to land use means recognizing culture and belief systems and how people relate to the land. That is, community solutions can be found best within the framework of the community's own local knowledge system.

CONCLUSION

The agroforestry studies conducted with the Winnebago Tribe of Nebraska provided a unique window of opportunity to show the all-important interconnectiveness of people and land. The studies promoted a Native American group's social and economic development program by providing an avenue for many of its tribal activities and rites, by enhancing its pride in managing the land in accordance with its philosophy of land stewardship, and by providing economic and other social benefits in the process. It involves different sections of the tribe, making it a more dynamic land use planning process, reflecting a truly tribal activity for the Winnebagos. For the most part, it does not introduce any product or resource that is alien to the land or to the landowners. Instead, it provides an option to make use of what they have, but in new combinations to meet their needs. For the outsiders (i.e., researchers), this provided a unique learning experience of finding solutions to a problem in the natural resource arena and getting the individuals concerned (the tribe) involved in the whole land use planning process.

REFERENCES

Adamowicz, W., Beckley, T., MacDonald, D. H., Just, L., Luckert, M., Murray, E., and Phillips, W., In search of forest resource values of aboriginal peoples, the applicability of nonmarket valuation techniques, in *Rural Economy,* Staff Paper 94-08, University of Alberta, Edmonton, Canada, 1994.

Anonymous, *History of the Winnebago Tribe of* Nebraska, Winnebago 130th Homecoming Celebration Program, July 25–28, 1996, Winnebago, Nebraska, 1996, p. 12.

Canham, H. O., Decision matrices and weighting summation valuation in forest land planning, *North. J. Appl. For.,* 7, 77, 1990.

Chambers, R., *Whose Reality Counts? Putting the First Last,* International Institute for Environment & Development, London, 1997.

Gregersen, H. M., People, trees, and rural development: the role of social forestry, *J. For.,* 86(1), 22, 1988.

Johnson-Trussell Company, *The Winnebago Tribe Land Use Plan,* Prepared for the Winnebago Tribe of Nebraska. Albuquerque, NM. 1989, 23.

Mercer, D. E., *A Framework for Analyzing the Socioeconomic Impacts of Agroforestry Projects,* Forest Private Enterprise Initiative Working Paper No. 52, Research Triangle Park. NC: USDA Forest Services Southeast Forest Experiment Station, 1993.

Messerschmidt, D. A., *Rapid Rural Appraisal for Community Forestry: The RA Process and Rapid Diagnostic Tools,* Technical Paper No. TP 91/2, Institute of Forestry, Nepal, 1991.

Molnar, A., *Community Forestry: Rapid Appraisal,* Forestry Note 3, Food and Agriculture Organization, Rome, 1989.

Office of Native American Programs, *HUD, Brief statistical information.* Available: http://www.codetalk.fed.us/winebago.html, Accessed: December 18, 1998.

Pretty, J. N. and Scoones, I., *Rapid Rural Appraisal for Economics: Exploring Incentives for Tree Management in Sudan,* International Institute for Environment & Development, London, 1989.

Rosacker, J., Colletti, J, Rule, L., Faltonson, R., Skadberg, A., Stubben, J., and Wray, P., Winnebago Forest and Agriculture Alternatives Feasibility Study: Final Report, Iowa State University. Ames, IA, 1992.

Rule, L. C., Colletti, J. P., Liu, T. P., Jungst, S. E., Mize, C. W., an Schultz, R. C., Agroforestry and forestry-related practices in the Midwestern United States. *Agrofor. Syst.,* 27, 1994a, 79.

Rule, L., Colletti, J. Faltonson, R., Rosacker, J., and Ausborn, D., Evaluating conversion of cropland to forests., *J. For. Econ.,* 1(3), 1995, 329.

Rule, L. C., Colletti, J. P., Faltonson, R. R., Rosacker, J., and Ausborn, D., Resource analysis research to compare the economic, environmental, and social benefits of agroforestry options, in *Proceedings of workshop on Potentials for Agroforestry in Northern Mexico,* Saltillo, Mexico, April 12–15, 1994b, p. 28.

Schultz, R. C., Colletti, J. P., and Faltonson, R. R., Agroforestry opportunities for the United States of America, *Agrofor. Syst.,* 31, 117, 1995.

Scoones, I. and McCracken, J., *Participatory Rapid Rural Appraisal in Wollo: Peasant Association Planning for Natural Resource Management,* International Institute for Environment & Development, London, 1989.

Smith, D. H., The issue of compatibility between cultural integrity and economic development among Native American tribes, *Am. Indian Cult. Res. J.,* 18, 177–205, 1994.

Smith, D. L., Ho-Chunk tribal history: the history of the Ho-Chunk people from the Mound-Building Era to the present day, *Winnebago Tribe Neb.,* 1996, p. 88.

Sultzman, L., *Winnebago history, Web Site, Lee Sultzman's Compact Tribal History.* Available: Hyperlink http://www.dickshovel.com/win.html), Accessed: December 18, 1998.

Szymanski, M., Whitewing, L., and Colletti, J., The use of participatory rural appraisal methodologies to link indigenous knowledge and land use decisions among the Winnebago Tribe of Nebraska, *Indigen. Knowle. Deve. Monitor,* 6(2), 3, 1998.

Whitewing, L., Personal communication with Winnebago Land Management, *1991 Winnebago Land Use Plan,* Winnebago Land Management Department, Winnebago, NE, 1997.

Winnebago 130th Homecoming Celebration Program, Winnebago, Nebraska, July 25–28, 1996.

York, E. T. Jr. 1988. Improving sustainability with agricultural research. *Environment,* 30 (9), 1988, 18.

14 Innovation in Indigenous Production Systems to Maintain Tradition[1]

Maria de Lourdes Barón and David Barkin[2]

CONTENTS

Sustainable Management, Indigenous Population, and Tradition 212
Avocado Production and Waste 213
The Medicinal Effects of the Avocado in Diets for Humans and Hogs 214
Hog Raising as Part of Purhe'pecha Traditions 215
An Alternative Strategy for the Rural Economy 217
Building the Basis for Sustainability: The Introduction
and Adaptation of New Technologies 218
References 219

For centuries, backyard animal raising has been a central element in a diversified strategy for community consolidation in peasant societies around the world. Transnational corporations have systematically undermined this strategy by imposing new technologies that make small-scale family units uncompetitive and unviable (Suárez and Barkin, 1990). With the introduction of new genetic stock better suited to intensive feeding and factory-like reproduction and fattening, new poultry and hog production technologies, including new genetics, are displacing the local races of those animals that are more efficient in processing household and small-farm waste streams and require more time before they can be marketed.

In the authors' search for strategies to promote sustainable regional resource management, they found that hogs fattened with avocados have lower blood-serum cholesterol levels and produce meat with a low fat content. By introducing small modifications in traditional diets for backyard animals, backyard hog raising is being

[1] This project was financed by the Mexican Council on Science and Technology (CONACYT), in collaboration with the School of Veterinary Medicine at the Universidad Micoacana de San Nicolás de Hidalgo and the Hospital Civil Miguel Silva in Morelia Michoacam. Lic Nora Vargas Contreras participated in the field work for this study. Dr. Mario Alvizouri directed the medical and dietary research.

[2] Maria de Lourdes Barón is Professor of Rural Development at the Universidad Autónoma de Chapingo, Morelia Campus, and David Barkin Professor of Economics at the Universidad Autónoma Metropolitana, Xochimilco Campus. For comments, access barkin@cueyatl.uam.mx.

encouraged as a complementary and profitable activity that would strengthen the regional economy and the role of women as a new social force.

To implement the project, the authors began working with an umbrella group that encompasses approximately 350 communities and more than one half million people in west central Mexico who share a common ethnic heritage (Purhe'pecha, or Tarascan as they were called by the colonial settlers). The research and community work proposes to use local agricultural wastes that lower costs to create a quality product ("lite" pork), for which a premium price can be obtained. As part of the project, the communities are responding by undertaking a broad series of complementary environmental cleanup measures to improve living conditions in the region.

SUSTAINABLE MANAGEMENT, INDIGENOUS POPULATION, AND TRADITION

Students of sustainability critically maintain that while generating wealth for a few, existing models of development are creating poverty among the masses. This process undermines the viability of rural communities, with their rich social and cultural traditions that developed productive systems to ensure provision for their basic needs. In rural areas, poverty forces people to abandon centuries-old traditions of ecosystem maintenance, because their search for employment often compels them to migrate from their communities. Now there is evidence that if successful rural management strategies are able to ensure better living conditions and higher incomes, the rural poor not only will care for the environment, but will undertake the tasks needed to protect their scarce natural resources (Barkin, 1998a; Toledo, 1995).

Today's problems have their roots in settlement patterns created during the colonial period (16th to 19th centuries). As the invaders expropriated the best lands, indigenous populations found themselves relegated to increasingly marginal ecosystems. These areas frequently were very different from their original places of settlement, and the natives were obliged to pay tribute to their conquerors when they were not enslaved. These changes were not new, however, because commerce and war were a common element in even the most ancient of societies (Wolf, 1982). After independence, the indigenous groups continued to be pushed to increasing inhospitable and fragile areas, just as colonization schemes transferred peasants to the tropical rainforests.

Indigenous communities continued under increasing pressure. Their living conditions deteriorated as their production systems demanded more from the land. They produced crops for human consumption on their rainfed lands, developed handicrafts and other artisan products, and raised animals and horticultural products, including hogs, chickens, fruits and herbs, in their backyards. The most fortunate among them were able to protect their access to other natural resources, such as a lake or river for their water needs, and fishing, and to a forest for wood or hunting. Over the decades, they accumulated a rich experience in managing these resources, developing sophisticated management systems that were integrated gradually into their customary practices. They continued trading activities among themselves and with others, maintaining and modifying their traditions, adapting them to changing conditions, strengthening their communities and their identity, and choosing to protect their most cherished values and practices in each historical moment.

This process is crucial because it incorporates innovation as a permanent part of social practice, as a means to maintain and even to reinforce tradition. One recent example of a change in productive activities to protect a valued tradition among people in the ethnic group with which the authors are working is the case of women who decorate cloth (unraveling the threads in attractive ways) to make *huanengos,* a traditional blouse open on the sides. They recently modified their techniques to produce blouses and dresses for visitors, because they noted the demand for their skills in decorating Western styles of clothing, without necessarily modifying the way they dress in their own communities.

Nowadays, the Purhe'pecha people, like other indigenous groups throughout Mexico, are attempting to exert greater control over their natural resources as well as their economic and political life. As they acquire a greater capacity for self-governance, their social and political organizations have begun to develop strategies to support demands for more local autonomy and productive diversification. The innovation in hog raising reported in this chapter exemplifies a means of implementing change to maintain and strengthen tradition.

AVOCADO PRODUCTION AND WASTE

Avocados are endemic to Mexico, one of the world's most important centers of avocado production. The high plateau of Michoacan is the main avocado-producing area in Mexico, accounting for approximately 80% of national production. The dominant commercial variety, Hass, a hybrid introduced into the region at the end of the 1950s, has been growing rapidly ever since. This variety is tasty, with high oil content and a thick skin that facilitates its transport and marketing. In 1997, almost 900,000 tons were produced (Stanford, 1999). Despite Mexico's predominant position in world markets, supplying about 45% of all avocado exports, they represent less than 10% of domestic avocado production. (Figure 14.1). Mexicans consume almost eight times as much of the fruit as do people in the wealthy countries.

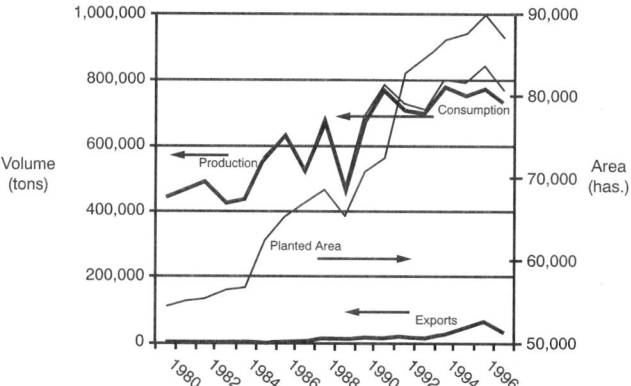

FIGURE 14.1   Avocado Product, Consumption, Planted Area and Exports, Mexico, 1980-1997. [From the Center for Economic Statistics, Secretary of Agriculture, Livestock, and Rural Development (SAGAR), *Consumos Aparentes de la Producción Agricola,* Mexico City, Mexico: SAGAR, electronic database].

Not everyone can participate in the export bonanza, however. Although export prices are often substantially higher than local prices, the financial risks and higher costs required to comply with strict export standards often are beyond the means of many growers, especially those with smaller orchards. These orchards range in size from 1 to 5 hectares, in contrast to commercial orchards that can exceed 500 hectares. There are about 6000 producers growing the fruit on approximately 90,000 hectares.

Because most local production is for the domestic market, considerable imbalances exist between demand and supply. As a result, substantial volumes of the fruit are never shipped. In the packinghouses, where fruit is crated and prepared for local and export markets, large volumes are rejected or spoil. Historically, much of this fruit was left to rot in the fields, but with increasingly strict phytosanitary regulations, this is no longer possible for those growers interested in the lucrative export market. More recently, the packing houses have resorted to dumping the unsalable fruit in nearby ravines. Adding to the environmental burden are the leftovers from several agroindustries in the region that use low-quality fruit for guacamole and oil for industrial, medicinal, and cosmetic purposes. Local experts estimate, conservatively, that more than 10% of avocado production could be used as animal feed.

Although avocado trees produce fruit only for several months, the variety of altitudes and microclimates in the region allows for a harvest during 10 or 11 months each year. The harvest, the processing, and the shipping schedules now are virtually permanent activities because the growers have been able to adjust their harvests and relocate production to ensure a continuing flow of fruit (Tuminaro, 1998).

Avocado production has brought an important increase in water and agrochemical use, competing with the region's traditional pine forests. Heightening environmental pressures, the industry also has generated an important increase in demands for wood to make the packing crates in which the fruit is shipped. Clandestine logging became a serious problem that was resolved partly by introducing plastic crates and cardboard boxes for export. The impacts on the water table have been so great as to create water shortages in some communities, including San Lorenzo, one of the villages selected for the pilot study. Many of the larger orchards are introducing drip irrigation to reduce their demands in response to rising water charges and to exhortations for a more careful use of the resource. As a move to improve economic conditions in the poorer communities, the use of surplus avocados for fattening hogs is expected to encourage the adoption of complementary ecosystem management activities and other changes in community environmental practices that will advance the process of developing a sustainable resource management program in the region.

THE MEDICINAL EFFECTS OF THE AVOCADO IN DIETS FOR HUMANS AND HOGS

On the basis of extensive trials, Dr. Mario Alvizouri, a member of the research team, has been offering clinical services to people with hypercholesterolemia for more than 25 years (Alvizouri et. al., 1992). His early observation that pigs fed with avocados did not store body fat in solid form (lard) led to a long-term research and treatment program. Because the animal fat maintained a viscous quality and had a chemical structure similar to the oils produced from olives, with a high content of Omega 3, he deduced that avocados could become an effective food in a treatment program. Although he had never systematically studied the precise way in which the avocados affected the pigs, this unusual response was common knowledge in the region, with butchers actually imposing a penalty on people selling animals fattened on the fruit because of their lack of lard, a valuable subproduct.

The initial project design was based on the use of this vernacular knowledge to develop more precise dietary guidelines that would allow local families to raise pigs profitably by using waste avocados and other local products as fodder in place of grains that had to be purchased. Because there already was considerable evidence that hogs fed on avocados experienced substantial changes in their metabolism and fat content, the initial project was designed to determine the optimal diets and to explore the best means for introducing these changes into the communities. With this knowledge and the scientific evidence that the meat was lower in cholesterol than commercial pork, it was clear that the product could be marketed at a premium price to a rapidly growing segment of the market willing to pay for healthier foods.

HOG RAISING AS PART OF PURHE'PECHA TRADITIONS

Hog raising was introduced into the region by the Spanish conquerors during the 16th century. The Purhe'pechas had been forced to introduce this activity into their communities as part of the tribute they paid to the colonial governors. The pigs were sent to feed the slaves in the mines and the Spaniards themselves.

Although there is no precise history concerning the introduction of pork into local diets, it is clear that it happened quickly. Pork and almost every other part of the hog became important parts of regional cuisines throughout Mexico, and to this day, Michoacan is justly renown for its *pozole* (a soup prepared with hominy corn, pork, greens, and seasoning) and *carnitas* (pork fried in its own fat), two succulent dishes traditionally reserved for feasts, especially at Christmas and the *Candelaria,* as well as during the corn harvest. Lard is also important because it is used to cook *mole* and *corrundas,* two other highly cherished dishes in ceremonial meals that now are eaten quite frequently.

Even today, hog raising continues to be an important part in the organization of the corn harvest, which generally lasts for several days in each household. In communities where traditional leaders still play an important role in social and productive organization, such as San Lorenzo, the family usually offers pork as the central part of the main meal of the day to the people who help in harvesting the corn. The workers also receive a sack of corn as a symbol of gratitude in lieu of money payments, which are not common in corn harvesting.

The continued importance of these traditions among the Purhe'pecha is quite remarkable. Since the 1940s, repeated attempts by successive governments to integrate indigenous peoples into the national culture have eroded some traditions. Nevertheless, in Michoacan, many natives have preserved successfully what they found to be important, such as their mother tongue, traditional markets, productive activities, clothing (particularly women's clothing), and even artisan products that are useful in local homes and attractive as souvenirs for visitors. Along with the products, these traditions and the inherited local governance structures have survived to varying degrees in the different communities.

Families usually keep one or two animals per year, sometimes including a sow to produce piglets. There are two approaches to raising hogs in the region. In many communities, the pigs are left free to rummage in the streets and fields of the communities, eating whatever they find. After harvest, the pigs, together with large animals, may be allowed to scavenge in the fields among the cornstalks and other leftovers. This has been a common approach, especially when the community was too poor to fatten the animals with corn and household wastes, or when no drinking water and drainage were available. In this setting, the pigs could usually be induced to return to their sties regularly at night by being offered some succulent rations.

A second approach involves maintaining the animals in the household compound, usually confined in stalls. With this technique, they generally will fatten more rapidly because their activity is limited, but the cost is greater because the pigs usually are fed corn along with cooking and harvesting wastes, and some form of green fodder when the corn is scarce.

When new breeds of hogs were introduced in the late 1960s, however, the industry changed dramatically, affecting local groups in Michoacan and elsewhere in Mexico. The new breeds of animals thrived on grains, usually enriched with nutritional additives and antibiotics as a prophylactic measure. Agribusinesses were established to construct large-scale intensive feed barns where the hogs were fattened under a very strict regime to minimize the time required and to maximize the rates of conversion (the increase in live weight per pound of grain fed the pig). A new crop, sorghum, was introduced in Mexico as the basis of this diet because of its high yields, ready mechanization, lower price, and lack of government control (Barkin and DeWalt, 1988). As long as grain prices remained low, this new technology allowed meat prices to fall, and small-scale farmers could continue to raise hogs, replacing their locally bred animals with pureblood lines. When the prices of grain and other inputs rose, however, along with the cost of the piglets, and when feed-lot technology generated greater productivity, the small-scale traditional producers were squeezed from the market. They could not compete with the economies of scale that lowered costs substantially (Suárez and Barkin, 1990).

The introduction of new genetic lines and feeding practices sounded the death knell for backyard hog raising for the market. In the process, the peasant economy itself was weakened, as yet another source of income disappeared. In small rural communities throughout the country, families were forced to search for new sources of income, as the diversified family farm and subsistence production became untenable strategies for survival. Urban-based jobs in construction and services and migratory farm work in export agriculture elsewhere in Mexico were attractive alternatives, along with the search for work in the United States.

This new strategy became particularly evident in another of the authors' study communities, Caltzontzin, a new town created for the people displaced from a village that had been destroyed by the eruption of the Paricutín volcano in 1943. Traditional community institutions and leadership came under severe pressure, and the people rapidly made the transition to a semi-urbanized economy, in which agriculture and livestock production were relegated to a marginal status. Here, the urgent need for new sources of income induced people to become teachers and look for any available employment. Despite these pressures, however, many people in the new town reintroduced traditional systems of cultivation, and today are exploring the possibility of participating in the project with waste avocados from nearby packinghouses.

As national policies to accelerate the pace of modernization affected an ever-widening sphere of activities, and as policies to speed the pace of international economic integration replaced the strategy of import substituting industrialization, small peasant villages were particularly affected (Barkin, 1990). The pillars of their economy for the previous half century were dramatically undermined, placing great pressures on their community organizations and social structure. Throughout the country, the village economy was impoverished, and the growing gap between urban and rural populations was producing a profound transformation in all dimensions of life in Mexico.

AN ALTERNATIVE STRATEGY FOR THE RURAL ECONOMY

The authors' research in the avocado area of Michoacan pointed to the possibility of collaborating with Purhe'pecha communities in their efforts to stem the decline in their economy and sense of ethnic identity. By the mid-1990s it was clear that the national strategy of regional integration was incapable of offering attractive opportunities to many poor people in Mexico. Throughout the country, these social groups were actively searching for alternatives that would allow them to preserve many features of their traditional organization and the quality of life that small villages offered.

This search for alternatives to preserve the rural community became so powerful that it was transformed into a major new social movement (Barkin, 1998b). This movement assumed many forms, ranging from the dramatic uprising of indigenous groups in Chiapas in January 1994 to the less spectacular but widespread efforts by peasant communities to assume greater control of their local economies. They began organizing their own enterprises and developing new mechanisms to strengthen the rural economy. These initiatives were particularly striking in areas with large indigenous populations, such as the Purhe'pecha region of Michoacan, where a regional organization emerged in an attempt to implement new approaches for economic improvement and institutional consolidation.

In this setting, the possibility of introducing a productive innovation that would strengthen a traditional pillar of the family economy seemed particularly promising. Hogs were still significant in the local economy, both because of their importance in the corn harvest and ceremonial life and because of the ease with which they could be raised and sold in times of dire need. Despite the marginal contribution of hogs to monetary incomes, many families still raised them as a source of meat or as one part of a complex diversification strategy that has always been central to the logic of the peasant community.

The authors approached local leaders and community members with the proposal to introduce modifications in the diet of pigs raised in confinement. A strategy was offered that would lower costs while producing a higher-quality product that could command a premium price in the market. The proposal was designed to improve the local economy by recycling waste products and making traditional activities profitable. By focusing on fattening pigs, an activity that had become central to communal organization, the authors expected that the proposal would contribute to the strengthening of local cultures. Because backyard activities were the domain of women, the authors also expected that the proposal would contribute to increasing their participation in community management and economic life. Furthermore, because they were proposing that the hogs should continue to be fattened within the family unit to reduce possible adverse environmental impacts, it would be highly decentralized and require broad participation while avoiding contamination. Despite the proposal's apparent attractiveness and simplicity, however, significant resistance was anticipated because of the important changes involved.

BUILDING THE BASIS FOR SUSTAINABILITY: THE INTRODUCTION AND ADAPTATION OF NEW TECHNOLOGIES

The initial effort to develop optimal diets for raising hogs with low cholesterol levels proved successful, and the enthusiasm for the new technology exceeded expectations. Although the project is still in the initial trial stages at this writing, it is clear that the approach has been accepted, and that the main obstacle to its full implementation will be the need for people from the region to advise and supervise the quality of the diets and the conditions in which the pigs are raised in the backyard stalls.

Traditional hog raising, still an important part in local Purhe'pecha communities such as San Lorenzo, is being reintroduced in more acculturated villages such as Caltzontzin. There are a few well-trained people (including veterinarians) to promote its growth, and the community appears ready to welcome this innovation, precisely because it allows them to retain control over the production process.

In retrospect, the proposed innovation is proving relatively easy to implement because of a design that fits it into the existing structure of village life and political organization. Although based on a declining activity, the proposed changes are clear to all participants and the commercial logic compelling, especially within today's precarious rural economy. Because of the focus on an activity that women historically have managed and the declining presence of men who must seek work elsewhere, the project has struck a particularly responsive chord. Furthermore, with a growing awareness of the need to improve sanitary conditions resulting from improved channels of information and concern about health, the project also has created an opportunity to discuss environmental issues, such as water quality and sewage disposal and treatment.

As the production of "low-fat pork" moves from the experimental to the implementation stages, the authors expect to find a growing demand among villagers to participate in the new industry. From the perspective of sustainable resource management and popular participation, another attractive feature of the program is its limited scale: the volume of production is inherently limited by the supply of waste avocados. It would not be advisable or profitable to use commercial grade fruit as fodder for the hogs. In this way the authors hope to avoid the health and environmental problems that usually are associated with large-scale hog raising elsewhere. They anticipate that there will be sufficient production to merit the construction of a small, certified butchering facility operated by the organization of the Purhe'pecha communities, providing an opportunity to raise the quality of meat available in the region while producing the low-fat pork products for the specialized markets that they are developing.

REFERENCES

Alvizouri Muñoz, M., Carranza Madrigal, J., Herrera Abarca, J. E., Chavez Carvajal, F., and Amezcua Gastelum, J. L., Effects of avocado as a source of monounsaturated fatty acids on plasma lipid levels, *Arch. Med. Res.*, 23(4), 163, 1992.

Barkin, D., *Distorted Development: Mexico in the World Economy.* Westview Press, Boulder, CO, 1990.

Barkin, D., *Mexican Peasant Strategies: Alternatives in the Face of Globalization.* Presented at the XXI International Congress of the Latin America Studies Association, Chicago, IL, September, 1998b.

Barkin, D., *Wealth, Poverty and Sustainable Development,* Editorial Jus, Mexico City Mexico, 1998a.

Barkin, D. and DeWalt, B. Sorghum and the Mexican food crisis, *Lat. Am. Res. Rev.,* 23(3), 30, 1988.

Stanford, L., Dimensiones sociales de la 'organización agrícola: la producción de aguacate en Michoacán, in *Agricultura de Exportación en Tiempos de Globalización: El caso de las hortalizas, frutas y flores,* de Grammont, H. C., Gómez Cruz, M. A., González, H., and Schwentesius R., Eds., CIESTAAM y Juan Pablos Editor, México, 1999, p. 211.

Suárez, B. and Barkin, D., *Porcicultura: La Producción de Traspatio-Otra Alternativa,* Oceano Editores, Mexico City México, 1990.

Toledo, V. M., *Peasantry, Agroindustraility, Sustainability: The Ecological and Historical Basis of Rural Development,* Morelia, International Council of Sustainable Agriculture, Mexico, 1995.

Tuminaro, A., *Los Aguacates Michoacanos, Ubicación, Empacadoras y Plantas de Procesamiento* (unpublished manuscript prepared for this project), Uruapan, Mexico, 1998.

Wolf, E., *Europe and the People without History,* University of California, Berkeley, 1982.

15 Ethnicity, Multiple Communities, and the Promotion of Conservation: Strawberries in California

Daniel C. Mountjoy

CONTENTS

Multiple Communities and the Changing Agroecosystem 222
Community Conflicts, Alliances, and Strategies 224
Issue Identification and Agency Response 225
Targeting an Outreach and Technology Delivery Program to Diverse Farmers . .226
    Knowledge and Perceptions: Human Capital 226
    Information Networks: Social Capital 227
    Production Loans and Investment Strategies: Financial Capital 227
  Outreach Strategies: Providing Relevant Service 229
  Technology Selection: Providing Relevant Products 230
  Community Education and Planning Process 232
  Interagency Coordination Strategy 234
    Regulatory Coordination 234
    Coordinated Farming Assistance 235
Shaping the Future of the Land and the Community 235
References 236

Winter storms blowing off the Pacific Ocean into the Elkhorn Slough watershed of Monterey County, California, rarely deliver more than 2 in. of rain in a day (Monterey County, 1998). Yet, when this precipitation falls on a steep, newly planted strawberry field, a 3-foot-deep gully can easily form in the sandy soils in less than 1 hour. Erosion rates in the area are some of the highest in the nation at 33 tons of soil lost per acre per year (Soil Conservation Service, 1984). The mechanics of the erosion process are well understood, and technical solutions have existed for years, but the motivations of the farming community and the constraints on it have been overlooked

0-8493-0917-4/01/$0.00+$.50
© 2001 by CRC Press LLC

221

in the effort to promote sustainable management of the agroecosystem. This case study demonstrates how recognition and assessment of multiple social communities can result in an improved use of public technical assistance funds.

An underlying assumption about the general farming community pervades public policy discussion of technical and economic incentives to promote change in the agricultural sector. The assumption is rooted in the diffusion of innovation model and assumes that members of the farming community will respond to incentives in similar and predictable ways. They may not all respond in the same way at the same time, but eventually "they will come around." This view assumes that all members of the farming community are motivated by similar aspirations and have access to equivalent information and resources. The complex structure of agriculture communities in the coastal hills of central California reveals the problems of this assumption and has forced local residents, politicians, and government agencies to rethink the meaning of community.

MULTIPLE COMMUNITIES AND THE CHANGING AGROECOSYSTEM

The California agricultural community, if it can be referred to as a community, is the product of successive waves of immigration over the past 200 years (Iwata, 1962; Saloutos, 1975; Wells 1988, 1991). Each wave of immigrants brought new ideas and technologies for managing the natural environment. The arrival of the Spanish missions and cattle ranching in the late 1700s began to alter the traditional American Indian management of the coastal oak woodlands and prairies surrounding Monterey Bay. When California boomed during the gold rush of the 1850s, European immigrants introduced commercial agriculture in the fertile Pajaro and Salinas valleys that drain into the bay. Later, Portuguese immigrants used the less fertile hill lands for dairying. Advances in irrigation, processing, and transportation technologies resulted in an expansion of the local agricultural industry in the late 1800s, but growth was constrained by a shortage of farm labor. Periodic labor shortages continue to this day and have been met through recruitment of foreign labor. Initially, Chinese and Japanese laborers were imported to tend the vegetable row crops of the Anglo landowners. Successive sources of labor have come from the Philippines, Dust Bowl refugees, and, since the 1950s, from Mexico.

The strawberry industry, which today dominates the agricultural landscape of the hills around Monterey Bay, had its beginnings a century ago alongside the vegetable crops in the fertile valleys. The Anglo strawberry farmers relied on the efficient and reliable labor of the Japanese immigrants, but were alarmed by their ability to accumulate savings and begin buying land of their own (Saloutos, 1975). The Alien Land Law of 1913 put an end to this competitive threat and returned the control of limited valley land to the hands of Anglo landowners. However, Japanese farmers continued to expand the strawberry industry as tenants and sharecroppers. When the Alien Land Laws were repealed in 1952, the experienced Japanese farmers began to acquire small acreages in the available lower-priced coastal bluffs and hills. Former grazing lands were converted to intensively managed strawberry farms.

As new strawberry varieties and production methods were adopted, yields increased, and Anglo and Japanese farms demanded more labor than the family could provide. The federally managed Bracero program of the 1950s met the seasonal demand for farm workers by encouraging the migration of Mexican laborers. As these government-sanctioned laborers and later undocumented migrants, learned strawberry production methods from their employers, they did as the Japanese had done before them and began to go into business on their own as sharecroppers and tenants. They also sought low-cost hill land to farm and found landowners interested in converting steeper undeveloped grazing lands and woodlands in exchange for strawberry rents. The expansion to increasingly steeper lands was made possible by the development of drip irrigation and the presence of marketing brokers who sought to expand their own share of the growing industry by encouraging the development of new acreage. These brokers played a key role by offering production financing to Mexican laborers, who otherwise lacked social ties or collateral to obtain commercial loans to begin farming independently. However, many of the new farmers had learned to farm on the flatter valley bottomlands and were unprepared to manage the highly erosive soils of the hills.

The dynamics of the strawberry industry and the aspirations of the ethnic immigrants led to the rapid change in the use of hillside lands of the Monterey Bay area. The advantages of farming in the hills proved great as farmers discovered they could "farm the climate," taking advantage of local microclimates and achieving earlier production than on the valley floor. They also could enter the fields to weed or spray sooner after a rain because of the well-drained sandy soils. These advantages allowed two groups of ethnic newcomers—the Japanese and Mexicans—to achieve economic success and establish themselves as major sectors of the local community.

The impact of these new farming communities on the natural ecosystem has also been great. The cultivation of the hills has created many negative downstream consequences, including burial of roads and choking of streams with eroded sediment, flooding of homes, loss of riparian habitat, and the accumulation of agricultural chemicals in the sensitive wetland habitats of Elkhorn and Watsonville Sloughs.

Farming in the hills also has its costs to the farmer resulting from the higher production costs associated with preventing or cleaning up erosion damage. Two generations of Japanese farmers in the hills have had time to develop a stable farming system in which the higher costs of managing erodible soils are compensated for by lower rents. Only a few of the newer Mexican farmers have been able to earn enough to begin investing in sustained management of their lands. For some, strawberry growing is a relationship of accumulating indebtedness to their lender-shippers. The long-term aspirations of these farmers focus on remaining viable as independent farmers even though some earn no more than the laborers they employ. Their planning horizon rarely exceeds 3 years or the length of the lease they hold. As a result, the concept of investing in agroecosystem sustainability is abstract and irrelevant.

From the perspective of Anglo ranching family descendants, these immigrant newcomers are mining the soil for profit. For the long-time ranch families who have lived on the land for several generations, the Mexican farmers do not appear to manage or value the land as their home. The fact that two thirds of the Mexican farmers are tenants adds to this perception of transiency, although it is applied equally to most of the Mexican landowners. Most of the Mexican farmers have "homes" in the towns of Watsonville or Salinas, but frequently relocate their farming operations as leases expire and they encounter better rental terms on other parcels. Owning or holding a lease on farmable land is fundamental to a farm business and perhaps even more so to new immigrant farmers who previously have been unable to control the means of production. However, the value of the land to most of the hillside farmers is instrumental rather than intrinsic. This contrast in the value of land highlights the difference between the two communities as more than one of ethnicity. The long-time, land-owning residents form a "community of place," whereas the farmers who live in town participate in the watershed area as a "community of economic interest" physically located in the hills only as long as farming remains a viable source of income.

Several other communities of interest have increasingly played a role in the changing landscape of coastal watersheds. These include property-based interests, such as the environmental organizations and agencies that manage tracts of public and private lands protected by conservation easements, as well as a growing population of new rural residents who have elected to live in the area because of its rural character. One environmental group, the Elkhorn Slough Foundation, has been very effective in creating an extended community of place centered on the stewardship of the 1440-acre Elkhorn Slough Estuarine Reserve through an education and docent program for its members, most of whom live in outlying local cities.

Non–property-based interests also seek to have an influence on the way the region is managed even though they do not directly manage land resources. These include regulatory government agencies that control land management behavior through creation and enforcement of environmental regulations such as land use zoning, water quality protection, or endangered species protection. Members of this community claim legitimacy as guardians of the public trust through legislative mandate. Members of advocacy environmental organizations also fall into this category of non–property-based interests. They base their legitimacy in local resource management debates on a moral prerogative to protect interests that have no established political or economic voice. They are active in issues such as protecting school children from pesticide exposure or wildlife from development encroachment.

COMMUNITY CONFLICTS, ALLIANCES, AND STRATEGIES

The boundaries between these various communities are somewhat arbitrary and vary situationally. Community identification surfaces most obviously when resource management objectives and actions conflict. The most notable and ongoing conflict centers on the impacts of agricultural activities on the other land uses within the watersheds. Lowland ranchers and rural residents literally find common ground when agricultural soils from upland farms wash down over their properties, obstruct stream flow and cause flooding. Water quality threats from agricultural runoff to wetland habitats are a rallying issue for guardians of public lands and agency regulators of water quality. New rural residents are often alarmed by agricultural practices such as pesticide spraying, the traffic caused by farm laborers, and aesthetic changes resulting from development of new farm acreage. These nonfarming communities have a history of joining forces in response to ongoing or worsening farm management. Their interests overlap around the goal of preventing further natural resource degradation. Often, this collaboration takes the form of formal complaints to local policy makers to demand strengthening or enforcement of regulations.

Regulatory enforcement and fines are not the only strategies that local communities have used to control farming, but these strategies are appealing, because they shift the cost to the public agencies. In an era of shrinking public budgets and increasing resource problems, some community groups have given up on the slow pace of enforcement as a tool to direct change. A more immediate, and apparently more effective, tactic has been the use of private law suits, whether threatened or real. Groups of rural residents or downstream farmers have used law suits effectively to motivate upland farmers to reduce runoff and sedimentation. Unfortunately, these approaches have done nothing to improve communication among the watershed communities or secure long-term farmer commitment to the health of the watershed.

An alternative approach advocated by some is the use of financial incentives, education, and technical assistance to farmers to improve their farming practices voluntarily. Advocates of this approach argue that farmers will adopt better farming practices if given the technical information and if cost constraints are reduced. These are the primary techniques of the University of California Cooperative Extension and the U.S. Department of Agriculture (USDA) Natural Resources Conservation Service. Once implemented, these more sustainable farming practices will be better maintained because they benefit the farmer, not because they are imposed by regulation or threat of fines.

Advocates of the disincentive approaches (regulatory enforcement, law suits) and the positive incentive approaches (technical and financial) assume that, if their approaches are applied evenly to the farming population, all farmers eventually will come into compliance or modify production practices. The focus of both regulatory and technical assistance programs is on changing farming behavior, yet rarely are the diverse social and economic causes of existing behavior considered. In coastal California, historically derived conditions have led to several distinct farming communities rather than a single organized farming community that will respond in predictable ways. Programs, whether regulatory or incentive-based, applied in this setting will produce different results depending on the farming community. The Anglo farming community is the most integrated with the public policy arena, and public programs usually are designed with this group in mind. In contrast, the Mexican farming community historically has been overlooked, has less representation in the policy-making process, and often is not even aware of regulations or assistance programs intended to modify farming behavior.

ISSUE IDENTIFICATION AND AGENCY RESPONSE

In the early 1990s, ongoing environmental degradation from farmland erosion and runoff in the 70-square-mile Elkhorn Slough watershed galvanized the interests of the agency regulator community, the environmental community, and the rural residential community. In response, the Resource Conservation District of Monterey County together with the USDA Natural Resources Conservation Service (NRCS) proposed an 8-year watershed project to provide technical assistance to farmers and ranchers to reduce sediment delivery to Elkhorn Slough by 50%. Funding for technical staffing of the project was obtained in 1994 through the NRCS Small Watersheds (PL-566) Program. Instead of funds being appropriated to construct large sediment trapping structures at the base of the watershed, the emphasis was placed on hiring a multi-disciplinary staff to develop outreach techniques and technical products that were appropriate to the varied farming communities within the watershed. By recognizing the multiple communities involved, the project has been able to promote sustainable agriculture management.

Since 1994, the Elkhorn Slough Watershed Project has employed three inter-related approaches to work most effectively with the communities involved: (1) a farmer outreach and technical assistance program that is responsive to the needs of all the ethnic and economic farming groups (communities of interest), (2) a community education and planning process to work with ranchers and rural residents (communities of place), and (3) interagency coordination activities to pool the resources and interests of other agencies and organizational partners toward the diverse needs of the watershed communities. Each of these approaches is described in the following discussion.

TARGETING AN OUTREACH AND TECHNOLOGY DELIVERY PROGRAM TO DIVERSE FARMERS

The Elkhorn Slough Watershed Project is built on a socioeconomic approach to promoting sustainable farm management. The farming population in the watershed is economically and ethnically diverse, and no single outreach method or conservation technology will work for all the farmers. The first task in developing an adaptable outreach and technology delivery program was to assess the social and economic conditions of the targeted farming communities and understand how these factors influenced resource management. The results of an in-depth social assessment (Mountjoy, 1995, 1996) provided staff with key indicators of farmers' ability and interest in adopting natural resource conservation practices.

Of the strawberry farmers in the Elkhorn Slough watershed, 80% are of Mexican descent, and 14% are of Japanese ancestry. The remaining 6% are of European origin ("Anglos"). The melting pot of American culture has not blended these cultures despite many years of side-by-side farming. The ethnically distinct farming communities can be differentiated based on three distinguishing categories of characteristics.

Knowledge and Perceptions: Human Capital

The three ethnic groups are historically distinguished by the length of time they have lived and farmed in the Monterey Bay area and the knowledge they have acquired during that time. Virtually all of the current Anglo and Japanese farmers grew up in the area, whereas the Mexican farmers have arrived in the area as farm workers within the last 30 years. Only half of the Mexican immigrants speak English, and few have completed primary school, in contrast to the Japanese and Anglo farmers, some of whom have college degrees. Very few of the Mexicans ever received any form of environmental or science education, nor have they been exposed to the English language media coverage of environmental issues. As a result, few of the Mexican farmers are aware of the downstream impacts of their farming operations, and therefore are not motivated to control runoff because they perceive it to be "a natural consequence of farming."

The average Mexican farmer began farming independently in the mid-1980s, a period characterized by low rainfall. This limited historical perspective taught that preparation for rain-induced erosion was an unnecessary expense. Weather patterns have changed since then, and the 1990s have brought 3 years of 100-year-frequency storm events, but the farmers continue to manage their farms as if the drought years are expected to return. This lack of preparation is compounded by the tendency of Mexican farmers to change leases every 6 years on the average. Rapid turnover of ranches prevents farmers from getting to know the unique problems of the land and developing site-specific solutions.

Information Networks: Social Capital

The ethnic farming communities are further distinguished by the information networks they tap into for access to farming and personal advice. Anglos are the only group to make regular use of government-supplied farming information such as university research and USDA farming advice. Japanese farmers, although more integrated with the mainstream Anglo agricultural network than the Mexicans, also maintain uniquely Japanese social ties through religious and business association membership. As newcomers that do not speak English, many of the Mexican farmers have limited access to established public resources, and rely instead on personal information networks that link extended family members together to share the successes and setbacks of their common farming experiences. Most Mexicans are unaware of the services provided by public agency farm advisors and often mistrust any public officials. Representatives from the berry shipping companies and other agricultural product vendors recognize the financial opportunities of working with the Mexican farmers and serve as a limited conduit for outside farming information to this ethnic community. These sources have a financial interest in preserving their roles as information and product brokers, and therefore tend not to provide information on alternatives to chemical use, or on new crops and marketing strategies. As a result, reliance on the familial networks serves to reinforce existing knowledge.

Production Loans and Investment Strategies: Financial Capital

In a farm business context, information and access to knowledgeable advisors must be paired with financial resources. The ethnic farming communities differ in the way that they secure financial capital for production. Anglo farmers are well established in the area and often run corporate farms that can secure commercial bank loans at favorable interest rates. Japanese farmers on smaller acreages can sometimes self-finance from savings, but also borrow from commercial banks or personal friends. The Mexican growers lack financial experience or collateral to obtain commercial loans and usually are in business through a loan from a strawberry shipping company. The terms of these loans include higher interest rates and the provision that the farmer must sell the entire crop through the shipper. This arrangement prevents the farmer from shopping around for buyers of his crop and has led farmers to suspect that shippers are fixing the prices to make additional profit. Many Mexican growers in this financial arrangement fail to make a profit after deducting interest on the loan and a sales commission to the shipper for marketing the crop. As lenders, the shippers also dictate the production methods that a farmer must use. Many growers feel that they have become farm managers, or worse, "slaves" of the shippers because of these financial contracts. It is no surprise that these farmers have no cash surplus to invest in conservation practices or alternative production methods.

Together, the influence of different levels of human, social, and financial capital predicts variation in management styles within the farming community (Mountjoy, 1996). Management style refers to the "standardized cognitive precedents" (Bennett, 1978) that groups of farmers use to evaluate management decisions. Four farm management styles evolved in the Elkhorn Slough Watershed in response to the social and economic circumstances of distinct ethnic groups:

1. *Farming as a business:* Virtually all of the well-educated Anglo farmers have easy access to the latest production technology and research, and can secure loans for their large farms at favorable rates. These farmers have a management style that emphasizes and trusts the competitive, market-driven business of farming.
2. *Managing the details:* This management style is characteristic of established Japanese farmers on small acreages. Financial success has been achieved through careful management of inputs to achieve efficiency. This strategy eliminates the risk of depending on large production loans and is based on adequate information and the support of fellow farmers.
3. *Scraping by:* The newest farmers to the strawberry industry tend to be Mexican immigrants with limited education or access to information and a reliance on lender-shippers. These farmers are always short on cash and usually feel powerless to change their condition. Their management style encourages avoidance of risk, and change is resisted even though the current system is only marginally successful.
4. *Taking chances:* A number of more established Mexican growers have achieved economic success by taking risks to expand acreage or find new markets for their crops. Many of these farmers immigrated earlier or were born in California, and thus have larger social networks and better understanding of economic opportunities than the "scraping by" farmers.

Individuals with similar socioeconomic backgrounds have developed similar farm management styles, and individuals with similar strategies for adapting to the local human and natural environment tend to socialize with people who share common experiences and values. This has been the basis for identification with a common community. This assessment of management styles helped the Elkhorn Slough Watershed Project team to identify the predominant farming communities in the area and anticipate their attitude toward resource conservation.

When the watershed project began in 1994, several public meetings were announced and convened to collect public comment on the best strategies for preventing agricultural impacts to water quality. These meetings were well attended by agencies and a few rural residents, but representatives from the various farming communities were noticeably absent. Judging by the assessment of the farming communities, this lack of response was not surprising, because few of the farmers recognized off-farm environmental impacts as an issue worthy of attending a meeting. This lack of awareness of downstream effects is especially true of the "scraping by" Mexican farmers, who attempt to farm the steepest, most erodible slopes with the least financial resources available to control the resulting erosion problems.

OUTREACH STRATEGIES: PROVIDING RELEVANT SERVICE

The first challenge to the newly formed watershed project team was to promote the vision of protecting off-farm resources, as articulated by the regulatory agency and environmental community, but without the buy-in of the targeted farming communities. The social assessment provided clear guidance, however, on how to communicate with the farmers. An outreach campaign that emphasized environmental protection would not pique the interests of the farmers. Instead, to gain farmer involvement, the project has focused on saving the farmers money by reducing crop damage from erosion, eliminating gullies that blocked farm road access, and avoiding costly fines or lawsuits from neighboring property owners. A secondary goal, once the project had provided a useful money-saving service, has been to educate the growers about the environmental protection benefit of their efforts. One Mexican farmer who successfully reduced his erosion problem is now the first to point out the environmental benefits of his conservation system proudly when groups of other farmers are brought to tour the farm.

Another challenge that surfaced through the assessment of the farming communities was that there was no single farm forum that the project team could visit to promote the services and technologies the team had to offer. All of the Anglo and most of the Japanese farmers are members of the Farm Bureau and Strawberry Commission, but only 20% of the Mexican growers are members, and they virtually never attend meetings and rarely read the information mailed to members (Mountjoy, 1995). The first goal of the project was to establish communication with the Mexican farmers.

A variety of strategies have been used to establish communication with the Mexican farming communities. The long-term goal was to overcome the mistrust or lack of awareness of government services and build personal relationships with local farmers so that NRCS would be recognized as a source of relevant information and technical advice. To build this trust, the team worked with a few willing farmers to set up working demonstrations of affordable conservation systems. Through the use of bilingual field days, tours, and farmer referrals, additional farmers were exposed to the technologies and services available. In addition, the primary shippers in the area were contacted, and workshops were conducted for their farmers. The support of the shippers has been an important means of spreading information, because the shippers serve as a network hub through almost daily contact with their growers. New ideas often are reinforced internally because growers tend to interact more with other growers in their shipping group than with others.

Motivating the interest of the "scraping by" farmers has been the greatest challenge. This group is the most likely to lease land and change leases frequently, thus eliminating the chance to improve the farming system incrementally. One strategy to overcome this problem has been an effort to work directly with landowners who lease land to small-scale producers. It is in the best interest of the landowner to protect the soil resources on the land to ensure future rental income. Several landowners now are playing a key role in pushing their tenants to work with the NRCS as a condition of the lease.

The financial cost and risk of implementing a conservation system has been partially offset by the availability of federal cost-share contracts under the USDA Environmental Quality Incentives Program. A farmer can be reimbursed for up to 75% of the cost of a conservation system and up to $10,000 per year. Unfortunately, the farmer must pay for the costs up front, and the poorest farmers often cannot afford to wait for the reimbursements. The project team is working with the shippers, encouraging them to offer financing for conservation projects along with other direct production costs, because conservation investments protect productivity and profits.

After 4 years, outreach is no longer a priority. Most of the farmers in the watershed are aware of the services offered by NRCS, and requests for assistance now exceed the availability of staffing to respond. Five full-time staff members now provide assistance to approximately 50 farmers per year.

TECHNOLOGY SELECTION: PROVIDING RELEVANT PRODUCTS

The ultimate measure of a conservation project's success is not whether people *want* to improve their farming systems, but whether they *do* actually change their practices. The ability to match available technology to the natural resource setting and to the interests and management capability of the farmer is critical to the long-term adoption of a new practice. Instead of "selling" farmers the most effective "best" management practice, the Elkhorn project team encourages farmers to consider alternatives, tour existing systems, talk to other farmers within their community, and try, on a limited scale, practices that suit their land, financial capacity, and planning horizon. A high-cost pipeline or sediment basin that is highly effective at conserving soil over a 15-year life span is not the "best" management practice for a tenant farmer who is just scraping by. Instead, assistance in laying out the slope of the furrows and seeding an annual erosion control grass may be the most relevant technology for the situation.

The notion of incremental or partial adoption is a key in working toward sustainable farming systems. Most farmers are unwilling to invest the money and take the risk to convert the whole farm from familiar production methods to some new approach without first trying the ideas on a limited scale. As technical advisors, the NRCS staff welcomes any change the farmer is willing to make and helps the grower assess the benefits of the modified practice in comparison with conventional techniques. This flexibility and patience ultimately leads to more complete adoption of new farming practices.

The conservation practices recommended by the NRCS have been tested extensively and conform to rigid standards for materials and installation to ensure reliability. Unfortunately, farmers do not always want to implement the most reliable practice, especially when it costs more to follow the NRCS standards. Often, the grower has some old irrigation pipe, a 50-gallon drum, or other construction materials lying around the farm that can be incorporated into the design to save costs. The project team has remained flexible to the ideas of the farmers and continues to test new ideas and materials through research projects or field trials.

Another technical issue affecting adoption of more sustainable technologies is the farmers' concerns about the unintended side effects of an otherwise useful practice. One example has been the reluctance of farmers to plant erosion-control grasses and shrubs around the bare areas of the farm out of concern that these plants will attract insect pests. To achieve adoption of the erosion-control vegetation, the NRCS team worked with University of California Santa Cruz entomologists to show through monitoring that there are no new pest buildups. Better yet, this collaborative research showed that selected erosion control plants actually attract beneficial insects that can reduce pest populations as well as pesticide applications and costs.

By remaining aware of farmers' practical skepticism and responding to it, promoters of sustainable farming systems become more involved and trusted agents of change, providing better, more integrated farming technology. This approach in the Elkhorn Slough Watershed has resulted in an exponential increase in the number of farmers served and the amount of soil erosion prevented. In the first 4 years of the project, conservation systems were implemented on 38 farms comprising almost 1000 acres of farmland. The cumulative effect of the conservation practices implemented to date has been to prevent 20,800 tons of soil annually from eroding from the farms and into the streams and wetlands of the watershed (USDA, 1998).

Perhaps more noteworthy than the increase in the use of conservation practices is the fact that nearly 60% of the implemented projects occurred on farms operated by Mexican farmers. Only 5 years earlier, when the social assessment was conducted, none of the Mexican farmers who were interviewed identified a government agency as a trusted source of information. Through the use of personal contacts, community leaders, shipper networks, and promotion of relevant and affordable technologies, Mexican farmers now make use of NRCS services in proportion to their Anglo and Japanese counterparts. It must be noted, however, that the Mexican farmers from the "scraping by" community have been the least likely to participate, as was predicted. Reaching this group remains a challenge that will require the partnership of other organizations that can provide more immediately needed assistance such as alternative marketing, financing, and basic farming skills. This is discussed in the Interagency Coordination Strategy section.

COMMUNITY EDUCATION AND PLANNING PROCESS

A second approach that the Elkhorn Slough Watershed Project has used to promote enhanced management of natural resources is a neighborhood-based education and planning process involving groups of rural residents, ranchers, and lowland farmers. This approach is suited to working with communities of place to address common resource problems. Initially, the project team hoped to develop a sense of stewardship by engaging all rural residents and farmers in discussions of watershed-wide resource management issues such as erosion, flooding, and natural habitat protection. A Watershed Coordinator/Educator position was established in 1996 by the Resource Conservation District of Monterey County to facilitate these types of discussions and extend the services of the NRCS technical team. It was soon apparent that the diverse communities of the area did not have a common set of issues around which to organize. Instead, more manageable units for discussion and planning turned out to be subwatersheds or neighborhoods either hydrologically or socially defined.

Nature provided a catalyzing event for expanded communication with residents and landowners along one stream in the watershed during the heavy winter storms of 1996–1997. Carneros Creek became plugged by eroded sediment from surrounding farmland and caused flooding of homes and farms along the creek. The obvious flooding problem mobilized the energy of dozens of residents, and the Watershed Coordinator seized the opportunity to facilitate discussion of the issue and bring residents together as a community of common interest.

At the early meetings, participants focused on placing blame for the flooding problem on farmers for unmanaged upland erosion and on the county for inadequate maintenance of the Carneros "ditch." The primary goal for the assembled community was to dredge the ditch. There was confusion and anger at the changing governmental policy about ditch management. As long-time residents, many remembered the days when public agencies paid for dredging of the ditch, first for "land reclamation," and then regularly for "mosquito abatement" and "flood control." Now they were being told that environmental protection legislation introduced in the 1970s prohibited the cutting of trees to dredge the channel because of "riparian corridor protection" and "endangered species habitat."

The NRCS and the Watershed Coordinator met with the residents and representatives from numerous local government agencies to begin an education process to introduce other related natural resource management issues before the group began clearing the creek. Specialists were invited to address the group and explain topics such as sediment transport, stream flow dynamics, water quality laws and regulations, the effect of riparian vegetation, and groundwater recharge. The purpose of these presentations was to provide the residents with adequate information for evaluating the full ramifications of management decisions they might make for the creek. Gradually, the definition of the problem shifted away from personal complaints about flooding to recognition of watershed management as an issue of community concern.

The group also came to the decision that the best way to resolve the many interrelated problems was to work together as an association. A core group of multigenerational resident families had worked together before, but many new relationships were established. Despite their common sense of place, many residents had never met their farming neighbors, and therefore had not functioned as a community. They discovered the power of numbers when attempting to get the county's attention. They formally organized themselves into the Carneros Creek Association and registered as a California Coordinated Resource Management Planning (CRMP) group under a memorandum of understanding with numerous federal, state, and local government agencies. The CRMP status conveys the understanding that the group is involving all stakeholders in the planning process to address multiple resource issues and also brings the support of the signatory agencies.

The combination of the watershed perspective and the community planning process led to the creation of a mission statement and the identification of multiple goals for the area. In addition to the obvious goals of reducing upland erosion and preventing flood damage, the group added improved water quality, habitat protection, and the preservation of the rural character of their valley. Dredging remained a priority objective where homes or structures were threatened with flooding, but other landowners began to consider the option of creating floodwater holding areas to increase wildlife habitat and groundwater recharge, and to reduce flood levels and sediment delivery downstream. Even the dredging effort was refined to include streambank restoration with native riparian species.

To maintain the group's momentum, implementation of the community's plan began even as it was being written. The association organized volunteer creek cleanup days, with the assistance of the Watershed Coordinator, in preparation for future creek restoration. The first restoration work was done in the fall of 1998 to remove accumulated sand from the channel near flood-prone homes. Members of the association have begun pushing the county to increase enforcement of erosion control ordinances to eliminate the source of eroded soil. Others have met with local farmers to exert peer pressure as neighbors, and to instill a sense of community responsibility.

The Watershed Coordinator describes his job as one of making "stone soup" when seeking funding and materials to implement the community's objectives. Once that first contribution (the stone) to the community pot is made, other resources start to be offered from local agencies, private contractors, volunteers from the environmental community, residents, and farmers. For every dollar spent on the coordinator's salary, he has leveraged three additional dollars' worth of services and products for the community.

What is unique about this accomplishment is that no formal political process was involved. No municipal structure or special district needed to be formed. Instead, a grass-roots community effort began to produce significant results in the management of the riparian corridor and flood plain. Moreover, the concept is reproducible. Several other subwatershed neighborhood groups have requested assistance in forming their own associations to deal with their unique problems. As these groups form, the local communities of place are redefined and reinforced. A common theme of the groups has been the desire to preserve the rural quality of life that the residents value. The education and planning process provides the communities with the tools to protect and shape their future rural quality. These communities provide hope for the establishment of a land stewardship ethic.

INTERAGENCY COORDINATION STRATEGY

The basic premise of the Elkhorn Sloough Watershed Project is that natural resource problems must be addressed at a watershed scale. This perspective directs attention to understanding the upstream causes and downstream effects of land management decisions. It also expands the definition of potential participants who can help to develop management solutions. These stakeholders include not only all the farmers and residents, but also the government agencies, nonprofit organizations, and businesses that operate within the watershed.

Coordination of these multiple partners' efforts is difficult to achieve for several reasons. Citizen requests for coordination of public agency services is confused by the fact that citizens rarely speak as one voice. Multiple-interest groups or communities rarely organize themselves for mutual benefit, much less to coordinate assistance from various agencies. Public agencies and private organizations also have difficulty organizing themselves to achieve a mutual goal, because of prescribed jurisdictions or missions and the lack of funding commitment for hiring a coordinator to serve multiple organizations.

Informal coordination frequently occurs as representatives attend the meetings of each other, share information, and discuss collaboration. The fact that so many agencies and organizations share an interest in the protection and enhancement of natural resources creates a community of interest around these issues. Unfortunately, from the public's perspective and perhaps in reality, there is a great deal of redundancy in environmental regulatory review, and there are gaps in service to certain silent minorities such as small-scale ethnic farmers. The Elkhorn Project has sought to play a role in increasing organizational coordination within these two areas of the agency community.

Regulatory Coordination

The Elkhorn Slough Watershed Project, in partnership with Sustainable Conservation, a nonprofit environmental organization, worked to design an innovative program to offer "one-stop regulatory shopping" to land managers willing to implement conservation practices. This program was developed to overcome the frustration, time, and cost to farmers of obtaining individual permits from as many as six regulatory agencies for conservation work intended to reduce agricultural runoff and to protect and enhance the natural environment. Most farmers will continue with current land use practices if the time and financial costs of seeking governmental approvals exceed the perceived benefits of engaging in conservation activities. Six federal, state, and local signatory agencies participated because they realized that current agency review processes, intended to protect natural values, often act as disincentives to voluntary initiatives to enhance and sustain management of agricultural and natural resources.

The strategy for developing the streamlined permit process involved gathering innovative representatives from each of the agencies to identify overlapping issues for regulatory review, and to develop acceptable implementation standards for conservation activities. After 2 years of discussions and negotiations, the permit agreement was made available to farmers and ranchers who work with the NRCS in the Elkhorn Slough Watershed. In the first 2 years of the agreement, 22 projects have been implemented.

Coordinated Farming Assistance

More widespread implementation of sustainable farming practices has been limited, in part, by the problem of limited financial resources for certain sectors of the farming community. As mentioned earlier, farmers who have had to adopt the "scraping by" management style are constrained by scarce production financing, poor production practices, and limited marketing opportunities. To increase adoption of conservation, these related conditions must be simultaneously addressed. However, many of these topics lie outside the mission and expertise of the NRCS.

To overcome these barriers, the NRCS has worked with other agency and non-profit partners to provide more holistic farming assistance. One example is the formation of an interagency outreach committee sponsored by the Cooperative Extension to coordinate technical outreach and training programs targeted to Mexican small-scale farmers. Topics have included information on alternative marketing, new crop varieties, organic production, and farm record-keeping.

Another example is a partnership with the nonprofit Rural Development Center in Salinas to obtain land and funding for a Farmer Training and Demonstration Center in the Elkhorn watershed. The center will provide training and offer plots of land to existing farmers, allowing them to experiment with alternative production methods without jeopardizing their leases, mortgages, or marketing contract if the crop yield declines. The goal of the center will be to coordinate and focus the technical support from a variety of participating research and extension organizations, and to provide a place for farmers to gather and discuss issues. A permanent center may serve as the locus for farmers to begin coming together as a community of place. The environmental community is financially supporting the purchase of the land because they view it as an opportunity to influence the future form of agriculture in the watershed.

SHAPING THE FUTURE OF THE LAND AND THE COMMUNITY

The history of land management in the Elkhorn Slough Watershed illustrates the role that human communities play in shaping the use of productive resources. Over the past 200 years, the coastal woodlands and bluffs were transformed into one of the most intensively managed crop production areas of the world. A combination of new technologies and hard-working immigrant communities made this change possible. But even as the resulting communities prospered, the environmental consequences of this transformation began to degrade downstream resources and the livelihood and interests of other communities. As resource and social conflicts increased between the rural communities, new alliances formed, and government intervention was requested. Environmental protection groups were formed to speak for the wetland resources, whereas the Elkhorn Slough Watershed Project, together with other organizations, has sought to resolve the upland resource degradation problems while respecting and preserving the diverse rural communities.

The mobility of human populations continues, and the current trend is toward a transformation of the watershed from a productive land-based rural community into a bedroom community for Silicon Valley commuters. This latest wave of immigrants brings with it a new culture and a different vision for the watershed. The land is viewed as an aesthetic amenity rather than a productive resource.

Existing communities are once again seeking alliances in anticipation of this land use change. They are attempting to define their vision for the future of the area before it is shaped by the newcomers. The regulatory, environmental, and long-time rural resident communities share an interest in preventing excessive development densities and the resulting fragmentation of natural habitats and potential impacts on riparian and wetland systems. Despite the associated problems with agriculture in the hills, the preservation of agriculture for its open space value is now being advocated by some through agricultural conservation easements.

As contact increases between agricultural and expanding rural residential land uses, the need for improved communication between communities grows. The opportunity for conflicts continues to grow because of the diverse values represented by these two communities. There is a fundamental difference between a wealthy homeowner's view of land management and that of a struggling farmer. Yet, through facilitated dialogue and acquisition of knowledge about one another, a new integrated vision for the watershed is beginning to evolve. Alternative farming methods that use fewer chemical inputs and invest in the health of the soil are more compatible with adjacent residential uses. These methods can help protect the livelihood of the farmer, preserve open space and rural character, and protect the downstream environment. The achievement of any future vision must build on the past. The diversity of communities involved in the Elkhorn Slough watershed requires a multifaceted approach that respects and seeks to address the priority concerns of each community. The ongoing efforts of the public agencies and the nonprofit organization community can assist in building further community support for this future.

REFERENCES

Bennett, J. W., A rational choice model of agricultural resource utilization and conservation, in *Social and Technological Management in Dry Lands: Past and Present, Indigenous and Imposed,* Gonzalez, N., Ed., Westview Press, Boulder, 1978, 151.

Iwata, M., The Japanese immigrants in California agriculture, *Agric. Hist.,* 36(1), 25, 1962.

Monterey County Water Resources Agency, *Rainfall Data for Salinas Municipal Airport,* Salinas, CA, 1998.

Mountjoy, D., *Culture, Capital, and Contours: Ethnic Diversity and the Adoption of Soil Conservation in the Strawberry Hills of Monterey, California,* Ph.D. dissertation, University of California, Davis, 1995.

Mountjoy, D., Ethnic diversity and the patterned adoption of soil conservation in the strawberry hills of Monterey, California, *Soci. Nat. Resourc.,* 9, 339, 1996.

Saloutos, T., The immigrant in Pacific Coast agriculture: 1800–1900, *Agric. Hist.,* 49(1), 182, 1975.

Soil Conservation Service, *Strawberry Hills Target Area Study,* USDA, Davis, CA., 1984.

U.S. Department of Agriculture (USDA), *Elkhorn Slough Watershed Project 1997 Report,* USDA Natural Resources Conservation Service, Salinas, CA, 1998.

Wells, M., Commodity systems and family farms, in *Food and Farm: Current Debates and Policy*, Gladwin, C. and Truman, K., Eds., Westview Press, Boulder, 1988.

Wells, M., Ethnic groups and knowledge systems in agriculture, *Econ. Dev. Cult.* Change, 39(4), 739, 1991.

16 Ecobelts: Reconnecting Agriculture and Communities

Michele M. Schoeneberger, Gary Bentrup, and Charles A. Francis

CONTENTS

Introduction 239
   Introduction to Ecobelts 239
   We–They: Defining the Challenge 241
Potentials of Woody Buffers 244
Potential Role of Ecobelts 249
Designing Ecobelts 252
   Planning Framework 252
   Conceptual Ecobelt Systems 254
Conclusion 258
References 259

INTRODUCTION

"Why do you insist on working the land on Sunday and creating all this noise and dust that disturbs our barbeque party? And the smell from your cattle and hogs is really gross! I'm sorry I live by a farm!" says an irate city resident who lives near the city limits.

"That's part of farming, and we were here first, of course. I sure wish you would keep your dogs out of our pasture and quit throwing your grass clippings and leaves over the fence. This really causes us problems with our livestock and machinery! I'm sorry the city grew out this way," replies the farmer across the property line.

INTRODUCTION TO ECOBELTS

The interface between agriculture and city is one site of major conflict between urban and rural residents. This physical place often is occupied by people with two completely different sets of goals, lifestyles, and daily activities that can lead to problems. In almost every situation, farms were there first, and farmers can rationally argue that their longevity, ownership, and land use take priority over those of people who arrived later on the scene. Homeowners contend that the growth of cities is inevitable, and that progress is measured by housing starts, local population, economic growth, and the infrastructure that comes with urbanization. "If this is zoned residential, then I have every right to expect a comfortable place to live without interference from the problems of a nearby farm," a resident on the edge of the city might argue.

Historically, landscapes were graded from an urban center to more scattered villages, to a diverse mosaic of farmlands and woodlands. This gradient allowed both a visual and physical land use transition while ecologic, economic and social connections were maintained within the larger landscape. Today, massive urban developments abut equally massive agricultural enterprises, creating not only an abrupt and sometimes harsh visual and physical interface, but also one that is highly charged politically (Bull et al., 1984). Many times, the human interactions and ecologic consequences at the interface are at odds with each other and the larger watershed (Moll et al., 1995). Continuing urban growth, dramatic increases in large lot developments for people escaping the city, and ease of transportation for those who want to live in the country and work in the city are accelerating the conversion of agricultural lands into urban or semi-urban environments. Over the past two decades, land put into development per new resident has been at least twice as much as the land per person used for development before that time (Olson and Lyson, 1999).

Despite people's close proximity, part of the conflict in this zone of rural and urban boundaries grows from an urban population that has become increasingly distanced socially from their agrarian neighbors. In the past, many urban residents had agrarian relatives who provided a tangible connection to agriculture. However, with increasing job specialization and agricultural production efficiency, fewer and fewer people have this familial connection. In addition, the escalating concentration of U.S. farming on basic crop commodities and the dominance of processing and advertising by vertically integrated major food companies have accented this distance from field to households, resulting in even less involvement and limited concern by urban neighbors about where and how food is produced. There is growing dissatisfaction in city populations that see tax revenues spent for expensive federal farm programs when they also see food surpluses, cheap products in the supermarket, and food of all types available every day of the year. The predominantly urban population is fast becoming distant from the natural environment also, as people become accustomed and adapted to a built-up cityscape.

The focus of most initiatives to address this "zone of tension" has been a we-or-they approach, with projects designed to meet the objectives of one or the other, but not both. Urban objectives for this interface often are met by creating vegetative barriers or greenbelts that are protected from further encroachment of the city, and which can mask the effects of farming from adjacent housing. Approaches to protect agricultural interests in this zone include special zoning or tax codes that provide exemption for farmers if they continue to make productive use of the land instead of selling it for development. In each case, the area between farm and city is viewed as one of conflict, of competition for space and resources, and of no-win compromise solutions that neither side may view as optimum from its point of view. The social and increasingly important ecologic needs to reconnect these two sectors demand a more proactive planning approach for the interface that should link rather than separate these two land uses and peoples.

The use of tree-based buffers, linear arrangements of "working trees," in the landscape is not a new concept. From the ancient hedgerows in Europe and the shelterbelts in the Great Plains created to provide services in the agroecosystem, to the greenways or linear parkways in the center of urban establishments, these tree-based plantings have been used to meet objectives of rural and urban residents. The U.S. Department of Agriculture (USDA) National Agroforestry Center, located in Lincoln, Nebraska, promotes a number of tree-based buffer programs: Working Trees for Agriculture, Working Trees for Communities, Working Trees for Wildlife, Working Trees for Livestock, Working Trees for Treating Waste, and Working Trees for Snow Management. Working trees are the right trees planted in the right place for a specific purpose. With the ability to conserve and develop natural resources while increasing economic diversity at both the site and community levels, these multipurpose greenways can create an appealing entity in which the rural and urban neighbors can interact physically, and which can foster the information-sharing and consensus-building needed to rebuild the connection between these two groups. Through the planned use of tree-based buffer practices in this interface, the authors propose a redefinition and redesign of this zone of conflict into one of shared ownership and use (Francis and Schoeneberger, 1998), in which both groups see the area as one of positive social, economic, and ecologic interaction (Figure 16.1). Although only a few examples of this approach exist, the authors consider it a valuable model for the future and one that should be explored for the interface between farm and city. They define this as the concept of "ecobelts."

WE–THEY: DEFINING THE CHALLENGE

The most obvious problems between farm families and neighboring city homeowners at the rural–urban boundary revolve around their differences in goals, life experiences, expectations, and tolerance. Many activities, and even discomforts, on the farm are an accepted part of that way of life for farm families. These same situations may be highly uncomfortable and unexpected by a family that has always lived in an urban setting before moving to the city limits. Likewise, many challenges faced by city dwellers may be an accepted part of their environment, but completely foreign and out of step with people in the countryside. These challenges may provide a source

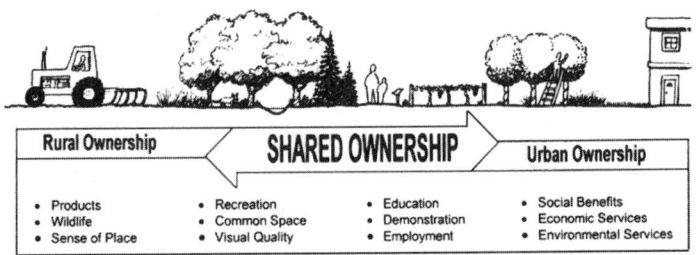

FIGURE 16.1 The rural–urban interface: a zone of shared ownership

TABLE 16.1
Problems and Different Perspectives at the Rural–Urban Interface

Problem	Farm Family Perspective	City Family Perspective
Agricultural-Induced		
Livestock odors	Natural part of farm environment	Unnatural and disgusting
Dust from fields	Normal to farm operations	Hazard to motorists
Noise from equipment, livestock	Accepted part of farming	Disturbs outdoor activities
Slow-moving equipment	Essential to reach fields	Road hazard, slows traffic
Insects from livestock	Normal farm environment	Nuisance to families
Herbicide spray drift	Hard to eliminate or control	Kills yard plants, lawns
Insecticide spray drift	Unfortunate, part of farming	Dangerous to people and pets
Urban-Induced		
High-speed traffic	Dangers to children, animals, tractors	Normal commuter challenge
Dogs in fields	Harmful to livestock	Dogs need open space to run
Snowmobiles, hiking	Compact soil, harm crops, open gates	People need recreational space
Garbage in/near fields	Interferes with operations	Over-the-fence, out-of-mind
Need to lock doors, equipment	Complication that costs time	Accepted way of life
Complaints to authorities	Interrupts farm operations	Normal approach to solving problems

Source: Stokes, S. N., Watson, A. E., and Mastran, S. S., *Saving America's Countryside,* 2nd edition, Johns Hopkins University Press, Baltimore, 1997.

of annoyance for farm families on the other side of the fence. Social tensions at the boundary have been described in other publications (Stokes et al., 1997).

Among the key problems that confront the urban immigrants to the border are odors, dust, noise, and insects (Table 16.1). Although the farm family may accept these minor discomforts as necessary parts of their way of life, they are often new and unacceptable nuisances to the city family next door. Herbicide spray drift may cause the loss of bedding plants or new trees, and insecticide drift is dangerous for people and pets. Equipment that must be run at night or on weekends, that also generates dust that flows across the boundary, may seem inconsiderate of other people's comfort or need to sleep. That same, perhaps wide, equipment driving down the road impedes traffic as it moves from one field to another, which is a source of frustration for the impatient driver trying to commute to a distant office. For a city family, these may be negative components of moving to the edge of the community and an endless source of confusion, danger, and anger.

Across the fence is a farm family coping with all the risks and stress of a difficult occupation and uncertain business. Some of their problems are listed in Table 16.1. Their operations are more dangerous because of increased traffic and higher speeds of vehicles on the road, perhaps driven by people not used to moving over on the shoulder to avoid a tandem disk or wide planter. Dogs may cross the fence to disrupt plantings or chase livestock. People accustomed to using nearby open space feel free to hike, ski, or drive a snowmobile across fields. Garbage tossed over the fence can hamper field operations by clogging equipment or damaging crops. Farmers now need to lock their doors and secure equipment, adding cost and interruption to their operations. Farm families accustomed to solving their differences by talking with neighbors may face legal challenges from city dwellers who are accustomed to dealing with grievances in a more impersonal way. Add to this the uncertainty about continuing to farm on land whose development value has inflated far beyond what is feasible for farming, and a sense of total frustration may emerge on the farm side of the fence.

In addition to social tensions between neighbors, there is an ecologic disconnectedness between these two sectors, compromising the health of the watershed on which both of these groups depend. Despite the abrupt land use change at the interface that seems to separate them physically, the urban and rural biophysical elements are still intimately tied and interact to determine the health and sustainability of the lands (Moll et al., 1995). Actions in one sector will affect the environment in the other and, regarding water, this is fast becoming a bone of contention between these two groups. For instance, in the interest of getting excess water off the land more quickly, vegetation is removed, and channels, storm sewers, tiling, and other drainage systems are put into place. Urban development exchanges large expanses of land that would otherwise retain most of the rain where it fell for new, predominantly impervious surfaces such as parking lots, roads, and buildings. As a consequence, the occurrence and intensity of stormwater flooding has dramatically increased, as has the physical damage to adjoining properties. In addition, this increased runoff or discharge carries with it chemicals, sediment, and other pollutants from both lands, impairing water quality for consumption, recreation and other ecosystem support (i.e., aquatic habitat). With little thought to the role that natural terrain plays in managing ecosystem health, agricultural and urban development has had numerous adverse impacts on the environment ranging from water, soil, and air quality to animal populations and beyond. Neither group necessarily feels the responsibility to bear the burden of correcting these problems. Furthermore, proposed solutions may then place the burden on their neighbors rather than on themselves (Box 16.1).

Box 16.1 Stormwater Flooding: Who's to Blame? Who's to Fix It?

The Beal Slough has become a zone of tension between urban and rural residents in Lancaster County, Nebraska. The 8-mile-long slough and tributary of the Salt Creek suffers from significant flooding and erosion problems due to urban development. Stormwater volumes have increased with the escalating urban development in the sub-watershed and now exceed what the slough was expected to carry, resulting in several flooding events that have damaged homes adjacent to the slough. Initial proposals to address the problems from the perspective of the homeowners called for construction of flood-control structures upstream in the basin on rural lands that would have had deleterious consequences to the rural landowners. One rural resident would have had her land essentially covered by a large flood control pond. Luckily, the conflict that ensued has brought the two impacted groups together and has resulted in the unanimously approved Beal Slough Stormwater Master Plan that better meets both their objectives.

Source: Hain, J. Christopher, "Beal Slough plan wins final OK" p.3A *Lincoln Journal Star* (6/6/00).

It is easy to see how a zone of frustration can be created at the boundary of a city, with families on each side of the boundary embracing different objectives and expectations.To complicate the situation, the boundary most frequently is not fixed, and there often is discord in the farming community between those willing to sell their land and others determined to continue farming in the same place. The boundaries often are highly irregular, with leapfrog development of parcels not contiguous with the existing city and the sale of small parcels for acreage development. Such development drastically increases the linear boundaries between farms and small properties for home sites, and further complicates the relationships between rural and urban people. The complexity of farmland loss and the magnitude of this change over the past several decades are described well in the recent book by Olson and Lyson (1999).

This confrontation situation is repeated thousands of times in different ways across the United States, where communities along the major highway routes are expanding into the adjacent rural countryside. Is this a temporary challenge that will be solved for the current protagonists when development moves one more mile out from the city center, and a whole new cast of players meets across the fence on the new front line, or can more permanent solutions be found with the establishment of firm boundaries and ecobelts?

POTENTIALS OF WOODY BUFFERS

Tree-based buffers provide more than just shade and beauty, bike trails and linking parks. By adding structural diversity to the landscape, these tree-based linear plantings perform ecologic functions that can have significance far greater than the relatively small amount of land they occupy (Box 16.2). These ecologic functions are described in greater detail elsewhere (Dramstad et al. 1996; Forman and Godron 1986; Johnson et al., 2000). The five functions described in Box 16.2 operate simultaneously and fluctuate with time, season, and weather. By manipulating the composition, arrangement, and placement of these plantings within the landscape, we can alter the level of expression of these ecologic functions in an attempt to attain the environmental outcomes we desire.

Of the five basic agroforestry practices (tree-based buffer practices recognized for their deliberate integration of trees and cropping or livestock production systems), four have applications in the rural–urban interface:

Box 16.2 Ecologic Functions Created by Tree-Based Buffers

- *Habitat:* provides resources (e.g., food, shelter, reproductive cover) to support an organism's needs.
- *Conduit:* conveys energy, water, nutrients, genes, seeds, organisms, and other elements.
- *Filter/Barrier:* intercepts wind, wind-blown particles, surface/subsurface water, nutrients, genes, and animals.
- *Sink:* receives and retains objects and substances that originate in the adjacent matrix of land.
- *Source:* releases objects and substances into the adjacent matrix of land.

- *Riparian forest buffers* are natural or planted streamside plantings composed of trees, shrubs, and grasses that buffer nonpoint source pollution of waterways from adjacent land use. They also provide bank protection, protect aquatic environments, improve wildlife habitat, and increase biodiversity.
- *Windbreaks* are planted strips of one to multiple rows of vegetation. Normally serving as upland buffers, these strips intercept the wind, creating a modified microclimate downwind. Windbreaks are planted to prevent soil erosion and to protect crops, livestock, buildings, work and recreation areas, roads, or communities.
- *Forest farming* is the cultivation of high-value specialty crops under a forest canopy that has been modified to provide the correct light conditions for the crops. Crops such as ginseng, shiitake mushrooms, and decorative ferns are sold for medicinal, culinary, or ornamental uses. Forest farming provides an added income while trees are grown for high-quality wood products or to provide an aesthetically pleasing site.
- *Special applications* is a catchall category for different practices that can address the many opportunities to use trees and shrubs for specific agricultural or community concerns, such as disposal of animal or municipal wastes and irrigation tail water filtration, while producing a short- or long-rotation woody crop.

The predominantly linear arrangement of these systems provides many of their services by establishing a screen or barrier. By creating barriers to the wind and reducing windspeeds, windbreak practices can increase crop and livestock production, improve irrigation efficiencies, enable the production of wind-sensitive row, cereal, vegetable, orchard and, vine crops that otherwise would not survive, and reduce energy costs (i.e., expense for heating buildings). Windbreaks in urban areas can modify environments around hospitals, schools, homes, recreation areas, parking lots, and industrial parks, creating more pleasant living and working areas. This wind breaking function also can be used to alter snow deposition in targeted areas. Strategically placed near access roads and emergency routes, these plantings work as living snow fences to reduce dangerous crosswinds, trap blowing snow, lower snow removal costs, and increase driving safety. In fields, they can be designed to enhance the deposition of snow by capturing moisture either through a more uniform distribution across a crop field or as a concentrated collection for filling ponds. Windbreaks serve as living barriers to screen and buffer residential areas from unsightly or loud areas and from dust associated with roads, industry, organized sports, businesses, landfills, or farm operations. They can filter and trap particulates generated from upwind areas, enhancing air quality.

Riparian plantings also are predominantly linear plantings used to intercept, filter, and trap sediment and chemical runoff from adjacent upland sites. By reducing the speed and energy of the water flows, and by increasing the retention of water in these areas, they provide valuable flood control for areas further downstream. A properly designed waterbreak can provide numerous benefits during flood conditions by trapping debris, reducing sand deposition and scouring, increasing bank stability, protecting levee systems, and reducing damage to roads and ditches. At the same time, during nonflood conditions, they can provide additional benefits such as timber and nontimber products, hunting, and other recreational opportunities.

Special applications include tree plantings used to capture excess nutrients produced by rural and urban operations. This natural alternative for using nutrients from livestock and farm operations, municipalities, and industries is able to turn waste into a product by applying it to the trees rather than processing it through expensive waste treatment systems. A direct economic opportunity from these systems involves the wood products from short-rotation systems, which can provide wood chips, fuelwood, and mulch to long-rotation systems that can provide veneer, lumber, paneling, molding and other specialty products.

These "working" tree plantings can provide a wide array of ecologic, economic, and social benefits (Table 16.2), which are necessary to meet the multiple and diverse objectives demanded of these private lands by the landowner and society. The structural diversity created by these linear plantings automatically creates additional habitats and corridors for wildlife. The plantings generally are more aesthetically pleasing and provide better recreational opportunities than the more developed rural and urban systems. The products produced from these tree plantings may include those used by the landowner for personal enjoyment, or they may provide a significant alternative income to help diversify the landowner's income and risk. For instance, a riparian buffer on a farm may produce specialty products, protect stream banks, and provide an aesthetically pleasing view or hunting opportunity to the landowner or to others. In addition to generating recreational/hunting fees for the landowner, these same plantings may reduce nonpoint source pollution of the water by filtering the runoff, stabilizing the stream banks, and altering the energy of water flow, thus protecting the lands and water resources for consumptive and recreational use by communities downstream. By providing additional social and economic returns to the private landowner, along with ecologic services, working tree buffers can create a win-win situation for the private landowner who must try to balance productivity and profitability with environmental stewardship (Boxes 16.3 and 16.4).

Many ecologic processes that contribute to the sustainability of the land, such as water quality and wildlife habitat, become fully expressed only at the landscape level. The actions of each landowner determine not only the health of his or her own land, but also that of the adjacent lands, the larger ecosystems, and the surrounding watershed. Although conservation practices on private lands tend to be applied in a piecemeal fashion, the cumulative functions of the activities of all the "neighbors" living on the landscape really determine the ultimate health of that system. This demands an "all lands" approach in land use planning.

The performance of a linked network of upland and riparian tree-based buffers will be optimized when buffers are planned and designed on a landscape scale. Therefore, designing these systems will be a task of creating strategic configurations across ownerships. What better place to start than at the rural–urban interface to restore and reconnect ecologic processes, and to create education and demonstration opportunities in which both groups can see their connections to the watershed?

TABLE 16.2
Environmental Functions Provided by Selected Working Tree Practices for Use in the Rural–Urban Interface and the Resulting Potential Co-benefits

Working Trees Practice	Environmental Function[a]	Urban Co-benefit	Agricultural Co-benefit
Windbreak	Modify microclimate	• Filter dust, agrichemical drift, odors • Create more favorable microclimate for homes, schools, recreational areas • Keep roads, emergency routes, parking lots open • Reduce home energy costs	• Protect crop • Enhance yield • Protect livestock • Reduce wind-blown soil erosion • Enhance irrigation efficiency • Reduce home energy costs • Increased capture of moisture for crops and livestock • Keep roads, emergency routes, parking lots open
Riparian buffer	Modify hydrology	• Reduce bank destabilization • Reduce stormwater volume • Reduce stormwater damage • Filter urban-generated pollutants (i.e., lawn chemicals, petroleum deposits) • Enhance aquatic habitat	• Reduce bank destabilization • Enhance aquatic habitat • Filter/trap/process field runoff • Reduce flooding damage to adjacent lands
Forest farming	Modify light environment	• Produce nontimber and timber products	• Produce nontimber and timber products
Special applications	Modify air quality Modify nutrient cycling Modify habitat	• Treat municipal waters • Enhance aesthetic value of view • Enhance wildlife • Provide recreational opportunities • Reduce noise	• Treat animal wastes • Enhance aesthetical value of view • Enhanced wildlife • Provide recreational opportunities

[a] Each of these practices will have all the major ecologic functions going on simultaneously but to different and varying degrees. They all therefore have the potential to impact more than just the targeted desired outcome and to serve multiple purposes (i.e., a windbreak designed for crop production [rural benefit] and to screen dust/chemical drift/visuals [urban benefit] also will impact wildlife populations by altering the habitat and conduit).

Box 16.3 Riparian Buffers for Agricultural and Urban Gain: South Carolina

One of the fastest growing counties in the country, Horry County in South Carolina, is home to Myrtle Beach and its massive recreation infrastructure, as well as vast acres of productive croplands, forests, and coastal wetlands. Between fall of 1997 and spring of 1998, over 1500 acres of buffer strips have been installed through the Conservation Reserve Program (CRP) for the purposes of filtering sediment, pesticides, and animal waste from crop- and pastureland. The fast growth in the area has already stressed water resources, and, given that the runoff from these areas drains to the popular beach area and valuable coastal wetlands, there's a tremendous amount of interest as to how these buffers can offset the problems created by the agricultural and urban development. Landowners who have installed these riparian buffers have already noted additional benefits in terms of enhanced wildlife, from turkey and deer to bobcat and fox, that use the buffers as habitat and corridors through the corn/soybean fields.

Source: USDA Natural Resources Conservation Service: South Carolina: economic benefits, riparian areas, in *Collection of Buffer Success Stories from the NRCS: Summer 1999,* Available: http://www.nhq.nrcs.usda.gov/ccs/Soutcar.html, Accessed: June 12, 2000.

Box 16.4 Riparian Buffers for Agricultural and Urban Gain: Illinois

High nitrate levels in the water from field runoff. Floods in Villa Grove, a town of 3000 people about 20 miles south of Champaign-Urbana, Illinois. Two seemingly unconnected problems with a common solution . . . How? These water-related problems in Champaign County are being co-addressed by the placement of riparian buffers along the Embarras River and its tributaries. Using Conservation Reserve Program (CRP) funds, farmers, like Don Koeberlein, are planting riparian forest buffers for the purposes of filtering field runoff before it can contaminate surface and ground water resources. These same buffers also serve to provide floodwater retention for downstream cities such as Villa Grove.

The mayor of Villa Grove notes a greater concern and involvement regarding the issues of flooding and water quality: "We have enough concerned farmers thinking conservation through the area. Mike Mooney [another producer planting riparian buffers in this area] wouldn't mind losing a couple of rows of corn for clean water, for wildlife, or for downstream. This younger generation seems more concerned about Villa Grove. Now, we both care about each other."

Source: USDA Natural Resources Conservation Service, Illinois: environmental and economic benefits, Conservation Reserve Program (CRP), in *Collection of Buffer Success Stories from the NRCS: Summer 1999,* Available: http://www.nhq.nrcs. usda.gov/ccs/ill 1 envir.html, Accessed: June 12, 2000.

POTENTIAL ROLE OF ECOBELTS

As described previously, with proper planting of trees and understory plants, it is possible to screen off some of what urban neighbors perceive as unpleasant consequences of farming shown in Table 16.1: dust from cultivation and harvest, odors from livestock, drifting chemicals from pesticide application, noise from equipment, or chemical runoff during a hard rain. Although this band-aid approach may solve some immediate problems, or at least put them out of sight or hearing, it could be argued that the placement of a woody barrier between urban and rural people creates additional types of distance: reduced human communications, partial solutions to serious differences, and reinforced "us versus them" opinions. With farmers increasingly dependent on votes and other support from consumers, it is essential to seek new and creative solutions to problems at the interface between these two cultures.

Given these challenges at the interface, how can buffers or other mixed types of plantings be envisioned and designed to help solve some of the obvious physical problems that create conflict? Can they in fact be designed to create positive linkages between urban and rural people? The first step has been accomplished: the definition of specific sources of disagreement listed and described earlier. In exploring a range of potential solutions, from isolation to barriers to greater physical separation of the activities, some options can be found that lead to win-win scenarios.

In conceptualizing the buffer area as one of shared ownership and concern, all parties must buy into the importance of this area to their economic well-being, property values, and quality of life (Box 16.5). Through education about the ecologic

---

Box 16.5 Where Private Forests Belong to All: Oslo, Norway

Common law in Norway that goes back more that 1000 years allows public access to private forests and other lands. Formalized into law in 1957, the *Allemansretten* (all man's right or law of access) says that any person is allowed to enter and roam freely on foot or skis through private forests. You may also picnic, camp out, ride a horse or cycle, pick berries, mushrooms, or flowers from this land, whether the owner knows you are there or not. You may not cut firewood or trees, or hunt animals; you also may not pick edible items from the forest within 100 m of a house. A person who enters the forest is expected to leave the place as he or she found it, and people in this culture respect the law and there is rarely any conflict.

Due to a strict zoning around cities that limits housing and commercial building from moving into either forest or agricultural land, it is obvious around Oslo and other cities that the boundary between community and rural areas is preserved and respected. This coupled with the public access to all private lands results in less pressure for people to want to acquire land and push up prices so they can move housing beyond the areas that are zoned for that purpose. The result is a planned culture that preserves wild areas near cities, ideal for hiking and skiing, and that at the same time preserves most of the ecosystem services the forest and agroecosystem lands provide to society.

Source: Francis C. and Meltzer, H. M., Case study 22: forests belong to everyone: public use of private lands in Norway, in *Under the Blade: The Conversion of Agricultural Landscapes,* Olson, R. K. and Lyson, T. A., Eds., Westview Press, Boulder, 1999, p. 450-452.

functions of buffers and how they can provide a series of recreation, ecosystem, economic, and social services, the interface or ecobelt can transform the area of current conflict into a zone that is mutually beneficial to everyone.

Some obvious uses of buffers are to alleviate or minimize many of the challenges described in a preceding section. They can be designed to perform these functions. But more importantly, they can be designed and used as places for recreation: hiking, biking, canoeing, picnicking, and observing birds, flowers, and other wildlife in a near-natural habitat. With the right guidelines and control, the area could be used also for limited economic activities.

Beyond visits to an area to enjoy the natural environment, there can be access to education about agriculture if the adjoining farmland is designed as a living laboratory for people to learn more about modern, ecologically safe farming practices. The design of integrated and diverse crop–animal systems can be demonstrated at this interface, bringing more knowledge and appreciation to the urban audience about the complexities and potentials of current and future food production systems.

The use of economically productive species could further enhance the value of the ecobelts and attract urban and rural people for a broader range of activities (Box 16.6).

Box 16.6 Community Resources' Urban Nontimber Forest Product Project

The "hidden bounty" of tree-based buffers in communities goes far beyond aesthetics and scenic bike and walking trails. They can provide a myriad of environmental services, from air and water quality to soil stabilization, climate modification, and wildlife habitat, and, as documented by Community Resources, simultaneously provide economic returns in the form of nontimber forest products. Alternative income opportunities from these tree-based linear buffers include aromatics, cooking wood (smoke/flavor wood), weaving and dyeing materials, decorative cones, Christmas trees and greens, medicinals (i.e., ginseng), edibles from fruit to nuts and fungi, floral products (i.e., pussy willows and ferns), and other ventures. From a 2-year study in the Baltimore urban forests, Community Resources found the following:

- Individuals and organizations currently collect over 103 products from 78 species.
- Collections were by a wide diversity of ethnic and socioeconomic groups.
- The net economic value of 60 products was calculated to be (1) direct net economic values ranging from $0.30 per pound for pokeweed to over $10 per pound for seeds and mushrooms, and (2) net annual per tree values ranging from $4 per year for an average mulberry tree to over $100 per year for mature Chinese chestnut, apricot, and peach trees.
- The potential value of these products was on par with the per acre values suggested for the environmental services such as energy savings and pollution prevention.

Source: Community Resources, *Exploring Urban Nontimber Forest Products: The Hidden Nutritional, Economic, Cultural, and Educational Resources of the Urban Environment,* Available: http://www.community resources.org/ntfp.htm, Accessed June 12, 2000.

Growing their own Christmas trees or other decorative woody species could provide people with limited economic return and further strengthen their's links with the areas. Agreements could be reached about harvest and replant of a tree each year. Economic species yielding such crops as fruits and nuts could be part of the buffer, and their benefits could accrue to the neighbors. Some young people could find jobs in managing, harvesting, and marketing these products. Many wood species can supply raw materials for crafts and projects, with limited harvest of plant material and great benefit to the neighbors. These activities of mutual interest and benefit would ensure that the ecobelts would be maintained and used, and that they would be renovated as necessary.

The case studies (Boxes 16.7 and 16.8) that follow describe several types of ecobelt systems, along with their places of application and the challenges they can help to solve. Although this concept has not been implemented entirely in any specific place, components of the system are used for similar objectives. The authors illustrate the use of ecobelts consisting of woody and mixed plantings to stimulate people in many communities to examine this option and look for solutions to their unique challenges at the local level. Through planning and design, ecobelts can address many of the problems and opportunities at the urban–rural interface.

Box 16.7 Maunulanpuisto Central Park in Helsinki: A Linear Analog to an Ecobelt

Imagine a linear forest and park running north–south for more than 10 miles through the center of an urban metropolis of one half million people. Then add hiking, jogging, and biking paths; a series of lighted cross-country ski trails; patches of small family garden plots; football practice fields; horse trails and stables; a creek and places to gather mushrooms; and even a pet cemetery embedded into the context of the landscape. Standing in the middle of this ecobelt, it is difficult to see the apartments or businesses that line the park boundary. Only the rumble of a distant train or muted sound of traffic disturbs this tranquil escape from the busy life of a surrounding city.

This is the Maunulanpuisto Central Park in Helsinki, an uninterrupted stretch of forest that connects the heart of downtown with distant suburbs, and a place that attracts grateful outdoors enthusiasts at all times of the year. Nordic peoples have lived close to the environment for centuries, and this linear park preserves the outdoor tradition. Ski trails are lighted for evening use because the sun sets between 3 and 4 p.m. during the winter months at 60 degrees north latitude. Garden plots are in great demand by apartment dwellers who have no space of their own at home. Even in the center of a large city in Finland it is safe for young people to go into these wild areas for the same adventure that helped many of us develop independence as children. Central Park provides a model of public ownership, multiple use, easy accessibility, and ecosystem services that can be copied in the development of ecobelts in our own cities. They bring together some of the components needed to establish zones of shared activity and responsibility that can define the boundary between urban and rural.

Box 16.8 The Town Forest in Weston, Massachusetts

A northern European heritage was obvious in the designation and management of town forests in many parts of New England in the early days of colonization. As part of this cultural tradition, town forests and their important wood resources were protected through the commons system for local citizens to use them for grazing, firewood, and timber rights. Everyone owned the forest, and there were careful regulations on who could cut trees and use them (Donahue, 1999). This pattern evolved into one of private ownership by the end of the seventeenth century, and by the mid-1800s most of this land had been cleared for farming. At this same time, there was a growing concern about loss of local forests and a move toward re-establishing them. A decline in agriculture throughout the region helped promote the reforestation and raised interest in community ownership.

The Town Forest in Weston is currently managed by Land's Sake, a nonprofit organization that is concerned with the cultural landscape as well as sustainable use of this renewable resource. Although there is continual debate about the level of harvesting that can be sustained, and in fact some people prefer to leave the forest untouched, there is community consensus that the forest belongs to all and that all should benefit from its services. This is a model for ecobelts, established for multiple purposes and managed for the benefit of all citizens. There is support from both a cultural history of common land and use, as well as a sustainability imperative that can be enhanced by community participation in planning and decision making in the town forest, or in future ecobelts.

Source: Donahue, B., *Reclaiming the Commons: Community Farms and Forests in a New England Town,* Yale University Press, New Haven, 1999, p. 217-277.

DESIGNING ECOBELTS

PLANNING FRAMEWORK

Agricultural and urban landscapes are complex assemblages of interactive components, which are continually being modified by humans to produce goods and services. The ecologic and social dimensions of landscape structure, function, and change demand a multiscale and interdisciplinary approach to the designing of ecobelts. The planning and designing of a comprehensive ecobelt network requires a flexible but holistic process that invites community participation. This section presents an open structured framework for designing ecobelt networks that accomplish multiple objectives and provide win-win solutions for both farm and city residents (Figure 16.2).

The framework is divided into three basic phases: setting goals, designing ecobelts, and implementing and managing ecobelts. Each phase is guided by a series of questions that assist rural and urban residents in creating a comprehensive ecobelt plan (Box 16.9). A question-based approach is used because questions are effective at providing specific but flexible guidance for analyzing resources and developing plans (e.g., Smith and Hellmund, 1993; Steinitz, 1990). This list of questions is by no

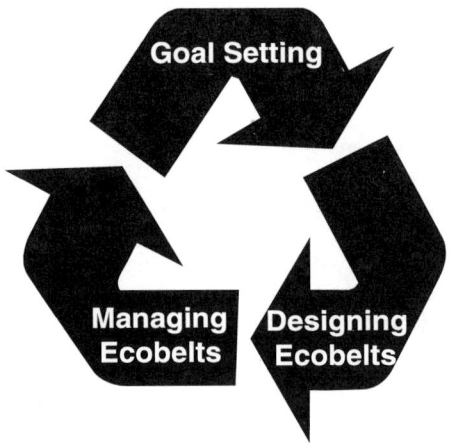

FIGURE 16.2   Ecobelt planning framework

means complete, but rather offers a starting point for ecobelt planning. In many cases, the questions in Box 16.9 will lead to other, more detailed questions that will need to be answered during the planning and design process. It is the responsibility of the design team to tailor the process to local ecologic and social conditions and requirements.

A key component of the process is community participation throughout the planning, design, implementation, and management stages. As described in earlier sections, the interface between farm and city is often one of conflict (Table 16.1). One valuable aspect of the ecobelt planning process is face-to-face dialogue between farm families and neighboring city homeowners to build understanding and trust. Rural and urban residents need to learn about the commonalities and differences in their goals, life experiences, expectations, and tolerances. Through this interaction, a shared vision, ownership, and management of the ecobelt network can be established. In essence, the design and implementation of ecobelts is as much about creating functional relationships between city and farm dwellers as about creating physical features in the landscape.

One of the best ways to initiate the planning process and dialogue between residents is through a quick watershed tour, known sometimes as rapid resource appraisal. The Social Science Institute of the Natural Resources Conservation Service (NRCS) has published a short handout on how to conduct a rapid resource appraisal with stakeholders (Box 16.10). A watershed tour or rapid resource appraisal is valuable because it removes the issues from an ambiguous context and places them in a real setting, allowing for discussion among rural and urban residents. From this foundation, the questions in phase 1: setting goals can be addressed.

For phases 2 and 3, emerging technologies offer many opportunities for planning and designing ecobelt networks, such as geographic information systems (GIS) and visualization programs. Computer-based, GIS facilitates inventory and analysis of resources and allows for what-if scenarios to be developed and evaluated. With the increasing availability of digital resource data, the use of GIS technology is a

---

Box 16.9 Ecobelt Planning Framework

Phase 1: Setting Goals

- Based on public perception, what are the key issues affecting the zone of tension between agriculture and urban areas?
- Are there other key issues the general public has not identified?
- How are the various resource issues interconnected?
- What ecologic and social processes are influencing the issues identified?
- How might these issues be rephrased as desire future conditions?
- How should the objectives of the ecobelt system vary with location across the planning area?

Phase 2: Designing Ecobelts

- Are there significant ecologic or cultural resources that should be protected, enhanced, or restored by a network of ecobelts?
- Where should ecobelts be proactively planned?
- Where can ecobelts be retrofitted into the landscape (i.e., along canals, right-of-ways, etc.)?
- What are the design characteristics necessary to achieve the desired future conditions?
- Where should ecobelts be located and designed to provide the ecologic and social functions of filter/barrier, sink, source, conduit, and habitat?
- Where can ecobelts be located to provide a means to educate residents about urban and agricultural land uses, impacts, and benefits?
- Are there priorities for developing different segments of the ecobelt network?

Phase 3: Implementing and Managing Ecobelts

- What are the potential mechanisms for shared ownership and management of ecobelts (i.e., acquisition, easements, incentive programs, etc.)?
- How will residents share implementation tasks?
- How will residents share management tasks?
- Based on monitoring and evaluation, do objectives, designs, or management practices need to be adjusted?

---

realistic option for ecobelt design. However, it is important to point out that ecobelts can be designed with limited information. A key concept of ecobelts is adaptive management, which allows for changes to be made as experience is derived from implementation and management of the ecobelts. To aid in communicating design ideas and alternatives, various visualization methods are available such as hand-drawn sketches, computer-produced illustrations, and photo simulations. For instance, inexpensive home landscaping software can be used to generate photo-realistic images illustrating vegetation types and composition as well as other structural features of the ecobelt.

CONCEPTUAL ECOBELT SYSTEMS

The following sketches (Figures 16.3 and 16.4) illustrate a conceptual ecobelt plan for a small mixed-use watershed surrounding and incorporating a community. This

---

Box 16.10 Planner's Toolbox

Conservation Corridor Planning at the Landscape Level: *Managing for Wildlife Habitat,* 2000. Published by the USDA Natural Resources Conservation Service (NRCS), National Biology Handbook 190-vi–Part 614.4. Available from NRCS State offices or for download at http://gneiss.geology.washington.edu/~nrcs-wsi/products.html

Landscape Ecology Principles in Landscape Architecture and Land-Use Planning, 1996. By W. Dramstad, J. Olson, and R. Forman. Published by Island Press. Available to order from http://www.islandpress.com/

Ecology of Greenways, 1993. Published by University of Minnesota Press. Editors D. S. Smith and R. C. Hellmund. Available to order from http://www.upress.umn.edu/

Partnership Handbook, 1996. Published by the Water Resources Research Center, College of Agriculture, University of Arizona, Tucson, AZ. Available for download at http://ag.arizona.edu/partners/

The Law of the Land [Legal Alternatives for Land Designation & Acquisition], 1999. By A. Olson. Published by Westview Press, Boulder, CO.

Exploring the Value of Urban Non-Timber Forest Products, 2000. A study of urban forest products published by Community Resources. Available to order from http://www.communityresources.org/index.shtml

TreePeople. An organization dedicated to urban watershed management. Information and resources for retrofitting urban landscapes available at http://www.treepeople.org/

USDA National Agroforestry Center. A multi-agency organization promoting agroforestry practices a variety of rural and urban landscapes. Resources available from http://www.unl.edu/nac/

Greenway: A Guide to Planning, Design, and Management, 1993. By C. Flink and R. Searns. Published by Island Press. Available to order from http://www.islandpress.com/

NRCS Social Science Institute. An agency organization providing social sciences technology resources to assist in equitable and environmentally sound use of natural resources. Resources available from http://people.nrcs.wisc.edu/SocSciInstitute/Default.html

Center for Watershed Protection. An organization dedicated to urban watershed management. Information and resources available at http://www.cwp.org/

Conservation Easement Handbook, 1996. By J. Diehl. Published by Land Trust Alliance. Available to order from http://www.lta.org/

Rural by Design, 1994. By R. Arendt. Published by the American Planning Association Planners Press. Available to order from http://www.planning.org/

American Farmland Trust. An organization dedicated to protecting farmland. Resources available from http://www.farmland.org/

Saving America's Countryside, 1997. By S. Stokes, A. E. Watson, and S. Mastran. Published by John Hopkins University Press. Available to order from http://www.press.jhu.edu/press/

Guide to Community Visioning, 1998. By S. Ames. Published by the American Planning Association Planners Press. Available to order from http://www.planning.org/

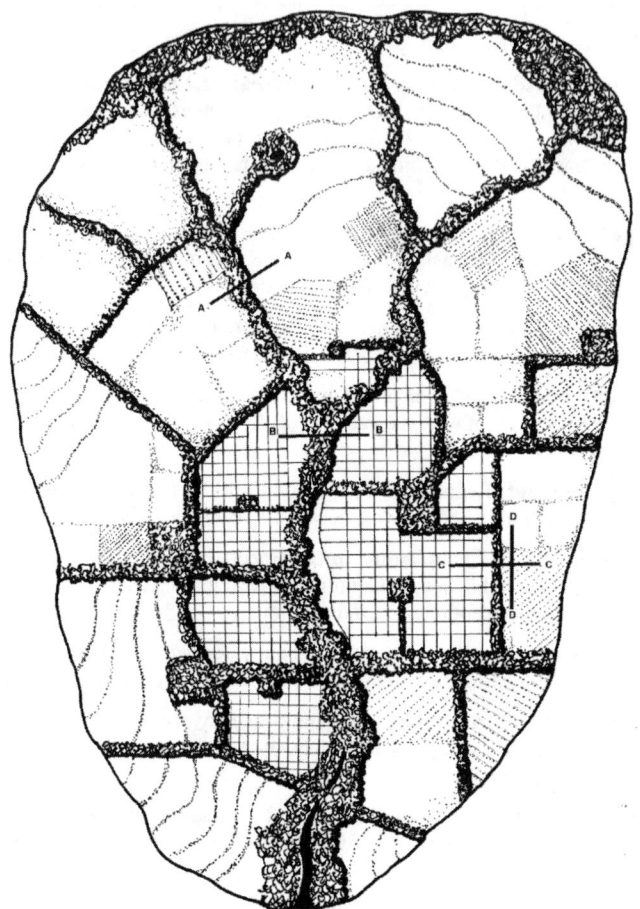

FIGURE 16.3   Conceptual ecobelt planning

example demonstrates how the location of the ecobelt within a watershed will play a
key role in determining the objectives and design parameters for a particular segment
of the ecobelt. For instance in Section A–A, the ecobelt is designed to address agricultural runoff by filtering runoff through a dense native vegetative buffer that also
provides a habitat and conduit for wildlife. This ecobelt also allows for passive recreation through a greenway trail, exposing urban residents to agricultural environments. In contrast, Section B–B illustrates an ecobelt in a more urbanized section of
the watershed. Because stormwater flow is concentrated, a constructed wetland is
designed in the ecobelt system to treat the stormwater before it flows into the stream.
More active recreation areas also are included in the ecobelt, providing a firebreak to
protect homes. Although wildlife may still benefit from this ecobelt, this objective
plays a lesser role than in Section A–A because of its landscape context.

An ecobelt between an agricultural field and residential area is presented in
Section C–C. This ecobelt, which serves primarily as a common garden for both rural

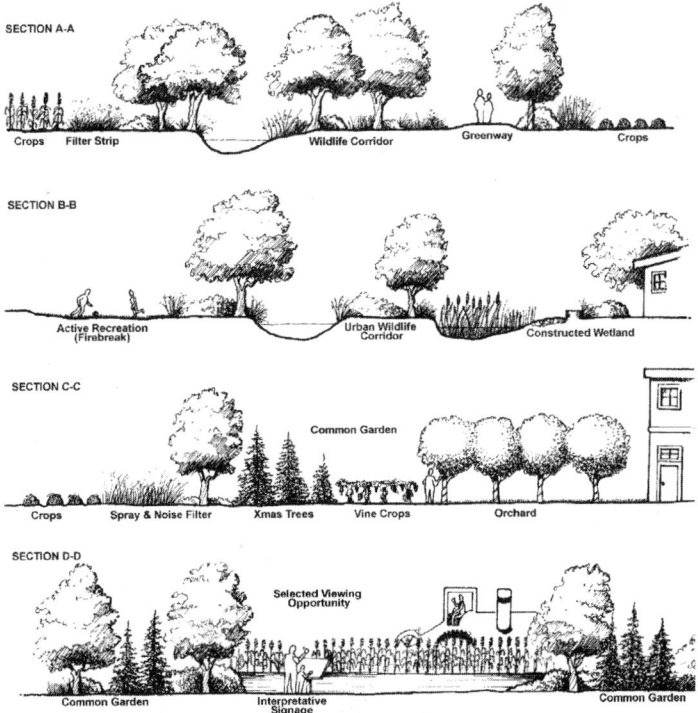

FIGURE 16.4   Ecobelt cross sections

and urban residents, is protected from noise and spray by a vegetative barrier. Products such as fruits, nuts, Christmas trees, and floral items can be harvested from the ecobelt, providing residents the valuable experience of maintaining and harvesting products. Section D–D illustrates how this same ecobelt can provide views between land uses at selected points.

In addition, interpretative signage has been incorporated into the ecobelt to educate residents about different land uses and benefits as well as conservation measures to protect natural resources. Although these conceptual illustrations are not drawn to scale, they clearly demonstrate how objectives and design characteristics change with location.

This brief description of the planning method provides a foundation on which to build an ecobelt system. Box 16.10 provides a list of valuable resources for planners and other individuals interested in planning and designing ecobelts for their watershed. There are many and varied options available for implementing an ecobelt plan including community land acquisition, conservation easements, federal and state programs, zoning, voluntary participation, and transfer of development rights, among others. The resources in Box 16.10 can provide more information on the many options available for urban and rural residents creating ecobelt plans.

CONCLUSION

Ecobelts as described in this chapter are not yet a reality. However, various components and applications of the concept are practiced in a number of U.S. communities. What we recommend is bringing these pieces of the puzzle together into a pleasing picture for the future, one that will help to meet the needs of both rural and urban citizens (Figure 16.5). The multiple-use ecobelts or linear areas through a community or connecting a community with nearby recreational areas outside its boundaries (see Afterword) are important, yet the most useful application of the concept comes at the interface between farming and urban residences.

The authors propose that the we–they mentality currently existing at the interface because of the many aforementioned conflicts can be converted into a consensus on land use and ownership that benefits people on both sides of the boundary. This will require the dedication of careful attention and considerable energy to a planning process that

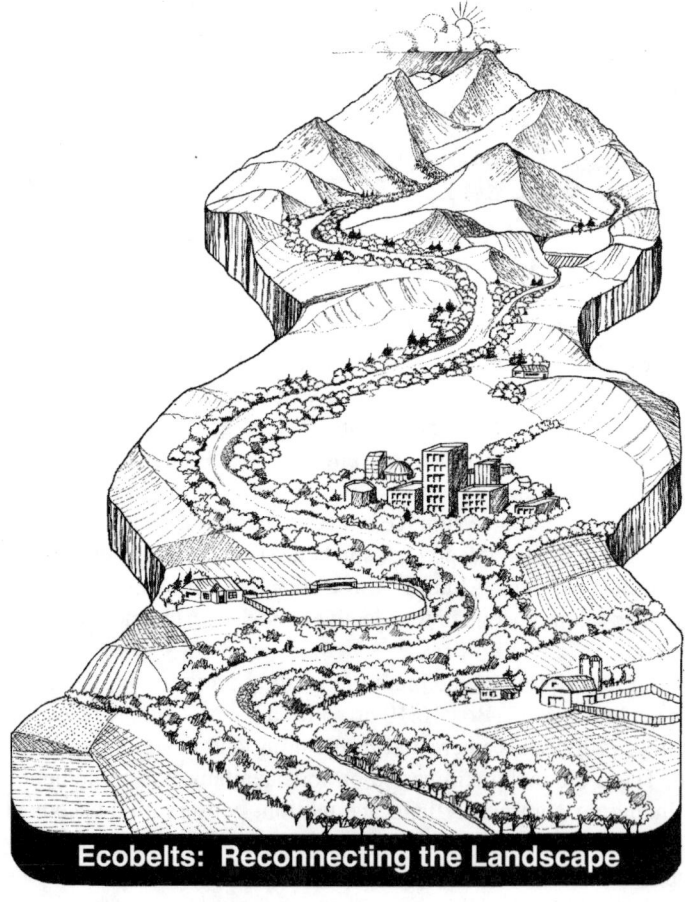

Ecobelts: Reconnecting the Landscape

FIGURE 16.5   Ecobelts: reconnecting the landscape

builds ownership of and identity with the program, and thus a lasting commitment to implementation of the plans. Location and design will require compromise, a difficult result to achieve in a culture so dedicated to individual needs and perceived desires as well as the feeling of independence that comes with ownership of land. Perhaps the best approach is to identify more successful models already in place across the country, and to use these as demonstration sites showing what is possible. This approach has been successful in selling the concept of bike trails through rural areas, where farmers and ranchers have often been opposed to the program until they have learned from other places that this can be successful and positive for them and for society.

The multiple functions of woody plantings have been realized in conservation plantings and other multiple-use areas in the rural landscape. They also are well-recognized as an integral part of our parks and recreation areas. It will be useful to bridge these already-accepted applications of buffer strips and tree plantings into the new concept of ecobelts. These areas can provide an effective barrier for solving some of the obvious problems at the rural–urban interface. They also can serve as a habitat for wildlife, a sink for carbon and a filter for undesirable materials moving from farm to community and vice versa, and a source of both recreation and limited economic activity. The educational benefits of learning from both farms and wooded areas adjacent to the city are difficult to quantify but important to establishing connections for the next generation of urban youth. For all these reasons, the authors consider ecobelts to be a viable concept for the future, one that will provide immense benefits for the co-owners and management of the areas while resolving conflicts between rural and urban people.

REFERENCES

Bull, C., Daniel, P., and Hopkinson, M., *The Geography of Rural Resources,* Oliver & Boyd, Edinburgh, 1984.

Community Resources, *Exploring Urban Nontimber Forest Products: The Hidden Nutritional, Economic, Cultural, and Educational Resources of the Urban Environment,* Available: http://www.communityresources.org/ntfp.htm, Verified: June 12, 2000.

Donahue, B., *Reclaiming the Commons: Community Farms and Forests in a New England Town,* Yale University Press, New Haven, 1999.

Dramstad, W. E., Olson, J. D., and Forman, R. T., *Landscape Ecology Principles in Landscape Architecture and Land-Use Planning,* Island Press, Washington, D.C. 1996.

Forman, R. T. and Godron, M., *Landscape Ecology,* John Wiley & Sons, New York, 1986.

Francis, C. and Meltzer, H. M., Case study 22: forests belong to everyone: public use of private lands in Norway, in *Under the Blade: The Conversion of Agricultural Landscapes,* Olson, R. K. and Lyson, T. A., Eds., Westview Press, Boulder, 1999, 450.

Francis, C. and Schoeneberger, M., Centers proposed education on conservation buffers: the rural/urban interface, in *Center for Sustainable Agriculture Newsletter,* July/August, 1, 5, 6, 1998.

Johnson, C. W., Bentrup, G., and Rol, D., Conservation corridor planning at the landscape level: managing for wildlife habitat, Section 614.3, in *USDA-NRCS National Biology Handbook Title 190,* U.S. Government Printing Office, Washington, D.C., 2000.

Moll, G., Macie, E., and Neville, B., Inside ecosystems, *Urban Forests,* 8, 1995.

Olson, R. K. and Lyson, T. A., *Under the Blade: The Conversion of Agricultural Landscapes,* Westview Press, Boulder, 1999.

Smith, D. S. and Hellmund, P. C., *Ecology of Greenways,* University of Minnesota Press, Minneapolis, 1993.

Steinitz, C., A framework for theory applicable to the education of landscape architects (and other design professionals, *Landscape J.*,9:2 (fall) 136–143, 1990.

Stokes, S. N., Watson, A. E., and Mastran, S. S., *Saving America's Countryside,* 2nd edition, John Hopkins University Press, Baltimore, 1997.

U.S. Department of Agriculture (USDA) Natural Resources Conservation Service, Illinois: environmental and economic benefits, Conservation Reserve Program (CRP), in *Collection of Buffer Success Stories from the NRCS: Summer 1999,* Available: http://www.nhq.nrcs.usda.gov/CCS/ill1envir.html, Accessed: June 12, 2000.

USDA Natural Resources Conservation Service, South Carolina: economic benefits, riparian areas, in *Collection of Buffer Success Stories from the NRCS: Summer 1999,* Available: http://www.nhq.nrcs.usda.gov/CCS/soutcar.html, Accessed: June 12, 2000.

17 Afterword: An Optimistic Future Scenario

Charles A. Francis

Inspired by Jackson, 1980; Piercy, 1976; and Thayer, 1994,
Setting: Lincoln, Nebraska,
A Saturday in late July, Year 2020.

"C'mon dad! If we wait much later it's gonna be too hot and the fish won't bite." Julie Thompson wheeled her 24-speed bike out of the shed. She had just finished lubricating the chain and hubs with soybean-based oil and bicycle grease and wiping down the frame with cloth made from milkweed, a perennial crop grown nearby.

"Don't forget the sandwiches and fishing gear. And be sure to get our bike helmets. Tell mom we're leaving. I'm just turning this last compost pile." Julie's dad, Brook, works with the regional office of State Farm Insurance, but today he's the 10-year-old's biking companion, and he's anxious to be outdoors. For now, the lush garden with ripening tomatoes and snap beans on tree-branch supports will be left behind. More mulching of vegetables can wait until later.

Their bikes roll down the path past butternut squash and green melon plants that replaced all the grass in their backyard. Protected by young trees and berry strips, these vegetables have given the family fresh produce and reintroduced Julie and her older brother Tim to the connections between food and environment. It's helped them discover some new roots!

"Bye Charlie," calls Julie to their resident cardinal who lives in the shrubs that line their lot. A complex mix of native shrubs and trees, these plantings shade the house and provide berries and cover for birds and small mammals, even right in the city of Lincoln. Their bikes turn onto the common path that goes through Tierra Park and gives access to a city network of trails that reach parks, schools, and the city center, never having to cross a city street.

Heading west on the bike trail, the bikers found the air to be cool, mainly because of the nearly closed tree canopy that shaded the path. "These trees were planted before you were born, Julie. Now you can enjoy them!" Brook called ahead to his daughter. As they rode along the level path, converted from an old railroad right-of-way, rabbits and squirrels darted across their route.

"Where do all these animals build their nests?" asked Julie. "I can see the squirrel nests up in the cottonwoods, but how about the rabbits, the possums, even the skunks?" She was mystified when these creatures appeared each spring, even though she knew about hibernation and thought they must stay somewhere nearby.

"They stay back in all this cover along the path," Brook said during this highly teachable moment. "When all these areas with trees, bushes, and tall grass are connected like they are along this old railroad route, the animals can move along with pretty good cover. That's why we've even seen a fox once in a while in the park near our house, and someone saw a couple of deer in Tierra Park last year."

Julie began to see the importance of connectedness of this habitat. It was not just a small patch every so often scattered through the city. "I guess all this woods really is all connected, even to those farms we pass and their windbreaks. Even all the way to the lake where we're going to fish! It's even connected for us bikers. Do you think the foxes go that far?" She imagined a mother fox making the 8-mile trip they were taking this morning to the lake, just to search for food and to keep track of her kits.

"I don't know about that," laughed Brook at his imaginative daughter. "Let's look up their range when we get back home. It's sure to be on the 'critter' web site on your computer." They were nearing the high pedestrian bridge that spanned 27th street and kept them well above and away from the traffic.

"This is the neatest bridge, but it looks sort of old to me," said Julie.

"You should have been here when it was put up in 1996. I was in high school then, and we loved to ride our bikes down from Antelope Park and cross these two bridges. We thought it was something out of *Star Wars* then," said her dad with tears of nostalgia in his eyes.

"Oh, dad! You always talk about those old movies. This looks ancient compared with the neat stuff in those ecotowns on *Bejor X* that we saw on interactive TV the other night. We could walk through those forests, make animal sounds and have them answer, and it was hard to believe that we were right in the middle of a town with houses and shops underground where we walked."

"Well Julie, some of those shows of yours are pretty far out. Do you even think people will live like that, really underground?" asked Brook.

"Of course they will, dad! You just don't have enough imagination anymore." They crossed 14th Street and a railroad on spanking new pedestrian/biking bridges and began the descent into Wilderness Park. Their destination was Wilderness Lake, another several miles south into the countryside.

"Dad, is it true that people didn't used to even care about other animals in their ecosystem? I can't believe that when you were young they were paving over all this land for malls and shopping centers, and everyone didn't even care what happened to rain and to everything that lived here. Was it really that way?"

Brook looked ahead on the path as they scared two large crows into the air and a covey of quail into the brush. "You may not believe it. People in the last century scarcely knew where their food was coming from, much less anything about the natural environment. Until we started teaching seriously about animals, soils, plants, cycles—all the stuff you take for granted now—most kids who weren't born on farms didn't know anything about the outdoors!"

"Gee, dad, you really are old. I can't imagine anybody not knowing about weather cycles, about why termites and ants are important, gosh, about how the soil is living—just like your compost pile with the nest of snakes underneath. Wow! Look at that coyote over near the corn in that field! I'll bet he can find some mice or rabbits over there." She was looking down the strips of corn where the coyote's ears were just visible above the green canopy of soybeans. "Why are those crops mixed up in the same field, dad?"

"Well, there were some people at the university and some creative farmers here in the area about 30 years ago who really believed that putting crops in strips was a good strategy. They looked at the yields and found that the taller corn could help the soybeans by blocking the wind, even if it did put some shade on the soybeans. Then some other people selected a soybean variety that did pretty well even with the shade from the corn. Now they have much higher yields in both crops, even in years when there's not much rain," said Brook.

"Wow! That's pretty amazing. I wondered why those crops were planted that way. I've seen a lot of fields like that. Look at those sheep eating grass under the trees. Why do they let them into the orchard, dad?"

"The technical name for that is a silvipastoral system. It just means putting trees and grazing animals together where they can help each other. See, the sheep eat forage and leave their manure to help fertilize the trees. All the flowering weeds and other plants around there attract beneficial insects that help keep the bad ones out of the fruit trees. Pretty good system, huh?"

But Julie was distracted. "Hey, there's the lake. Hope it's not too late to catch some big ones," enthused Julie.

Their bikes slowed as Julie and Brook wheeled through the buffer strips that lined the lake. The wetlands at the head end of the lake and the grasses and shrubs that lined the bank provided more habitats for wildlife, and the shade was good for keeping water cool and protected for aquatic life. Leaning their vehicles against a convenient tree, the pair began to unpack their fishing gear.

"How can we cast with so many trees around, dad?" queried Julie. "I'm always getting my line hung up. Why don't they cut down some of these trees?"

Brook laughed. "That makes it more of a challenge, Julie. You want to give the fish a chance, don't you? In fact they planted some of these bigger trees about 40 years ago when the dam was put in and the lake filled up. Most of the rest just came up from seeds blown in by wind or carried by birds or water. I'm glad they leave the natural cover. It really helps the fish to spawn."

"I saw an old picture of you out here on one of the lakes, dad. You were driving one of those great water ski things that threw up a big wake. Did you really do that out here?"

Looking a bit sheepish, Brook admitted, "That was in the old times before we knew about the damage that we did to the shore and the spawning areas. People back then were really thrilled with power, you know, motorcycles and fast cars. They even made those 'jet skis' here in Lincoln! We didn't really think about all the gas we were burning up and how many problems we caused in the environment. That's why they only allow canoes out here now, plus swimming and wading in part of the lake."

"But that's common sense, dad. What was wrong with you people back then? You really didn't think about the other animals that much?"

"Well, it seems pretty amazing today, I guess. You start caring for other critters from the time you hit the back yard, and especially in school where you all have to adopt an animal and take care of it. What have you had so far—a gerbil, the pair of parakeets, even some worms one year?"

Julie added, "Don't forget the year I was in charge of cockroaches. My teacher said they will probably outlive people. At least their children will live longer than most of us!"

"And that living space where you keep all the animals and plants, with nothing in cages. That's something new to me. What do you call it?"

"Oh dad. You always forget. It's our own class's 'Enviroom.' It's sort of a small copy of the biosphere we all read about in Arizona. It's pretty neat the way all the plants and animals are balanced in there. We see the frogs eating flies, and the snake even caught a small mouse the other day. It was kind of yucky, but the teacher said that's all part of life. I wish I could have a snake at home. That would really gross out the little kids next door!"

"Well Julie, why don't you adopt the nest of garter snakes that's under the compost? You could even give them names, check on them every so often, and see if the conditions are right for them. I'll bet we could find some more snake information on the *Econet* channel. Hey, let's catch some fish. Remember, the limit is two, and we have to put back any that are under 25 centimeters long."

Seated in the shade, the two continued their talk about the lake, the fish, and even about the way things used to be.

"Dad, is it true that you found a frog here one time that had four back legs? I can't imagine finding something like that. It would be totally weird! Did you really find something like that?"

"It was pretty scary," said Brook thoughtfully. "The little frog only had three back legs, but that was bad enough. He looked pretty normal otherwise. We put him in a jar and took it to biology class in high school. The biology teacher said there had been quite a few like that found up in Minnesota. They really never found out what caused that type of frog. There was a lot of talk about pesticides that were causing the genetic changes. At least we don't have to worry about that any more. All the frogs you've seen are normal, aren't they?"

"You bet! The only weird ones I've seen were in pictures in school. Did the chemicals really do stuff like that?"

"It's hard to say, Julie. There were so many pesticides used then to kill weeds and insects, even to fumigate the soil to kill nematodes. Quite a bit of the pesticides went into the water. When we started the big push back in 1999 to plant conservation buffers along all the streams, that helped to filter out some of the chemicals and kept the soil from washing completely out of the fields. We could trap them in these strips just like we have here around the lake. Then over the next 15 years, most of the chemicals were phased out of farming. Did you know that we used to put chemicals on our lawns and even on the gardens where our food came from?"

"Wow! That must have been just as scary as the frog with three legs? Did you wear all those masks and big suits like the astronauts?"

"Well, that was the big problem. Most people at home didn't really think much about what they were doing. They could just walk into the garden store and buy about whatever they wanted, then spray it on their yards to kill all the bugs," mused Brook.

"The farmers did a better job of it. At least they had some training. But now we're pretty well phased out of the worst of the chemical pesticides."

"But we know that most of the bugs are good. They're part of the system too. I can't believe you tried to kill them all?"

"That's the way things were. People thought they could control everything, even by spraying all the good insects to get rid of a few that made them uncomfortable. We even did that in our yard at home. I'm glad things have changed, Julie. You know we've started to see the number of people with cancer going down now, and it's probably because there are not so many chemicals in our food and all around us."

"I remember that's how grandma died," thought Julie sadly. "I really hope we never do that stuff again! Hey, I've got a fish. Hope it's big enough!"

With Brook's help she pulled it in and it was a keeper. Over the next couple of hours they filled their limit, carefully released several small fish, and prepared to head for home. Packing up their gear, Julie pondered what she had seen that morning.

"Dad, it's hard to believe that people really didn't think much about the other parts of the system back when you were a kid. How come they didn't know about chemicals and how the gas was running out and where food came from and all that stuff?"

Brook thought a while as they pushed their bikes up to the path. "I know it seems strange, but people back then—just 20 years ago—were still pretty disconnected from the environment. We really didn't have so much in school. Sure, we were recycling quite a bit. In fact it was a push from the kids in elementary schools that started people thinking about how much they were throwing away. Can you believe that our family had two garbage cans full on Mondays and Thursdays, and all that ended up in the landfill?"

"Gee, dad. We have only a small can once a month. I always thought it was only a few things that broke or really couldn't be burned at home or put in the compost that had to go in the trash. There's really not very much stuff we really call trash is there?"

"You're right! All the organic material—you know, kitchen trimmings, paper, cardboard, natural cloth—goes through the grinder into the compost. I think everyone does that now. And all the glass, tin, aluminum, newspaper, and plastic goes into the recycle bin. There's not much left is there? We even used to see a lot of stuff along the roads and sidewalks. I'm really glad that all we can see now is grass and wild flowers. It was pretty bad in some places. You know, Julie, the people who used bike paths and jogging paths were some of the first ones to show others that we really didn't have to trash the planet?"

"That's simple, dad. Who's going to carry a bunch of extra weight around on the bike, and throw it out so it looks bad along the path? Did people really throw things out of their cars? We'd really get fined if that happened today."

"You're right. That's the way to scare people. But what really works is when everyone knows inside that it's wrong to scatter things around. It's a lot better today, and I think it's because people care about the environment."

They passed a tree with birds singing in the July heat. "Julie, here's a good example. Think about Kerrey High School for a minute. You know we're riding by the edge of the school grounds and football field, but it's over there about 200 meters through the trees. You can't even see the high school over there when we ride by on the path. Would you believe that back in 1998 they tried to zone this all industrial and housing? There was a session with the city council when 150 people showed up to protest the mad rush to development. They all had green stickers that said, "Plan a Sustainable Wilderness Park." I guess that started everyone thinking about how we could keep this habitat for animals and still have schools and houses. That's when the real question started coming up about how big these city lots should be, and why people who weren't farming needed all that land. Now with the green belt around the city to protect farmland we're in better shape. Even our insurance rates at State Farm favor the homes with reasonably sized yards and native plantings."

"Why are those better, dad? I like all the different plants and our garden, but why is that really better for your company? There are still some of those green front yards around. You mean they have to pay more for insurance plus all they spend to take care of a big lawn?"

"We found out some years ago that the people who put in native shrubs and cover and gardens and didn't use chemicals on their yards were a much lower risk for the company. They were healthier, maybe because they spent more time outside working in the garden and took in more fresh air. We really don't know why, but it's a business deal for us. You always hear me say that it's the economics that makes people change, right?"

"Sure, dad. But there are lots of other things that are important, too. I really like the animals we see in the yard. It's good to know that we have water and some natural food for them all year long. And I like to go out in the park and play with my friends there and run around on the grass. I don't see why everyone thought they needed to have a big lawn back in the old days?"

"Well Julie, people's ideas change. We used to put a high value on that—a clean and manicured lawn that looked perfect. It used to make your house more valuable. Now people see it as a waste of time, besides being a really unnatural place to live as well as pretty expensive to keep up. You probably won't believe that people used to seed their lawns, put on fertilizer, water them all the time, and put on insecticides, then mow them and throw the clippings away! We'd get visitors from developing countries who were astounded that we could do things like that when we could be producing food right on our own lots. But people do change. That's what education's all about."

"Dad, can we go by the little farm market on the way home? It's not far out of the way. I really liked the mushrooms we got there last week. And maybe they'll have some more of those dolls made from the notch of a limb. You know, the ones with the big eyes painted on them? C'mon, dad, let's go see what's there today!"

"OK. We'll stop by, but we have to get these fish home to the cool cellar before they spoil in the heat."

The small Saturday farmer's market was one of Brook's favorite places too. Since the main market place in Haymarket had expanded to have six smaller satellites around the city, there were many more vendors bringing in their extra garden produce. Nearby farmers were producing specialty crops, including fruits and nuts, in their windbreaks and wildlife plantings. This was more than a Saturday social event. It provided much of the income for people who produced food, prepared canned jellies and jams, baked cookies and breads, and even made unique handicrafts from their natural resources at home. Brook and Julie turned from the main bike path, and in a couple of blocks they reached the market.

"Hey Cassie, come see the fish we've caught," called Julie to her best friend from school. "We've just gone eight miles on our bikes, and we're all sweaty, and we caught these fish for supper tonight. Come see them!"

"Wow, those are neat," said Cassie. "Wasn't it too hot to ride out to the lake? It looks like you're pretty tired. Are you still coming over this afternoon?"

"Sure! I'm not too tired. It's shady all the way, thanks to all the trees. We saw about 22 meadowlarks, and even a coyote in one of the cornfields, and a couple of hawks soaring around, and a possum, and a skunk crossed the path. You should come with us next time!"

"Hey, I'd like that. My bike is ready. Maybe we can go tomorrow?" asked Cassie. "Do you want to go swim out at the lake?"

"I'll ask my mom. She likes to bike and swim. And the lake's really clean now that we don't have chemicals on the crops and jet skis on the lake, and now that there are buffer strips all around to keep out the soil from the fields."

"Chemicals, jet skis, buffers? What's all that stuff?" asked Cassie. "What are you talking about?"

"That's a long story," said Julie sagely. "I'll tell you all about it tomorrow if we ride to the lake. Don't forget that I'm coming over later!" She and Brook rode off toward home, tired but happy after a morning in the fresh air.

"Lincoln's a good place to live, dad. I'm glad we can see so many things so close to home. I feel like I'm really a part of the system. And you should really think more about those underground houses in *Bejor X!* You might have to come visit me in one of those sometime!"

REFERENCES

Jackson, W. *New Roots for Agriculture,* Lincoln: University of Nebraska Press, 1980.

Piercy, M. *Woman on the Edge of Time,* New York, AA. Knopf, 1976.

Thayer, R. L., *Gray World Green Heart: Technology, Nature, & the Sustainable Landscape.* New York: J.Wiley, Ch.9, 296.

Index

A

Agroecosystem(s)
  changes, 1
  communities and, 54–55
  culture and, 1
  as ecosocial processes, 16
  environment and, 3
  farmer participation, 133
  hog-based, 50
  management, learning for, 138
  performance, 148–150
  social capital, 177–178
  sustainability, 1, 2
  transhumant, 107
  uses, 3
Agricultural Conservation Program, 195
Agriculture
  crop diseases, 81
  Fordist, 120
  interface between city and, 239
  irrigated, 69, 73, 77, 79
  market fluctuations, 76
  multifunctional nature, 159
  nature protection and, 90
  packing plants, 74, 75
  production contracts, 82
  public sector research, 70
  small farmer, 69
  success in, 119
Agroforestry, 188–208
  basic practices, 244
  development, 201–202
  as developmental tool, 191–192
  origins, 192
Alien Land Law (1913), 222
American Farmland Trust, 255
Anticorporate farming law, 34
Apiculture, 90
Apples, 85
  pile village, 87
  streuobst yields, 94
  varieties, 91
Argentina, 2. See Patagonia
As You Sow, 158
Avocados, 211
  medicinal effects, 214–215

Avocados (*Continued*)
  production, 214
    ecological benefits, 214

B

Banana industry
  forest fragmentation and, 20–21
  Panama disease epidemic and, 20
  shifting, 22
  small-scale growers, 24
Beal Slough, Nebraska, 243
Beans, 72, 76
Bureau of Indian Affairs, 190

C

California, 221–236
Capital
  business usage, 9
  defined, 9
  financial/built, 10, 11
  forms, 9
  human, 10, 11
  natural, 10
  social, 10, 11
Carneros Creek Association, 233
Center for Watershed Protection, 255
Charcoal makers, 59
Chemical waste, 122
Civil society, 7–8
  markets and, 8
Columbia City Market, 165
Commercial forestry, 59
Common property resources, 107
Community(ies)
  acroecosystems and, 54–55
  as ecosocial processes, 16
  interest, 5
  learning instrument, 143–151
  place, 5
  sustainability and, 175
  tree-based buffers, 241
  urban vs. rural, 240
  zone of tension, 240
Community-acroecosystem, 2
Community forestry, 59

Community(ies) (*Continued*)
  corruption, 64
  distribution of proceeds, 61
  subsidizing agricultural production via, 63
Community Resources' Urban Nontimber Forest
    Product Project, 250
Community supported agriculture, 162
Concentrated animal feeding operations
    (CAFO), 33
Conservation Corridor Planning at the Landscape
    Level, 255
Conservation Easement Handbook, 255
Conservation Reserve Program, 199, 203, 248
Contract farmers, 34
Convention of Biological Diversity, 95
Crianceros
  agricultural, 104
  communities, 112
    resistance, 113
  land access, 106
  negotiating power, 110
  public pastureland, 106
  sedentary, 104
  transcontinental corporation and, 111
  transhumance, 104 (*See also* Nomad breeders)
Crop diseases, 81, 83
Culture, 1, 158
Current resource flows, 121
Cuyamel Fruit Company, 25, 27

D

Davis Correctional Center, 41
Desertification, 108, 109
Development Support Services

E

Ecobelt(s), 249–259. *See also* Tree-based buffers
  conceptual planning, 254, 256
  designing, 252–257
    role of emerging technologies, 253
  implementing and managing, 254
  planners' resources, 255
  potential role, 249–252
Ecology of Greenways, 255
Elkhorn Slough Estuarine Reserve, 224
Elkhorn Slough Watershed Project, 226
  basic premise, 234
  Farmer Training and Demonstration Center,
    235
  land management, 235
  management styles associated with, 228–229
  neighborhood-based education, 232
  Sustainable Conservation partnership, 234

Environmental Easement Program, 199
Environmental Protection Agency, 50
Environmental Quality Incentive Program, 184,
    230
European Economic Community, 95
*Exploring the Value of Urban Non-Timber Forest
    Products,* 255

F

Farmer(s)
  brokers, 73, 75, 83
  contract, 34
  cooperatives, 81
  credit negotiations, 75–76
  federal support, 80
  Japanese, 223
  Mexican, 223
  organizations, 73–4
  packing plants, 74
  participation in agroecosystems, 133
  sharecropping, 76, 78
  as stewards of natural resources, 158
Farmer Training and Demonstration Center, 235
Farmland Industries, 34
Feed barns, 216
Food and Agriculture, Conservation, and Trade
    Act (1990), 195
Food Bank Farm, 162–164
  crops, 163
  labor, 163
  organic production practices, 163
Foodshed, 159–160
Fordist agriculture, 120
Forest farming, 245, 247
Forest fragmentation, 20–21
Forest Service National Agroforestry Center, 201
Forestry Incentive Program, 195
Forestry management regimes, 199
Free access resources, 107
Freedom to Farm Act, 123
Fruit(s)
  exports, 70
  production, 70, 85 (*See also* Streuobst)
    history, 86–87
    hobby, 93

G

Genetically modified seed, 160
German Federal Nature Protection Agency, 89, 95
German Society for Nature Protection, 95
Germany. *See* Streuobst
Greenway, 255
Groundwater, 37–38

Guide to Community Visioning, 255
Guymon, Oklahoma, 39
Guymon-Hugoton gas field, 39

H

Helsinki, Finland, 251
Hog business, 34, 215–217. *See also* Hughes
           County, Oklahoma; Texas County,
           Oklahoma
   groundwater and, 37–38
Holdenville, Oklahoma, 38–39
   water issues, 38
Horry County, South Carolina, 248
Hughes County, Oklahoma, 33, 38
   cultural heritage, 44
   demographics, 42, 46
   economic profile, 40
   farm size, 41
   farming acreage, 49
   job creation, 44
   largest employers, 41
   main crops, 38
   political affiliation, 49
   population, 40
   surface water, 38
   unemployment rate, 41, 42

I

International Support Group, 133, 138
Irrigation, 69, 73, 77, 79
   importance, 82

J

Johnson-Trussel Company, 193

K

Kenya, 133
King County Growth Management, 164

L

Lake Benton, Minnesota, 176
Lake Benton Watershed Holisstic Management
           Coalition, 180–183
Lake Improvement for Everyone, 177
Lancaster County, Nebraska, 243
Land, sustainability of. *See* Sustainability
Land use, 70, 71
   determining best, 193–201
      economic criteria, 199
      environmental criteria, 199

Land use, determining best (*Continued*)
      feasibility study for, 193–195
      political and institutional criteria, 199
      social criteria, 199
   transition, 240
   Winnebago Tribe, 190
Land's Sake, 252
*Landscape Ecology Principles in Landscape
           Architecture and Land-Use Planning,*
           255
Law of the Land, 255
Learning coalitions, 135
Learning instrument
   for clarifying characteristics of new partner-
           ships, 145–148
   for clarifying requirements, partnerships, and
           responsibilities, 143–145
   community(ies), 143–151
   for reflecting on agroecosystem performance,
           148–150
*Legal Alternatives for Land Designation &
           Acquisition,* 255
Local Agriculture Association, 73, 77
Logging business, 60

M

Maize, 72
Market(s)
   civil society and, 8
   defined, 6
   dynamism, 6
   ownership, 6
   purpose, 6
   relationship to state, 7
Mazard, 85
Melons, 76, 78
Mexico, 2, 55, 211–19
   agricultural community, 55
      crops, 55
         subsistence of, 56
   agriculture (*See* Agriculture)
   charcoal makers, 59
   commercial forestry, 59
   Council of Distinguished Men, 63
   fruit and vegetable production, 70
   land use, 70, 71
   Local Agriculture Association, 73
   logging business, 60
   migrations, 72
   Municipal Agricultural Association, 80–81
   National Confederation of Horticulture
           Producers, 74
   Purhe'pecha people, 213, 215

Mexico (*Continued*)
  reforestation, 63–64
  Regional Agrarian Unions, 74, 80
  subsistence agriculture, 57
  timber poaching, 59
  Union of Ejidos Emiliano Zapata, 80
Mezapa, Honduras, 28
Minnesota Department of Agriculture, 175
Multiculturalism, 48
Municipal Agricultural Association, 80–81
Mutable immobiles, 119

N

National Agroforestry Center, 241, 255
National Confederation of Horticulture
        Producers, 74
Natural Resource Conservation Service, 124, 226,
        229, 231, 232, 235
Nebraska, 188–208
Neighborhood-based education, 232
Nomad breeders, 104. *See also* Crianceros
  household production units, 105
Nomadic herding, 2
Non-governmental organizations, 62, 92, 143
  development projects, 134
  streuobst protection, 95, 98
North American Free Trade Agreement, 69

O

Omoa, Honduras, 15, 25
Onions, 80–82
Oslo, Norway, 249

P

Packing plants, 74, 75
Panama disease, 20
Partnership(s), 143–145
  building, 152
  performance, 150–151
Partnership Handbook, 255
Patagonia, 103–113
  common *vs.* government property, 106
  government *vs.* common property, 106
Pears, 85
  varieties, 91
Peru, 140, 142
Piatt County, Illinois, 116
  chemical farm waste, 122
  history, 116
  local community institutions, 118
Piatt County Service Company, 126
Pigs for Ancestors, 158

Plums, 85
  varieties, 91
Prospect and refuge theory, 92
Public works projects, 61–62
Purhe'pecha people, 213, 215

R

Reforestation, 63–64
Regional Agrarian Unions, 74, 80
Resource Conservation District of Monterey
        County, 226
Resource flows, 121, 126
Resource use, 158
Riparian forest buffers, 245, 247, 248
Rotation crops, 116
Row crops, 116
Rural by Design, 255
Rural community leadership, 175–184
  goals, 183
  implementation, 178
  strengths, 184
  workshops, 178–180
Rural-urban interface, 242–244, 247. *See also*
        Ecobelt(s); Tree-based buffers

S

Salinization, 160
Savannah type landscapes, 92
Saving America's Countryside, 255
Seaboard Farms, 33–34, 42–43
  economic and cultural differences, 39
  employees, 48
  officials, 35
  water issues, 39
Seattle University District Farmers' Market, 165
Sharecropping, 76, 78
Social Science Institute, 255
Society of Social Solidarity, 73, 75, 77
Soil compaction, 160
Soil erosion, 160
Sonaguera, Colon, 22–24
Sour cherry, 85
Standard Fruit Company, 22
State
  agencies, 7
  purpose, 7
  relationship to markets, 7
Stormwater flooding, 243
Strawberry industry, 22
Streuobst, 85–99
  aesthetic qualities, 93
  apiculture and, 90
  defined, 85

Streuobst (*Continued*)
  ecological significance, 86, 89–92
  history, 86–88
  profitability, 94
  protection, 95
  as savannah type landscapes, 92
Sustainability, 1, 2, 160
  basis, 218–219
  communities and, 175
  connectors, 161, 169–171
  water quality and, 246
  wildlife habitat and, 246
Sustainable Conservation, 234
Switzerland. *See* Streuobst

T

Tanzania, 133, 143, 144
Tela, Honduras, 28
Tela Railroad Company, 29
Texas County, Oklahoma, 36–38
  average wage, 41
  demographics, 46
  economic profile, 40
  ethnicity, 41
  political affiliation, 49
  population, 39
  school system, 36–38
  unemployment rate, 41
Thompson Farms, 161, 166–169
  crops, 166
Timber poaching, 59
  community dynamics, 60
  control, 64
Tolt Farm, 161, 164–166
  crops, 164
  fertilizer, 165
  weed control, 165
Town forests, 252
Transcontinental corporation, 111
Tree-based buffers, 241
  ecological functions, 244, 247
TreePeople, 255
Tribal Allotment Act (1887), 190
Tribal Operating Procedure for Land Acquisition, 201
Truxillo Railroad Company, 27
Tyson's Pork Group, 33, 41, 43–45
  corporate culture, 45
  economic and cultural differences, 39
  establishment, 44
  opposition, 44
  water issues, 39

U

Uganda, 133, 138, 139, 140
Undercropping, 98
Union of Ejidos Emiliano Zapata, 80
United Fruit Company, 15, 26
Urbanization, 5
  defined, 5

V

Vegetable(s), 70
  exports, 70
  production, 70
Villa Grove, Illinois, 248

W

Walnuts, 91
Water pollution, 160
Water quality, 246
Western Massachusetts Food Bank, 162
Western Sustainable Agriculture Research and Education Administrative Council, 169
Weston, Massachusetts, 252
Whole Farm Planning Program, 175
Wildlife habitat, 246
Windbreaks, 245, 247
Windward Islands, West Indies, 158
Winnebago Land management Department, 202
Winnebago Tribe
  agroforestry project, 188–208
    community participation, 202–204
    development, 201–202
    impact, 205–206
    participatory rural appraisal, 202
    planning, 204
    tribal participation, 207
  history, 190
  land use, 190
  location of tribe, 189
Woody buffers, 241
  ecological functions, 244
  potentials, 244–248
Working Trees for Agriculture, 241
Working Trees for Communities, 241
Working Trees for Livestock, 241
Working Trees for Snow Management, 241
Working Trees for Treating Waste, 241
Working Trees for Wildlife, 241
World Bank, 107